EQUATION-OF-STATE AND PHASE-TRANSITION ISSUES IN MODELS OF ORDINARY ASTROPHYSICAL MATTER

Related Titles from AIP Conference Proceedings

730 Atomic Processes in Plasmas: 14th APS Topical Conference on Atomic Processes in Plasmas
Edited by J. Cohen, S. Mazevet, and D. Kilcrease, October 2004, 0-7354-211-6

719 Physics of the Outer Heliosphere: Third International IGPP Conference
Edited by V. Florinski, N. V. Pogorelov, and G. P. Zank, September 2004, 0-7354-0199-3

703 Plasmas in the Laboratory and in the Universe: New Insights and New Challenges
Edited by Giuseppe Bertin, Daniela Farina, and Roberto Pozzoli, April, 2004,
0-7354-0176-4

679 Solar Wind Ten: Proceedings of the Tenth International Solar Wind Conference
Edited by Marco Velli, Roberto Bruno, and Francesco Malara, September 2003,
CD-ROM included, 0-7354-0148-9

669 Plasma Physics: 11th International Congress on Plasma Physics: ICPP2002
Edited by Ian S. Falconer, Robert L. Dewar, and Joe Khachan, June 2003,
Print: 0-7354-0133-0; CD-ROM: 0-7354-0134-9

637 Classical Nova Explosions: International Conference on Classical Nova Explosions
Edited by Margarita Hernanz and Jordi José, November 2002, 0-7354-0092-X

635 Atomic Processes in Plasmas: 13th APS Topical Conference on Atomic Processes in Plasmas
Edited by David R. Schultz, Fred W. Meyer, and Fay Ownby, October 2002, 0-7354-0090-3

634 Science of Superstrong Field Interactions: Seventh International Symposium of The Graduate University for Advanced Studies on Science of Superstrong Field Interactions
Edited by Kazuhisa Nakajima and Masayuki Deguchi, October 2002, CD-ROM included,
0-7354-0089-X

598 Solar and Galactic Composition: A Joint SOHO/ACE Workshop
Edited by Robert F. Wimmer-Schweingruber, December 2001,CD-ROM included,
0-7354-0042-3

565 Young Supernova Remnants: Eleventh Astrophysics Conference
Edited by Stephen S. Holt and Una Hwang, May 2001, 0-7354-0001-6

563 Plasma Physics: IX Latin American Workshop
Edited by Hernán Chuaqui and Mario Favre, May 2001, 1-56396-999-8

To learn more about these titles, or the AIP Conference Proceedings Series, please visit the webpage
http://proceedings/aip.org/proceedings

EQUATION-OF-STATE AND PHASE-TRANSITION ISSUES IN MODELS OF ORDINARY ASTROPHYSICAL MATTER

Lorentz Center, Leiden, The Netherlands   2 – 11 June 2004

EDITORS
Vladan Čelebonović
Institute of Physics
Belgrade, Serbia and Montenegro

Werner Däppen
University of Southern California
Los Angeles, California

Douglas Gough
Institute of Astronomy
Cambridge, United Kingdom

SPONSORING ORGANIZATIONS
FOM - Fundamenteel Onderzoek der Materie
   (Dutch Physics Funding Foundation)
NWO - Nederlandse Organisatie voor Wetenschappelijk Onderzoek
   (Netherlands Organisation for Scientific Research)
Leiden University

Melville, New York, 2004
AIP CONFERENCE PROCEEDINGS ■ VOLUME 731

Editors:

Vladan Čelebonović
Institute of Physics
Pregrevica 118
11080 Belgrade-Zemun
SERBIA AND MONTENEGRO
E-mail: vladan@phy.bg.ac.yu

Werner Däppen
Department of Physics and Astronomy
University of Southern California
Los Angeles, California 90089-1342
UNITED STATES
E-mail: dappen@usc.edu

Douglas Gough
Institute of Astronomy
Madingley Road
Cambridge CB3 0HA
UNITED KINGDOM
E-mail: douglas@ast.cam.ac.uk

Cover: Martinus Nijhoff, The Hague, Netherlands, "H. A. Lorentz, Collected Papers," Jan Veth del.1899, courtesy AIP Emilio Segrè Visual Archives.

Authorization to photocopy items for internal or personal use, beyond the free copying permitted under the 1978 U.S. Copyright Law (see statement below), is granted by the American Institute of Physics for users registered with the Copyright Clearance Center (CCC) Transactional Reporting Service, provided that the base fee of $22.00 per copy is paid directly to CCC, 222 Rosewood Drive, Danvers, MA 01923. For those organizations that have been granted a photocopy license by CCC, a separate system of payment has been arranged. The fee code for users of the Transactional Reporting Service is: 0-7354-0213-2/04/$22.00.

© 2004 American Institute of Physics

Individual readers of this volume and nonprofit libraries, acting for them, are permitted to make fair use of the material in it, such as copying an article for use in teaching or research. Permission is granted to quote from this volume in scientific work with the customary acknowledgment of the source. To reprint a figure, table, or other excerpt requires the consent of one of the original authors and notification to AIP. Republication or systematic or multiple reproduction of any material in this volume is permitted only under license from AIP. Address inquiries to Office of Rights and Permissions, Suite 1NO1, 2 Huntington Quadrangle, Melville, N.Y. 11747-4502; phone: 516-576-2268; fax: 516-576-2450; e-mail: rights@aip.org.

L.C. Catalog Card No. 2004114500
ISBN 0-7354-0213-2
ISSN 0094-243X
Printed in the United States of America

Contents

Preface..ix

PART I

THE SUN AND THE STARS: GENERAL ASPECTS

Equations of State for Solar and Stellar Modeling 3
  W. Däppen
An Introduction to Solar Oscillations and Helioseismology 18
  J. Christensen-Dalsgaard
Helioseismic Inversions and the Equation of State 47
  S. V. Vorontsov

PART II

PLASMAS OF STELLAR INTERIORS: BASIC PRINCIPLES

The Limit on Mean Field Theories and Nuclear Screening
in Stellar Plasmas 67
  G. Shaviv
Corrected EOS of Weakly-Nonideal Hydrogen Plasmas
without Mysteries 83
  A. N. Starostin and V. C. Roerich
Improved Phenomenological Equation of State in the
Chemical Picture 99
  R. Trampedach
Emulating the OPAL Equation of State in the
Chemical-Picture Formalism 106
  A. Liang

PART III

THE SUN AND THE STARS: DETAILED TECHNIQUES

The Power of Helioseismology to Address Issues of
Fundamental Physics 119
  D. Gough
The State of ^{7}Be in the Sun 139
  N. J. Shaviv and G. Shaviv

SAHA-S Model: Equation of State and Thermodynamic Functions of Solar Plasma 147
    V. K. Gryaznov, S. V. Ayukov, V. A. Baturin, I. L. Iosilevskiy,
    A. N. Starostin, and V. E. Fortov

Thermodynamics of the Convection Zone Through Adiabatic Exponent 162
    V. A. Baturin

Heavy Element Settling in the Sun and Equation of State 173
    S. V. Ayukov and V. A. Baturin

Solar Models Using the SAHA-S Equation of State 178
    S. V. Ayukov, V. A. Baturin, V. K. Gryaznov, I. L. Iosilevsky,
    and A. N. Starostin

The Effect of Using Different EOS in Modelling the α Centauri Binary System 187
    A. Miglio

Asteroseismic Helium Abundance Determination 193
    G. Houdek

Quantum Statistical Corrections on the Enhancement Factor in Solar Fusion Reactions 208
    A. Perez and G. Chabrier

Isolating the Effects of Chemical Composition in the Equation of State 219
    C.-H. Lin and W. Däppen

The Chemical Composition and Equation of State of the Sun Inferred from Seismic Models Through an Inversion Procedure 230
    C.-H. Lin and W. Däppen

PART IV

FROM WEAKLY TO STRONGLY COUPLED PLASMAS TO POSSIBLE PHASE TRANSITIONS

Phase Transitions in Dense Hydrogen-Helium Plasmas 239
    V. Filinov, P. Levashov, M. Bonitz, and V. Fortov

Quantum Statistical Approach to Dense, Weakly Coupled Plasmas 248
    J. Vorberger, M. Schlanges, and W.-D. Kraeft

Spinodal Decomposition of Metastable Melting in the Zero-Temperature Limit 255
    I. L. Iosilevski and A. Y. Chigvintsev

Anomalous Phase Diagrams in the Simplest Plasma Models 261
    I. L. Iosilevski and A. Y. Chigvintsev

PART V

FROM THE PLASMA TO THE SOLID STATE: ASTROPHYSICAL IMPLICATIONS

Selected Results and Open Problems in a Semiclassical Theory of Dense Matter 269
   V. Čelebonović

Basic Notions of Static High Pressure Experiments 280
   V. Čelebonović

A Model of the Internal Structure of Titan: First Results 288
   G. Pavičić and V. Čelebonović

The Narrow Line Region of an AGN Sample 291
   E. Bon, D. Ilić, L. Č. Popović, E. Mediavilla, V. Čelebonović, and G. Pavičić

On Phase Transitions in Bose Gases at Constant Density and Constant Pressure 295
   V. G. Ivanov and D. I. Uzunov

Phase Transitions to Spin-Triplet Ferromagnetic Superconductivity in Neutron Stars 302
   D. V. Shopova, T. E. Tsvetkov, and D. I. Uzunov

Author Index 311

PREFACE

The need for precise knowledge of the equation of state of hot dense matter, including the occurrence of phase transitions, arises frequently in physics and astrophysics. The need occurs in past and present laboratory experiments under high static and dynamic pressure, which will be supplemented in the future by large laser-fusion experiments at the National Ignition Facility at Livermore, USA, and Megajoule in Bordeaux, France. In astrophysics, accurate knowledge of the equation of state is required for understanding planets and stars, including brown and white dwarfs and especially the Sun. Although there has been some interaction between the plasma-physics and the astrophysics communities in the past, the subject has now reached a degree of sophistication that warrants much closer contact. Helioseismological diagnosis has enabled us to measure certain properties of the solar plasma to a degree of precision that has justified much more detailed theoretical analysis than had previously been thought necessary. This has stimulated the theoretical plasma physicists whose new findings have, in turn, triggered astrophysicists to refine their diagnostic tools. The time became ripe for bringing the two communities together to make each more aware of the knowledge and abilities of the other, and of the potential for further, perhaps joint, progress. Such a gathering occurred at a ten-day workshop in June 2004, at the Lorentz Center at the University of Leiden.

The story of the workshop begins and ends with the American Institute of Physics (AIP). In 1995 Vladan Čelebonović was studying the AIP journal donation program on behalf of the Institute of Physics in Belgrade, and noticed that someone called Wim van Saarloos was offering to donate Europhysics Letters. Unfortunately, by the time contact was made, the journals had been passed elsewhere. However, a year later Vladan wrote to Wim again, and this time he received some other journals for his institute. True contact had now been established. But only after a couple more years did Vladan come to realize that Wim was not only a senior professor in both Leiden and Paris, but he was also Director of the Lorentz Center. After the political changes in Yugoslavia in 2000, Wim proposed to Vladan a short visit to the Center, which duly occurred the following year. At the end of Vladan's seminar, which exposed the pregnant state of the subject, Wim proposed to Vladan that he organize a workshop; two nights later Vladan concluded that he would involve Werner Däppen, and Werner, in turn, proposed co-opting Douglas Gough. The rest is, as they say, history. We should at least add, however, that we were persuaded by the novelty and timeliness of the gathering that, contrary to Lorentz Center custom, it would be fruitful to publish proceedings, all contributions being refereed. The AIP welcomed the prospect, and so completed the near-decade-long story.

The workshop was found by all the participants to be both educational and inspirational. This was due to the enthusiasm of the diverse group of scientists from a broad range of countries, many of whom had never before met, to devote time to exchange their ideas and work towards a greater appreciation of what others were thinking. The well appointed working environment provided by the Lorentz Center was conducive not only to discussion but also to working seriously on ideas that had been generated by the discussions. Some of the interactions will lead to new and, it is to be hoped, fruitful collaborations. The success of the workshop is owed also to the superb organization of the Lorentz Center, and we thank the local program assistant Gerda Filippo, and the executive manager Dr. Martje Kruk - de Bruin for their professional support and hospitality. We are especially indebted to Professor Wim van Saarloos, whose continual support and encouragement were crucial at several stages of the preparation; we regret that owing to unexpected illness he was unable to be present for the latter half of the workshop. We wish him well.

Vladan Čelebonović

Werner Däppen

Douglas Gough

PART I

THE SUN AND THE STARS:
General Aspects

Equations of state for solar and stellar modeling

Werner Däppen

University of Southern California, Los Angeles, California, 90089-1342, U.S.A.

Abstract. Helioseismology has become the most successful diagnosis of the equation of state for the plasma of stellar interiors. Although in the solar interior the plasma is only weakly coupled and weakly degenerate, the great observational accuracy of the helioseismological observations puts nevertheless strong constraints on the nonideal part of the equation of state. For solar and stellar modeling, a high-quality equation of state is crucial. But the inverse is also true: the astrophysical data put constraints on the physical formalisms, making the Sun and the stars novel laboratories for plasma physics.

INTRODUCTION

Thermodynamic quantities of stellar matter are, together with opacity and nuclear reaction rates, the fundamental properties of stellar matter that enter stellar models (Christensen-Dalsgaard 2004). These quantities are responsible for the resulting overall parameters of stars and their differentiation according to chemical composition. These *material properties* determine the coefficients of the equations of stellar structure. Otherwise, the type of stellar matter would not enter these equations, which express the result of (*i*) a balance of forces, (*ii*) a balance between the energy loss at the stellar surface and energy generation in the core, and (*iii*) stationary energy transport between the core and the surface. Thermodynamics is most directly relevant for the balance of forces, which settles in the *hydrostatic equilibrium*, which is a relation between the local pressure gradient and gravitational acceleration. The force of gravity is determined by the density distribution in the star: thus stellar modeling requires a relation between density and pressure through the specific properties of the matter. More precisely, the relevant properties of stellar matter are expressed by the *equation of state*, connecting pressure p, density ρ, temperature T, and composition. A simple example is the perfect gas law for a fully ionized gas, which may be written as $p = \mathscr{R}\rho T/\mu$; here $\mathscr{R}$ is the gas constant, and μ is the mean molecular weight of the various species contained in the gas. Clearly μ depends on the composition, which is often characterized by the fractional mass abundances X, Y and Z of hydrogen, helium and heavier elements.

Present-day solar and stellar models are based on sophisticated new equations of state. Popular, for instance, are the ones underlying the two ongoing major opacity recomputation efforts. One of these efforts is the international Opacity Project (OP; see the books by Seaton 1995, Berrington 1997); it contains the so-called Mihalas-Hummer-Däppen equation of state (Hummer & Mihalas 1988, Mihalas *et al.* 1988, Däppen *et al.* 1988, Nayfonov *et al.* 1999, Trampedach 2004, Trampedach *et al.* 2004; hereinafter MHD) and it deals with *heuristic* concepts about the modification of atoms and ions in a plasma. The other effort is being pursued at Lawrence Livermore National Laboratory by the OPAL group (Iglesias and Rogers 1996, Rogers *et al.* 1996); its equation of state is based on a detailed *systematic* method to include density effects in a plasma.

The name "equation of state" is generally interpreted to encompass the thermodynamic quantities as well, in addition to the pressure-density-temperature relation. Knowledge of thermodynamic quantities is necessary for stellar models. For instance, the bulk of most convection zones is essentially adiabatically stratified, which means that the pressure-density-temperature relation is an adiabat. The adiabat is the integrated form of the local adiabatic gradients, which are purely thermodynamic quantities. One of the adiabatic gradients is related to adiabatic sound speed: $c^2 = (\partial p/\partial \rho)_s$ (s being specific entropy). This thermodynamic quantity has another important stellar application, because adiabatic sound speed is the key quantity that determines the frequency of acoustic oscillation modes; however, thermodynamics is also crucial when assessing the importance of non-adiabatic effects. This is understood by realizing that the temperature of the stellar interior is determined by the local energy balance of the outward flux of energy, which begins in the hot stellar core that is driven by nuclear reactions, and ends at the surface when the energy leaves the star in form of radiation.

THEORETICAL MODELS

A useful practical introduction to the equation of state is the book by Eliezer *et al.* (1986). On a higher theoretical and systematic level is the book about the physics of Coulomb systems by Kraeft *et al.* (1986). The most important classification when dealing with plasmas of stellar interiors is the one in ideal and nonideal plasmas. The simplest model is given by a mixture of nuclei and electrons, assumed fully ionized and obeying the classical perfect gas law. However, this is not the most general of an *ideal-gas* equation of state: it may include deviations from the perfect gas law, namely ionization or dissociation reactions, radiation and degeneracy of electrons, as long as the underlying microphysics of these additional effects is still ideal, that is, does not contain interactions. The "particles", however, can be be classical or quantum, material or photonic. In such an ideal framework, bound systems (molecules, atoms, ions) are allowed to have internal degrees of freedom (excited states, spin). All such ideal effects can be calculated as exactly as desired.

Nonideal effects in plasmas can be assessed by the so-called coupling parameter Γ. In a plasma of temperature T and density such that particles of charge e have an average distance $<r>$ from each other, one can define Γ as the ratio of average potential binding energy over mean kinetic energy $k_B T$ (k_B being the Boltzmann constant)

$$\Gamma = (e^2/<r>)/k_B T .$$

For simplicity, this expression is restricted to the case of hydrogen; generalizations to other elements are straightforward. Plasmas with $\Gamma \gg 1$ are *strongly* coupled, those with $\Gamma \ll 1$ *weakly* coupled. A famous example of a strongly coupled plasma is the electron gas in the interior of white dwarfs, where the coupling can become strong enough to force crystallization. Another example is given by the electrons in the conduction band of a metal at room temperature. Weakly coupled plasmas are, for instance, the interiors of stars with masses ranging from the slightly sub-solar ones to the largest.

Basic Concepts and Dimensional Parameters

To describe a degree of "nonideality" or "degeneracy" of the plasma, it is convenient to introduce a few dimensionless parameters to express the relative strength of the particular effects, such as the Coulomb interaction or quantum degeneracy. These parameters are defined as combinations of various "characteristic lengths" resulting from simple physical estimates.

The strength of the Coulomb interaction is given by the density-independent *Landau length* l_L

$$l_L = \frac{e_i e_j}{kT}.$$

Although the Coulomb potential has infinite range, the Landau length represents an effective radius of the interaction between particles i and j, i.e. for distances $\gg l_L$ the potential energy is negligible compared to the thermal energy kT.

The Landau length clearly reflects a degree of nonideality of the plasma and the case of an "ideal" plasma corresponds to $l_L = 0$.

Another purely classical characteristic length is the *mean distance* between two particles of the same species k (k = $\{e,i\}$)

$$d_k = \left(\frac{3}{4\pi n_k}\right)^{1/3}.$$

Here, n_k is the particle density N_k/V of species k. Note that d_k is the generalization of the quantity $<r>$ in Eq. (1). It does not depend on temperature.

The most evident quantum mechanical characteristic length is the *thermal de-Broglie wavelength*

$$\lambda_k = \frac{h}{\sqrt{2\pi m_k kT}}.$$

The limits of applicability of various theoretical models of the equation of state are usually described by different dimensionless parameters. The *degeneracy* of species k

$$n_k \lambda_k^3$$

is very closely related to the well-known *degeneracy parameter* η (sometimes also denoted ψ) by

$$F_{1/2}(\eta) = \frac{\sqrt{\pi}}{4} \frac{\lambda_e^3 N_e}{V},$$

where F_i is the Fermi integral

$$F_i(y) = \int_0^\infty x^i [1+\exp(y+x)]^{-1}\, dx.$$

In the case of a stellar plasma with heavy nuclei and light electrons, the heavy particles become degenerate only at much higher densities than electrons.

Screening of the Coulomb Potential

All characteristic lengths defined so far demonstrate either purely density- or purely temperature-dependent behavior. No collective behavior of particles has been considered. Here screening is discussed, which for normal stars is the most important deviation from ideality.

The seminal study of collective behavior in a plasma was the Debye-Hückel theory of electrolytical solutions. Despite its phenomenological approach this theory proved to be immensely successful not just for electrolytes, but also for plasma calculations.

Basically, one mixes two ideas. First, assuming nondegenerate electrons, and considering one particular ion being fixed, the mean electron distribution around that ion is given by the *Boltzmann factor*

$$n_e(r) = n_e \exp\left[\frac{+e\phi(r)}{kT}\right].$$

The mean ion density around the same ion is similarly given by

$$n_i(r) = n_i \exp\left[\frac{-e\phi(r)}{kT}\right].$$

Second, one assumes applicability of Poisson's equation for *mean charge distributions* (instead of point charges)

$$\Delta\phi(r) = -4\pi e[n_i(r) - n_e(r)].$$

The resulting system of equations is complicated and nonlinear (see Brüggen & Gough 2000); however, if the system is only slightly nonideal, that is, if $e\phi(r)/kT < 1$, the linearized system has the well-known solution of the static-screened Coulomb potential (SSCP)

$$\phi(r) = \frac{e}{r} e^{-\kappa r},$$

where κ is the reciprocal of the *Debye-length* r_D

$$r_D = \sqrt{\frac{kT}{4\pi(n_e e^2 + n_i e^2)}}.$$

This is the electrostatic and weak-screening approximation to a more general dynamic situation, which plays not only an important role in stellar thermodynamics, but equally in the determination of nuclear reaction rates (Shaviv & Shaviv 1996, Brüggen & Gough 1997; 2000, Shaviv 2004a;b, Perez & Chabrier 2004).

Chemical-Picture Models

With the exception of neutron stars, in stellar interiors the thermal de Broglie wavelength of nuclei, atoms, ions and molecules is always tiny and they can all be treated classically. Only electrons have to be treated according to quantum mechanics. Here, there is a bifurcation into two distinct classes of approach, the "chemical picture" and the "physical picture" (see the book by Kraeft et al. 1986). While in the more conventional chemical picture bound configurations (atoms, ions and molecules) are introduced and treated as new and independent species, only *fundamental* particles (electrons and nuclei) appear in the physical picture. In the chemical picture, reactions between the various species occur, and thus the thermodynamical equilibrium must be sought among the stoichiometrically allowed set of concentration variables by means of a maximum entropy (or minimum free-energy) principle. In contrast, the physical picture has the esthetic advantage that there is no need for a minimax principle; the question of bound states is dealt with implicitly through the Hamiltonian describing the interaction between the fundamental particles. See the section below on activity expansions.

The development of computational methods was strongly stimulated by the requirements for stellar evolution, where data for a wide range of parameters are needed. The chemical picture is very well suited for powerful, approximative models such as the free-energy-minimization method, which became computationally feasible since about 1960. Given a mathematical model for the Helmholtz free energy $F(T,V,\{N_i\})$, where $\{N_i\}$ is a set of particle numbers for all species i present in a plasma, one minimizes F subject to the stoichiometric relations governing possible reactions among the particle species in the plasma. The underlying principle is that nature adjusts the reaction equilibrium such that entropy is maximum (for given energy) or the free energy is minimum (for given temperature).

One starts from the canonical partition function. Consider a physical system (with Hamiltonian H) confined in a box of volume V in contact with a heat reservoir at temperature T. Then the canonical partition function is a trace (denoted by Tr)

$$\mathscr{Z} = \mathrm{Tr}\left(e^{-H/kT}\right).$$

From that, the free energy is obtained by the formula

$$F(T,V,\{N\}) = -kT \ln(\mathscr{Z}).$$

Here, the volume V is implicitly contained in the Hamiltonian operator. The free energy is calculated for all $\{N\}$ which are allowed by the *stoichiometric relations* and a given initial composition. Equilibrium concentrations are those $\{N\}$ that minimize $F(T,V,\{N\})$ for given T,V under the stoichiometric constraints. This gives ionization and dissociation equilibria. Numerically, this will be a task of finding a minimum of a nonlinear function $F(T,V,\{N\})$ in the $\{N\}$-space under the stoichiometric constraints.

The major assumption in any *practical* realization of the free-energy minimization method is that the total partition function $\mathscr{Z} = \mathscr{Z}_{\mathrm{total}}$ factorizes, that is,

$$\mathscr{Z}_{\mathrm{total}} = \mathscr{Z}_{\mathrm{e}} \mathscr{Z}_{\mathrm{trans}} \mathscr{Z}_{\mathrm{conf}} \mathscr{Z}_{\mathrm{int}},$$

where Z_e here stands for the electronic contribution, $\mathscr{Z}_{\text{trans}}$ is the result of the integration over momentum space for the heavy particles, $\mathscr{Z}_{\text{conf}}$ is the configurational integral over configuration space coordinates, and $\mathscr{Z}_{\text{int}}$ finally is

$$\mathscr{Z}_{\text{int}} = \Pi \mathscr{Z}_{\text{int}}^{(i)}, \qquad \mathscr{Z}_{\text{int}}^{(i)} = \sum_{j=1}^{\infty} g_{ij} \exp(-E_{ij}/kT),$$

where E_{ij} and g_{ij} denote energy and degeneracy of the state j of species i, respectively.

When going from the partition function to the free energy, the consequence of the assumed factorizability of the partition function is additivity, or modularity, of the free energy. This modularity accounts for the great appeal of the chemical approach from both a modeling and a computational point of view. Modularity basically tells us that all the interactions can be split into separate parts with clear physical meaning. And if, for example, one part needs to be modified to implement a higher order correction, it can be done without having to worry about consistency with the other parts. Terms can be introduced simply by adding or changing subroutines in a computer program, without a major overhaul of the previous work. That is why sometimes for the sake of modularity even relatively crude approximations are maintained, because a rigorous treatment, in the framework of the exact nonfactorizing partition function, would be entirely prohibitive.

The power of the free-energy-minimization method is that it allows the *consistent* inclusion of nonideal effects. The reason is that any modification to the thermodynamics is only made at one single place (the free energy). Since the ionization degree and all thermodynamic quantities follow from the free energy by purely mathematical steps, all these quantities are consistent with each other. For instance, they automatically obey all Maxwell relations. The corresponding shifts in the ionization equilibria are correct.

Achieving such consistency would be difficult in an approach where one considers the Saha equation and the corresponding thermodynamic quantities separately, all to be modified individually. There would be no general systematic procedure. The free-energy-minimization method, however, achieves it nicely and simply. It is therefore the natural nonideal extension of to incorporate interactions into simple equations of state which are based on the Saha equation.

As an example, to include screening at the level of the Debye-Hückel approximation, it suffices to add to the total free energy the "module" for the Debye-Hückel free energy

$$F_{\text{conf}} \equiv F^{\text{DH}} = -\frac{kTV}{12\pi r_D^3}.$$

For weakly-coupled plasmas, perturbational approaches are effective. An example is the Mihalas-Hummer-Däppen (MHD) equation of state mentioned in the introduction. It centers around a *heuristic* concept of modified atoms and ions due to the plasma environment. More specifically, the MHD equation of state is based on an occupation probability formalism. The internal partition functions Z_s^{int} of species s adopted by MHD are sums that are made convergent by weights

$$Z_{\text{int}}^s = \sum_i w_{is} g_{is} \exp\left(-\frac{E_{is}}{k_B T}\right).$$

Here, the index *is* labels the state i of species s. E_{is} and g_{is} are their energies (in MHD assumed to be unshifted) and statistical weights, respectively, k_B is the Boltzmann constant, T temperature, and the weights w_{is} are the occupation probabilities that take into account charged and neutral surrounding particles. In physical terms, w_{is} gives the fraction of all particles of species s that can exist in state i with an electron bound to the atom or ion, and $1 - w_{is}$ gives the fraction of those that are so heavily perturbed by nearby neighbors that their states are effectively destroyed.

Although the MHD equation of state was originally designed to provide the level populations for opacity calculations of stellar *envelopes*, the associated *thermodynamic quantities* of MHD can none the less be reliably used also for stellar cores. This is due to the fact that in the deeper interior the plasma becomes virtually fully ionized. This rather fortuitous circumstance is the reason why the MHD equation of state fares very well in models of the entire Sun and even in lower-mass models, down to 0.4 $M_\odot$ (for references, see Däppen & Guzik 2000).

Thermodynamics is more complicated for the interior of low-mass stars, because the kinetic energy of the particles does not dominate the electrostatic potential energies. In the deep interior of these stars, the plasma can become strongly coupled. In addition, temperature alone cannot ionize matter fully, and pressure ionization occurs, which require more elaborate modeling. The configurational free energy F_conf stems directly from the configuration integral $\mathscr{Z}_\text{conf}$. Besides containing the Coulomb interactions that lead in the classical weakly-coupled approximation to the Debye-Hückel free energy F_conf, it also contains excluded-volume effects of neutral particles. The well-known van-der-Waals equation of state is based on such an excluded volume effect, but there are more elaborate formalisms of hard-sphere or softer types. Such excluded volume effects become very important in the relatively cold plasmas of low-mass stars or planets (Saumon *et al.* 1995). Note that the sophisticate nature of hard-sphere potentials does not change the basically heuristic nature of the free-energy minimization methods. Only the physical picture can overcome this fundamental limitation (see the following section).

Physical-Picture Models

It is clear from the preceding sections that the advantages of the free-energy-minimization method and the chemical picture lie in the possibility to model complicated plasmas, and to obtain numerically smooth and consistent thermodynamical quantities. Nevertheless, the heuristic method of the separation of the atomic-physics problem from that of statistical mechanics is not satisfactory, and attempts have been made to avoid the concept of a perturbed atom in a plasma altogether. This has suggested an alternative description, the physical picture. In such an approach one expects that no assumptions about energy-level shifts or the convergence of internal partition functions have to be made. On the contrary, properties of energy levels and the partition functions should come out from the formalism.

Despite the impressive body of literature on the physical picture (Kraeft *et al.* 1986, Rogers 1986, Rogers *et al.* 1996; and references therein), the majority of work on the physical picture was not dedicated to the problem of obtaining a high-precision equation of state for stellar interiors. Only the OPAL group at Livermore has so far achieved such a goal.

To explain the advantages of this approach for partially ionized plasmas, it is instructive to discuss the activity expansion for gaseous hydrogen. The interactions in this case are all short ranged and pressure is determined from a self-consistent solution of the equations

$$\frac{p}{k_B T} = z + z^2 b_2 + z^3 b_3 + \ldots ,$$

$$\rho = \frac{z}{k_B T} \left(\frac{\partial p}{\partial z} \right) ,$$

where $z = \lambda_e^{-3} \exp(\mu/k_B T)$ is the activity [with λ_e being the thermal de-Broglie wavelength of electrons, see above], and μ is the chemical potential. The b_n are cluster coefficients such that b_2 includes all two particle states, b_3 includes all three particle states, etc.

In contrast to the chemical picture, which is plagued by divergent partition functions, the physical picture has the power to avoid them altogether. An important example of such a fictitious divergence is that associated with the atomic partition function. This divergence is fictitious in the sense that the bound-state part of b_2 is divergent but the scattering state part, which is omitted in the Saha approach, has a compensating divergence. Consequently the total b_2 does not contain a divergence of this type. A major advantage of the physical picture is that it incorporates this compensation at the outset. A further advantage is that no assumptions about energy-level shifts have to be made; it follows from the formalism that there are none.

As a result, the Boltzmann sum appearing in the atomic free energy for atoms and compound ions is replaced with the so-called Planck-Larkin partition function (PLPF), given by

$$\text{PLPF} = \sum_{nl} (2l+1) \left[\exp(-\frac{E_{nl}}{kT}) - 1 + \frac{E_{nl}}{k_B T} \right] .$$

The PLPF is convergent without additional cut-off criteria as are required in the chemical picture. However, despite its name the PLPF is not a partition function, but merely an auxiliary term in the second expansion (virial) coefficient of pressure.

So far, only one group (the OPAL group at Livermore) has realized a practically useful formalism that satisfies the exacting demands of stellar modelers. Since it contains an activity expansion, other than OPAL equation of state, it is alternatively called ACTEX. Because of its complexity, so far all OPAL results had only be published in the form of pre-computed tables, for certain chemical compositions of astrophysical relevance (Rogers *et al.* 1996). The most recent OPAL tables allow modeling of stars with masses down to 0.4 $M_\odot$ (Rogers & Nayfonov 2002). A new version that should go down to 0.1 $M_\odot$ will be announced later this year (however, models of brown dwarfs and giant planets might still need configurational-integral approaches, such as that of Saumon *et al.* 1995).

Relation between Chemical and Physical Picture

In theory, that is, if all expansions were to infinite order, the physical-picture formalism ought to be exact. The level of accuracy is determined by the order of expansion. No such statement can be made for the chemical picture. Its modular nature, hard-wired in through the imposed factorizability of the total partition function, inevitably brings in errors. We do not know if the resulting fundamental errors are so small that one can keep them always beyond reach of observational detection. Furthermore, practical chemical-picture formalisms suffer from other limitations, due, on the one hand, to badly chosen interaction terms of the free energy and, on the other hand, to the usual approximation of unperturbed continuum states for the free electrons (see below). In principle, one can reduce both these practical limitations with a suitable effort, analogous to that made in the physical picture when one goes to a higher order of of expansion. In principle, the heuristic nature of the chemical-picture formalism allows the introduction of *ad-hoc* terms that could correct for the fundamental flaw of factorizability, but it remains to be seen if such a strategy can be successful (Liang & Däppen 2004, Liang 2004).

For practical reasons, in the chemical picture, all equations of state have been using free (plane-wave) electron states in the continuum. Finding a more realistic model for the free electrons would be extremely difficult, since in the chemical picture, the charged background that determines the wave functions of the free electrons *is changing* with the degree of ionization of all elements. In contrast, in the physical picture, the systematic expansion from one-body to many-body terms allows to take into account the continuum corrections in a clear fashion, as as Coulomb functions for the two-body part for example.

In the physical picture, it is the continuum term that is responsible for the PLPF. The effect of the modification of the free states is opposite and equal to the one incurred by including high-lying excited states. The net result is a *cancellation*, which has two consequences. In the continuum part, there appears a specific form of a sum over scattering phases, and in the bound-state the PLPF. The physical explanation of this PLPF cancellation can be broadly be given by the following consideration. Electrons in low-lying bound states (such as the ground state) essentially "see" the central charge alone; neighboring charges are far away. Electrons in highly-excited bound states are subject both to the central charge and to neighboring charges, all in similar strength. Together, these charges more-or-less cancel (due to overall neutrality of the plasma), thus rendering the highly-excited electrons essentially free. That is why the highly-excited states have little effect in the bound-state partition function, a fact expressed precisely by the PLPF (which roughly corresponds to a truncation beginning at states whose binding energy is equal or less than $-kT$. Note that even after the PLPF cancellation is made, there still remains a continuum term. There is some freedom to modify this continuum term and the PLPF together, as long as the sum of their effects is the same. In other words, the PLPF is not unique.

HELIOSEISMIC EQUATION-OF-STATE STUDIES

Although simple ideal-gas-based models of the plasma of the solar interior were adequate before helioseismology, modern analyses require more sophisticated physical models. The need to go beyond the ideal-gas model for helioseismic applications had been recognized from the early 1980s (see, for instance, Berthomieu *et al.* 1980, Ulrich 1982, Noels *et al.* 1984). With the better data that became available towards the end of the 1980s, a clearer picture began to emerge. Christensen-Dalsgaard *et al.* (1988; 1996) demonstrated that the Coulomb correction had to be included in the equation of state. The relative Coulomb correction peaks in the outer part of the convection zone (about minus 8 per cent for pressure) and in

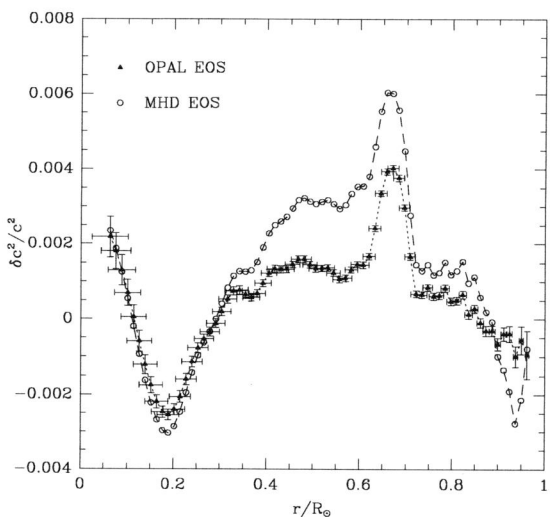

FIGURE 1. Difference (in the sense 'Sun − model') between squared sound speed from inversion and two solar models differing only by their equation of state (OPAL, MHD). Figure by S. Basu.

the solar core (about minus 1 per cent for pressure). For solar conditions, the Debye-Hückel (DH) theory is a good approximation for the leading term of the Coulomb correction.

Although approximate asymptotic techniques (see Christensen-Dalsgaard *et al.* 1985, Gough 1993, Christensen-Dalsgaard 2004) exist to invert solar oscillation frequencies for the internal sound speed, for an accurate analysis of the observations, a fully-fledged, non-asymptotic numerical treatment of the oscillations is mandatory (see Gough *et al.* 1996, Christensen-Dalsgaard 2004). Figure 1 is a typical result of such a numerical inversion (Basu & Christensen-Dalsgaard 1997). It shows the relative difference (in the sense Sun − model) between the squared sound speed obtained from inversion of oscillation data and that of a two standard solar models. The two solar models used are identical in all respects except for their equation of state, MHD (circles) and OPAL (triangles), respectively. For the present purpose, we can consider inversion results such as Figure 1 as the *data* of helioseismology, disregarding the procedure through which they were actually obtained from solar oscillation frequencies. It follows from Figure 1 that in most parts of the Sun the OPAL equation of state is a better approximation to reality than MHD, but OPAL needs to be improved as well. In the outermost layers of the Sun, however, the general trend might be reversed (see Figure 3).

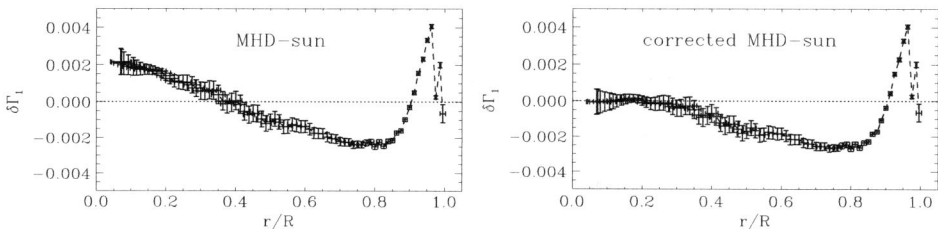

FIGURE 2. Difference between γ_1 (here denoted Γ_1) of a solar model and observation, for models with nonrelativistic electrons (left) and relativistic electrons (right), as determined by Elliott & Kosovichev (1998). Figures from Christensen-Dalsgaard (2002).

Effect of Relativistic Electrons

The strong constraints from helioseismology revealed the influence of *relativistic* electrons. The original versions of MHD and OPAL had not included relativistic electrons (although both did include degeneracy). In a recent helioseismic inversion for the adiabatic gradient $\gamma_1 = (\partial \ln p / \partial \ln \rho)_s$ (s being specific entropy), Elliott & Kosovichev (1998) found a discrepancy between, on the one hand, the observed structure of the Sun, and, on the other hand, models using the OPAL or MHD equation of state.

The top panel of Figure 2 shows this discrepancy for MHD. The relevant deviation occurs in the central 30% parts of the Sun. (A corresponding figure for OPAL would look essentially the same.) A *relativistic* treatment of the degenerate electrons in the solar model (bottom panel) removes the discrepancy nicely. As a result, both MHD (Gong *et al.* 2001a;b) and OPAL (Rogers & Nayfonov 2002) had since been upgraded to include relativistic electrons.

Effect of Excited States

Another effect beyond the Debye-Hückel correction is the signature of the internal partition functions. Nayfonov & Däppen (1998) discovered a "wiggle" in the thermodynamic quantities, located in the hydrogen and helium ionization zones. This effect, due to excited states, has probably already been observed in the Sun, because new observations (Basu *et al.* 1999) suggest that in the top 2% of the solar radius, MHD models can give a more accurate match with the data than OPAL models. Since it turns out that in this region, the discrepancy between MHD and OPAL is essentially reflected by the aforementioned wiggle (Nayfonov & Däppen 1998), the result of the inversion (Basu *et al.* 1999) could mean a validation of an MHD-like treatment (Hummer & Mihalas 1988) of exited states.

The main result of Basu *et al.* (1999) is shown in Figure 3. It is the result from an inversion of observed solar oscillation frequencies for the *intrinsic* γ_1 difference between the Sun and a solar model. The intrinsic difference is that part of the γ_1 difference which is due to the difference in the equation of state itself; there is a further component to the γ_1 difference caused by the change to the structure of the solar model resulting from the difference in the equation of state (Basu & Christensen-Dalsgaard 1997). The error bars shown in Figure 3 are based on combined errors of the inversion method and observational errors.

Figure 3 should not be over-interpreted, however. Present uncertainties in the inversion of the upper layers of the Sun, (*e.g.*, turbulent pressure, magnetic fields, nonlocal effects of radiation, uncertainties in the chemical composition) so far preclude a definitive interpretation, and further clarifying work is in progress. In the slightly deeper regions (below a depth of about 3% of the solar radius) the findings of the study (Basu *et al.* 1999) are more reliable, and they confirm the findings of Figure 1, that is, overall OPAL is a better equation of state than MHD.

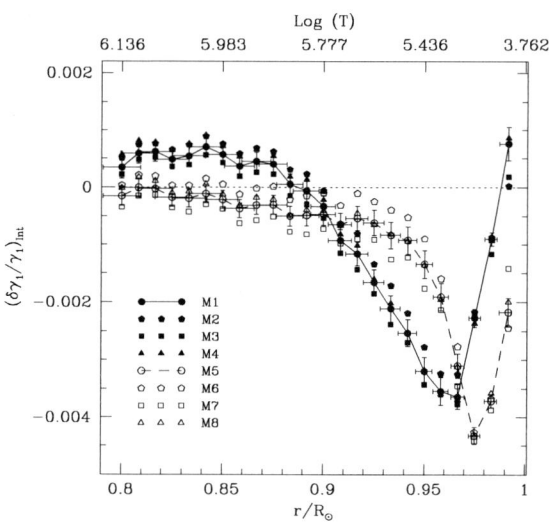

FIGURE 3. Intrinsic difference (see text) between γ_1 obtained from an inversion of helioseismological data (Basu *et al.* 1999) and 8 models [4 MHD models (M1-4, filled points); 4 OPAL models (M5-8, empty points)]. All results are in the sense 'Sun – model" (for more details about the models, see Basu *et al.* 1999).

However, should the results in the top 2% of the Sun remain in favor of MHD, they would demonstrate the significance of the different implementations of many-body interactions in the two formalisms. In principle, since density decreases in the upper part, OPAL, by its nature of a systematic expansion, inevitably becomes itself more accurate; but MHD might, by its heuristic approach (and by luck!), have incorporated even finer, higher-order effects. Since helioseismology gives localized information, it is natural that the various equations of state have their preferred regions in the Sun. One should, however, resist the temptation to produce a "combined" solar equation of state, with different pieces for different parts of the Sun. Such a hybrid solution is fraught with danger, because patching equations of state together can introduce spurious effects (Däppen *et al.* 1993). It is better to seek an improvement of individual equations of state, such as MHD and OPAL, in parallel and independently, guided by the progress of helioseismology.

GUIDELINES FOR FUTURE DEVELOPMENTS

- As evidenced by Figures 1 and 3, current equations of state (such as OPAL and MHD) appear to be quite satisfactory. However, improvements are still possible. One day, perhaps, further systematic expansions of a physical-picture formalism, such as OPAL, can reach agreement with helioseismological observations within the observational errors. However, such further expansions will be a formidable task, and the likelihood of success is difficult to estimate. The necessary effort has to be measured in terms of the complexity of current physical-picture formalisms. So far, only one group (the OPAL group at Livermore) has realized a practically useful formalism that satisfies the exacting demands of stellar modelers. However, even now, all OPAL results are still only available in the form of pre-computed tables, and they have been made for certain chemical compositions of astrophysical relevance (Rogers et al. 1996). And finally, the OPAL computer code is so far proprietary belonging exclusively to the Livermore group.

- Despite its limitations, OPAL fares better than MHD in most parts of the Sun. It makes therefore sense to retrofit chemical-picture formalisms, so that they better agree with OPAL. Liang & Däppen (2004) and Liang (2004) have successfully emulated the OPAL equation of state, for a simple hydrogen-only plasma under solar-interior conditions. Their work is based on the one hand on the PLPF and on the other hand on a continuum term borrowed from OPAL. Of course, such efforts need not stop there: because of their heuristic nature, chemical-picture formalisms can, in principle, be tuned to mimic any other formalism, or even observational results. There had been an earlier attempt the simulate certain aspects of the OPAL equation of state; it is the so-called SIREFF equation of state by Rogers, Swenson and Irwin, based on the Eggleton, Faulkner & Flannery (1973) (EFF) equation of state (for details see Däppen & Guzik 2000).

- In order to get an equation of state that can be superior to OPAL (but which will be less systematic and more intuitive), more generally, the chemical picture formalisms can be enriched with terms originating in the physical picture. The aforementioned OPAL emulator (Liang & Däppen 2004, Liang 2004) is an example for this strategy. Other efforts in this directions have also been made (Starostin et al. 2003, Starostin & Roerich 2004), and solar models have been computed. The results look promising (Gryaznov et al. 2004, Baturin 2004, Ayukov et al. 2004). Despite a lack of rigor in their physical foundation, such intuitive efforts are legitimate while one waits for a new generation of higher-order physical-picture equations of state.

- Among the rigorous efforts, there are pure density expansions, such as, for instance, the path-integral-based Feynman-Kac (FK) formalism of Alastuey & Perez (1992;1996) and Alastuey et al. (1994;1995) (see also Perez & Chabrier 2004). While density expansions converge only well in the case of nearly completely ionized plasmas, their systematic nature allows us, in principle at least, to find the higher-order terms needed to match the quality of the solar observations and inversions.

- Use of the PLPF has been very controversial for a long time (see Däppen et al. 1987). In the light of the helioseismic success of OPAL, it appears that the PLPF must be, in one or another form, the effective internal partition function for the calculation of pressure and all thermodynamic quantities (however, computing optical quantities is an altogether other matter, see Rogers 1986).

• It would be very helpful if one could find the scattering term for electrons *entirely from within the chemical picture*. If such a correction to the unperturbed free-electron states produced a PLPF-like effect, without recourse to the physical picture, it could convince the last remaining doubters of the PLPF (residing mainly in the astrophysical community). It could become universally accepted that "foreign" terms must be used in the chemical picture, that is terms, which can only be found and justified in the physical picture.

CONCLUSIONS

While currently available equations of state give reasonable accuracy for solar and stellar modelers, even the present observational data allow us to aim higher, that is, not only to strive for better solar models, but to advance the use of the Sun and stars as plasma-physics experiments, thus complementing on-going efforts in terrestrial laboratories.

ACKNOWLEDGMENTS

I am very grateful to my organizer colleagues of this workshop, Vladan Čelebonović and Douglas Gough, and to all participants, who collectively made such a fruitful and enjoyable program possible. I acknowledge the professional support and hospitality by the local program assistant Gerda Filippo and the executive manager Dr. Martje Kruk-de Bruin. But above all, had not Professor Wim van Saarloos approached Vladan Čelebonović three years ago, with the suggestion of such a workshop, it would not have happened. Wim's continuous encouragement and support had been crucial at several stages of the preparation of the workshop, and we all feel deeply indebted to him.
This work was supported by the grant AST-0307578 of the National Science Foundation.

REFERENCES

Alastuey, A. & Perez, A. 1992, *Europhys. Lett.*, **20**, 19
Alastuey, A. & Perez, A. 1996, *Phys. Rev. E*, **53**, 5714
Alastuey, A., Cornu, F. & Perez, A. 1994, *Phys. Rev. E*, **49**, 1077
Alastuey, A., Cornu, F. & Perez, A. 1995, *Phys. Rev. E*, **51**, 1725
Ayukov, S.V., Baturin, V.A., Gryaznov, V.K., Iosilevsky, I.L. & Starostin, A.N. 2004, Solar models using the SAHA-S equation of state, *these proceedings*.
Basu, S. & Christensen-Dalsgaard, J. 1997, *Astron. Astrophys.*, **322**, L5
Basu, S., Däppen, W. & Nayfonov, A. 1999, *Astrophys. J.*, **518**, 985
Baturin, V. 2004, Thermodynamics of the Convection Zone Through Adiabatic Exponent, *these proceedings*.
Berrington, K. A. 1997, *The Opacity Project*, vol. II, Institute of Physics Publishing. Bristol
Berthomieu, G., Cooper, A. J., Gough, D. O., Osaki, Y., Provost, J. & Rocca, A. 1980, in Hill, H. A., Dziembowski, W., eds., *Lecture Notes in Physics* **125**, Springer, Heidelberg, p. 307
Brüggen, M. & Gough, D.O. 1997, *Astrophys. J.*, **488**, 867
Brüggen, M. & Gough, D.O. 2000, *J. Math. Phys.*, **41**, 260-283
Christensen-Dalsgaard, J. 2002, *Rev. Mod. Phys.*, **74**, 1073

Christensen-Dalsgaard, J. 2004, An Introduction to Solar Oscillations and Helioseismology, *these proceedings*.
Christensen-Dalsgaard, J., Däppen, W. & Lebreton, L. 1988, *Nature*, **336**, 634
Christensen-Dalsgaard, J., Gough, D. O. & Toomre, J. 1985, *Science*, **229**, 923
Christensen-Dalsgaard, J., Däppen, W., and the GONG Team 1996, *Science*, **272**, 1286
Däppen, W., 1980, *Astron. Astrophys.*, **91**, 212-220
Däppen, W. & Guzik, J. A. 2000, in Ibanoglu C., ed., *Proc. Variable stars as essential astrophysical tools*, Kluwer, Dordrecht, p. 177
Däppen, W., Anderson, L., Mihalas, D., 1987, *Astrophys. J.*, **319**, 195-206
Däppen, W., Mihalas, D., Hummer, D. G. & Mihalas, B. W. 1988, *Astrophys. J.*, **332**, 261
Däppen, W., Gough, D. O., Kosovichev, A. G. & Rhodes, E. J., Jr. 1993, in Weiss W., Baglin A., eds., *Proc. IAU Symp. 137*, ASP Conf. Ser. **40** *Inside the Stars*, Astron. Soc. Pac., San Francisco, p. 304
Eggleton, P. P., Faulkner, J. & Flannery, B. P. 1973, *Astron. Astrophys.*, **23**, 325
Eliezer, S., Ghatak, A. & Hora, H. 1986, *An introduction to equations of state: theory and applications*, (Cambridge University Press)
Elliott, J. R. & Kosovichev, A. G. 1998, *Astrophys. J.*, **500**, L199
Gong, Z.-G., Däppen, W. & Zejda, L. 2001a, *Astrophys. J.*, **546**, 1178
Gong, Z.-G., Zejda, L., Däppen, W. & Aparicio J. M. 2001b, *Comp. Phys. Commun.*, **136**, 294
Gough, D. O. 1993, in Zahn, J.-P., Zinn-Justin, J., eds., *Astrophysical Fluid Dynamics*, North-Holland, Amsterdam, p. 399
Gough, D. O., Kosovichev, A. G., Toomre, J., and the GONG Team 1996, *Science*, **272**, 1296
Gryaznov, V.K., Ayukov, S.V., Baturin, V.A., Iosilevskiy, I.L., Starostin, A.N. & Fortov, V.E. 2004, SAHA-S model: Equation of State and Thermodynamic Functions of Solar Plasma, *these proceedings*.
Hummer, D. G. & Mihalas, D. 1988, *Astrophys. J.*, **331**, 794
Iglesias, C. A. and Rogers, F. J. 1996, *Astrophys. J.*, **464**, 943
Liang, A. 2004, Emulating the OPAL Equation of State in the Chemical-Picture Formalism, *these proceedings*.
Liang, A. & Däppen, W. 2004, in *"Helio- and Asteroseismology: Towards a Golden Future"*, SOHO14–GONG2004 Meeting held July 12-16 2004 at Yale University, New Haven, CT, USA (ESA Publications Division: Noordwijk, The Netherlands), in press.
Kraeft W. D., Kremp D., Ebeling W. & Röpke G. 1986 *Quantum Statistics of Charged Particle Systems*, (New York: Plenum)
Mihalas, D., Däppen, W. & Hummer, D. G. 1988, *Astrophys. J.*, **331**, 815
Nayfonov, A. & Däppen, W. 1998, *Astrophys. J.*, **499**, 489
Nayfonov, A., Däppen, W., Hummer, D.G. & Mihalas, D.M. 1999, *Astrophys. J.*, **526**, 451-464.
Noels, A., Scuflaire, R. & Gabriel, M. 1984, *Astron. Astrophys.*, **130**, 389
Perez, A. & Chabrier, G. 2004, Quantum statistical corrections on the enhancement factor in solar fusion reactions, *these proceedings*.
Rogers, F.J., 1986, *Astrophys. J.*, **310**, 723-728
Rogers, F.J. & Nayfonov, A. 2002, *Astrophys. J.*, **576**, 1064
Rogers, F. J., Swenson, F. J. & Iglesias, C. A. 1996, *Astrophys. J.*, **456**, 902
Saumon, D., Chabrier, G. & Van Horn, H.M. 1995, *Astrophys. J. Suppl. Ser.* **99**, 713
Seaton, M.J. 1995, *The Opacity Project* Vol. I, Institute of Physics Publishing. Bristol
Shaviv, G. 2004a, *Astron. Astrophys.*, **418**, 801
Shaviv, G. 2004b, The limit on mean field theories and nuclear screening in stellar plasmas, *these proceedings*.
Shaviv, N.J. & Shaviv, G. 1996, *Astrophys. J.*, **468**, 433
Starostin A.N. & Roerich, V.C. 2004, Corrected EOS of Weakly-Nonideal Hydrogen Plasmas without Mysteries, *these proceedings*.
Starostin A.N., Roerich, V.C. & More, R.M. 2003, *Contrib. Plasma Phys.*, **43**, 369
Trampedach, R. 2004, Improved phenomenological equation of state in the chemical picture, *these proceedings*.
Trampedach, R., Baturin, V.A. & Däppen, W. 2004, *Astrophys. J.*, in press.
Ulrich, R. K. 1982, *Astrophys. J.*, **258**, 404

An Introduction to Solar Oscillations and Helioseismology

Jørgen Christensen-Dalsgaard

Institut for Fysik og Astronomi, Aarhus Universitet, DK-8000 Aarhus C, Denmark, and High Altitude Observatory, National Center for Atmospheric Research, P.O. Box 3000, Boulder, CO 80307, USA

Abstract. Helioseismology offers unique possibilities for probing the detailed internal structure of a star and, in this way, constraining the physical properties of matter under stellar conditions. Here I provide a brief introduction to stellar structure and stellar oscillations, as well as to the techniques used in helioseismic analyses. In addition, I give a few examples of the results obtained from helioseismic investigations of solar structure.

1. INTRODUCTION

To learn about stellar interiors we need observables that reflect interior properties. Such information is provided, in a global sense, by the luminosity and surface temperature of a star, given that these quantities are determined by the stellar structure. Also, observations of the surface composition may provide some information about the internal properties, if the composition has been modified by mixing processes; in particular, surface abundances of lithium and beryllium may have been reduced by nuclear burning if sufficiently deep mixing has taken place during the evolution of the star. However, the relations between stellar structure, let alone the physics of the stellar interior, and these observables are rather indirect and themselves involve complex and somewhat uncertain physical processes. In contrast, frequencies of stellar pulsation are related in a simple way to the properties of the stellar interior; furthermore, they can be determined with extremely high precision. With a sufficiently rich spectrum of oscillations, which is available in several very different types of stars, it is possible to obtain information about subtle aspects of conditions in localized parts of the star, and hence about the physical processes responsible for these conditions.

The extreme example of this type of diagnostics is the Sun where thousands of frequencies of acoustic oscillations are known with very high accuracy. The modes have periods between around 3 and 15 minutes and spatial scales ranging from spherically symmetric oscillations to modes with surface wavelengths of a few thousand kilometers. The amplitudes of individual modes are tiny: up to around $20 \mathrm{cm s}^{-1}$ in velocity and a few parts per million in intensity. The variety of modes is such that the sound speed and other properties can be determined with high precision and resolution in much of the solar interior. Given the dependence of the sound speed on the thermodynamic state of the gas, particularly the adiabatic compressibility, this provides stringent constraints on the equation of state in the solar interior.

These aspects are discussed in detail in this volume. To set the scene the present paper provides an overview of solar modelling and of the observations and analysis that lead to the desired information about the thermodynamic properties of solar matter. Stellar modelling is discussed briefly in Section 2; for further details any of the large number of textbooks available, such as Kippenhahn & Weigert (1990), can be consulted. In Section 3 I present the basic properties of stellar acoustic oscillations, concentrating on how their frequencies are related to stellar properties; detailed descriptions of the theory of stellar oscillations was given by Unno *et al.* (1989) and Gough (1993). Observations of solar oscillations are considered briefly in Section 4, while Section 5 discusses a few of the results that have been obtained from helioseismic investigation of solar structure; there are several recent reviews on these issues (*e.g.* Christensen-Dalsgaard 2002). Finally, Section 6 discusses some of the problems and potentials of such investigations, including the extension to other stars with possibly more extreme physical conditions.

2. SOLAR MODELLING

It is obvious that stars, including the Sun, in principle are extremely complex physical systems, with a huge number of degrees of freedom. However, the description of stellar interiors are greatly simplified by the state of near equilibrium that is apparently satisfied at all levels, from the thermodynamic state to the overall energy budget of the star. The enormous disparity in scale between the size of the star and the mean free path of particles or photons in it ensures that local thermodynamical equilibrium is satisfied to a very high precision except in the stellar atmospheres, greatly simplifying the treatment of radiative transfer and nuclear reactions in stellar interiors and allowing the use of equilibrium thermodynamics. In general it enables a local description of the *microphysics*, *i.e.*, the properties of stellar matter and its interaction with radiation. On large scales, it seems that most stars are very nearly in hydrostatic and thermal equilibrium, with only slow changes with time as a result of the gradual depletion of nuclear fuel. These simplifications make it possible to construct simple, yet apparently fairly realistic, models of stellar interiors, an obvious prerequisite for the use of observations of stellar properties to study the underlying physics.

A remaining difficulty concerns possible hydrodynamical instabilities in stellar interiors, and other hydrodynamical effects such as might be induced, *e.g.*, by the rotation of stars. In most cases such effects are simply ignored, generally with little or no justification. An exception is convective instability, discussed below, which is certainly present in most stars with very significant effects on their structure and evolution. The commonly employed crude descriptions of the motion resulting from this instability is a serious uncertainty that must be taken into account in the interpretation and use of results of stellar modelling.

For future reference it is useful to discuss very briefly the relevant equations of stellar structure. I shall assume that the structure of the star is spherically symmetric and characterized by the distance r to the centre. This neglects the centrifugal force resulting from rotation, an excellent approximation for the slowly rotating Sun, and other effects that might lead to departures from spherical symmetry. Hydrostatic equilibrium requires a balance between the pressure gradient and gravity which may then be written as

$$\frac{dp}{dr} = -\frac{Gm\rho}{r^2}, \tag{1}$$

where p is pressure, ρ is density, m is the mass of the sphere contained within r, and G is the gravitational constant. Also, obviously,

$$\frac{dm}{dr} = 4\pi r^2 \rho. \tag{2}$$

The gradient of temperature T is determined by the requirements of energy transport, from the central regions where nuclear reactions take place to the surface where the energy is radiated. The temperature gradient is conventionally written as

$$\frac{dT}{dr} = \nabla \frac{T}{p} \frac{dp}{dr}, \tag{3}$$

where, obviously, $\nabla = d\ln T / d\ln p$. Its form depends on the mode of energy transport; for radiative transport

$$\nabla = \nabla_{\rm rad} \equiv \frac{3}{16\pi a \tilde{c} G} \frac{\kappa p}{T^4} \frac{L(r)}{m(r)}, \tag{4}$$

where κ is the opacity, L is the energy flow through the sphere of radius r, a is the radiation energy density constant and $\tilde{c}$ is the speed of light. The energy equation relates the energy generation to the energy flow and the change in the internal energy of the gas:

$$\frac{dL}{dr} = 4\pi r^2 \left[\rho\varepsilon - \rho \frac{d}{dt}\left(\frac{e}{\rho}\right) + \frac{p}{\rho}\frac{d\rho}{dt}\right]; \tag{5}$$

here ε is the rate of nuclear energy generation[1] per unit mass and unit time, e is the internal energy per unit volume and t is time.[2] Finally, we need to consider the rate of change of the composition, which controls stellar evolution. In a main-sequence star such as the Sun the dominant effect is the burning of hydrogen; however, we must also take into account the changes in composition resulting from diffusion and settling. The rate of change of the abundance X by mass of hydrogen is therefore given by

$$\frac{\partial X}{\partial t} = \mathscr{R}_{\rm H} + \frac{1}{r^2 \rho}\frac{\partial}{\partial r}\left[r^2 \rho \left(D_{\rm H} \frac{\partial X}{\partial r} + V_{\rm H} X \right) \right], \tag{6}$$

where $\mathscr{R}_{\rm H}$ is the rate of change resulting from nuclear reactions, $D_{\rm H}$ is the diffusion coefficient and $V_{\rm H}$ is the settling velocity.

To these basic equations we must add the description of the microphysics, most naturally, given the form of the equations, in terms of $(p, T, \{X_i\})$, where X_i are the

---

[1] reduced for the emission of neutrinos which escape the star and hence do not contribute to the energy budget.
[2] For a star evolving in near thermal equilibrium the terms in the time derivatives are small.

abundances of the relevant elements. The determination of ρ, u and other required thermodynamical variables follows from the equation of state, extensively discussed elsewhere in this volume. The opacity (describing the absorption and scattering of radiation by the constituents of the gas) is most naturally obtained as $\kappa = \kappa(\rho, T, \{X_i\})$. Evidently the calculation of the opacity requires knowledge of the relevant atomic parameters; however, the opacity also depends crucially on the thermodynamic state of the gas, through the occupation of atomic ionization and excitation levels and the perturbation of atomic states by neighbouring particles. Similarly, the rates of nuclear reactions depend not only on the nuclear parameters but also, as discussed by Shaviv in this volume, on the interactions between the particles in the gas which leads to partial screening of the repulsive Coulomb potential between the nuclei. Also, calculation of the diffusion and settling coefficients depends on the thermodynamic state.

I have so far ignored the convective instability. This sets in if the density decreases more slowly with position than for an adiabatic change, *i.e.*,

$$\frac{d\ln\rho}{d\ln p} < \frac{1}{\gamma_1}, \qquad (7)$$

where $\gamma_1 = (\partial \ln p / \partial \ln \rho)_{ad}$, the derivative being taken for an adiabatic change. In stellar modelling this condition is often replaced by

$$\frac{d\ln T}{d\ln p} \equiv \nabla > \nabla_{ad} \equiv \left(\frac{d\ln T}{d\ln p}\right)_{ad} \qquad (8)$$

which is equivalent in the case of a uniform composition.[3] Thus a layer is convectively unstable if the radiative gradient ∇_{rad} (cf. Eq. 4) exceeds ∇_{ad}. In this case convective motion sets in, with hotter gas rising and cooler gas sinking, both contributing to the energy transport towards the surface. The condition that the combined radiative and convective energy transport through a surface of radius r match the luminosity then in principle defines a condition for the average temperature gradient which can be written as

$$\nabla = \nabla_{conv}(\rho, T, L, \ldots). \qquad (9)$$

In practice this relation depends on the details of the convective flow which is very likely turbulent and represents conditions over a range of positions in the star; also, motion is inevitably induced outside the immediate unstable region.

Substantial progress has been made towards the modelling of time-dependent convection in the near-surface regions (*e.g.* Stein & Nordlund 1989, 1998; Robinson *et al.* 2003). However, for the general treatment of convection in stellar modelling a simpler prescription is needed. In the so-called mixing-length models this is typically based on rough models of the convective eddies, characterized by a length scale ℓ usually taken as a multiple $\alpha_{ML} H_p$ of the pressure scale height H_p (*e.g.* Böhm-Vitense 1958). This can be generalized to include non-local effects (*e.g.* Spiegel 1963; Gough 1977; Balmforth

---

[3] For the complications arising when composition is not uniform, see for example Kippenhahn & Weigert (1990).

1992a); treatments that take some aspects of the spectrum of turbulence into account have also been developed (*e.g.* Canuto & Mazzitelli 1991). A general feature of these treatments of convection, confirmed by the hydrodynamical simulations, is that except in a thin region near the surface the resulting temperature gradient exceeds the adiabatic value by a very small amount,[4]

$$\nabla_{\text{conv}} \simeq \nabla_{\text{ad}} = \nabla_{\text{ad}}(p, T, \{X_i\}) \,, \tag{10}$$

emphasizing that ∇_{ad} is a thermodynamical quantity.

The Sun is unique amongst stars in that we have fairly accurate measurements of its mass, radius and energy output, as well as a precise age inferred from the radioactively determined ages of meteorites. Furthermore spectral analysis has provided determinations of the solar surface abundances, although, as indicated in Section 5, spectroscopic analysis of the solar atmospheric abundance is complicated by uncertainties in the modelling of the solar atmosphere; additional information about abundances of refractory elements comes from the analysis of the composition of meteorites (*e.g.* Anders & Grevesse 1989). An important exception is helium: despite the fact that helium was first detected in spectral lines arising in the higher parts of the solar atmosphere, conditions in this region are so uncertain that no precise measurement of the helium abundance can be made from spectroscopy. Since hydrogen and helium are by far the most abundant elements the composition of stellar matter is typically characterized by the abundances by mass of hydrogen, denoted X, helium (Y) and elements heavier than helium (Z) with, obviously, $X + Y + Z = 1$. However, as discussed later the relative composition of these 'heavy' elements is also important, particularly for the determination of the opacity.

Models of the Sun should obviously be consistent with the measurements of surface radius, luminosity and composition. In practice this is achieved by evolving a model of solar mass (assuming that mass loss since the formation of the Sun has been negligible; see, however, Sackmann & Boothroyd 2003) to the present age of the Sun. To obtain the right radius, luminosity and present surface composition three parameters are adjusted. One of them, *e.g.* the constant α_{ML} in the mixing-length description of convection, determines the efficacy of convection and hence the superadiabatic gradient $\nabla - \nabla_{\text{ad}}$ required for energy transport in the thin significantly superadiabatic gradient of the convection zone. The initial helium abundance Y_0, which is not constrained from spectroscopic measurements, is adjusted to match the observed surface luminosity, and the initial heavy-element abundance Z_0 is chosen to obtain the observed ratio $(Z/X)_s$ between the surface heavy-element and hydrogen abundances.

The calibration of the convective treatment serves to determine the adiabat of the convection zone which in turn ensures that the model has the correct radius. It follows that, after calibration, models computed with different treatments of convection yield essentially the same structure except in the thin region of significant superadiabaticity. This is illustrated in Fig. 1. The fact that the adiabat of the bulk of the solar convection

---

[4] Crudely speaking this follows from the fact that the internal energy density is so high, except near the surface, that only a very small excess in the energy density and hence in temperature above the adiabatic value suffices for the required energy transport.

A second condition is formally obtained by assuming that there is no pressure perturbation at the perturbed surface of the star, *i.e.*,

$$\delta p = p' + \xi_r \frac{dp}{dr} = 0 \quad \text{at } r = R, \tag{33}$$

In practice, the 'surface' must be defined as the outermost point of the model, which is obviously somewhat arbitrary. Thus more sophisticated conditions are used in calculations of stellar oscillations. These show that trapped modes are restricted to have frequencies below the acoustical cut-off frequency ω_{ac} (*cf.* Eq. 40); at frequencies well below this value Eq. (33) is still approximately satisfied. In any case, as discussed below, the properties of the oscillations in the superficial layers remain a serious problem in the computation of solar oscillation frequencies.

The system of differential equations together with the boundary conditions has nontrivial solutions only for discrete eigenvalues. The equations and boundary conditions are independent of the azimuthal order m and so, therefore are the eigenfrequencies. This follows obviously from the assumed spherical symmetry of the equilibrium structure: the definition of m depends on the the orientation of the axis of the spherical harmonics which can have no physical effect for a spherically symmetric star.[5] Computed frequencies for a solar model, as functions of the degree l are illustrated in Fig. 2. The observed frequencies in the Sun, greater than around $1000\,\mu$Hz, clearly correspond to acoustic modes of fairly high order or high degree or, at high degree, to surface gravity (f) modes.

From Eqs (33), (1) and (25), neglecting Φ',[6] it follows that

$$\frac{\xi_h(R)}{\xi_r(R)} = \frac{g_s}{R\omega^2}, \tag{34}$$

where g_s is the surface gravity. For modes of high frequency, typical of those observed in the Sun, and low or moderate degree, this shows that the motion in the atmosphere is predominantly in the radial direction.

3.2. Excitation of Solar Oscillations

The description of solar oscillations made so far evidently provides no information about the origin of the observed solar oscillations: given the adiabatic approximation and suitable outer boundary conditions the modes preserve energy and hence neither grow nor decay in amplitude. A more complete treatment would include the energy equation and would then determine whether the modes are stable or unstable; formally,

---

[5] Rotation introduces a frequency splitting according to m which has been used to obtain detailed information about solar internal rotation and its variation with time. For a recent review, see Thompson *et al.* (2003).

[6] This so-called *Cowling approximation* (Cowling 1941) is quite reasonable for high-order or high-degree modes, such as observed in the Sun.

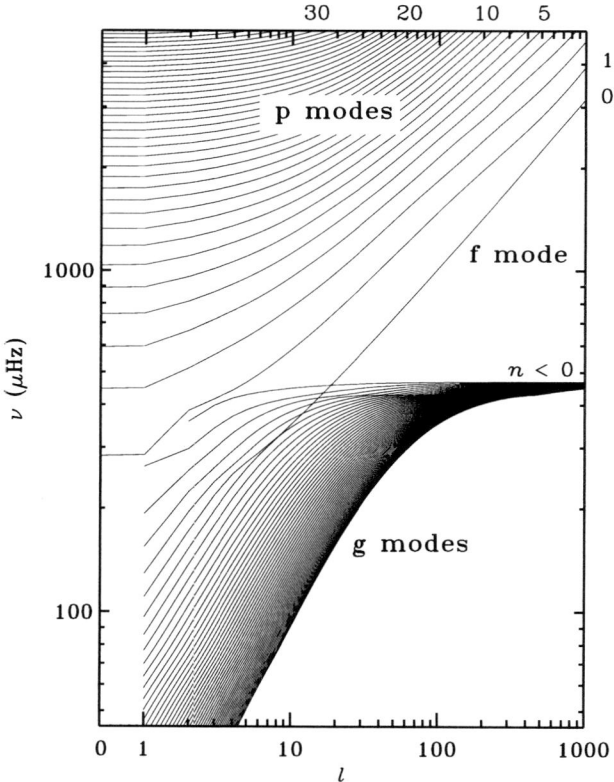

FIGURE 2. Computed frequencies for a model of the present Sun, as functions of degree l. For clarity, the points corresponding individual degrees have been connected; although the abscissa is generally logarithmic, the left-hand edge corresponds to radial modes, with $l = 0$. The radial order n is indicated for some of the modes; conventionally $n > 0$ is used for acoustic (or p) modes, $n = 0$ for the so-called f modes which at moderate and high degree have the character of surface gravity waves, and $n < 0$ for internal gravity (or g) modes. The order is defined such that, in most cases, $|n|$ gives the number of nodes in the radial direction.

the eigenfrequencies ω would then be complex, with positive or negative imaginary parts corresponding to growing or decaying modes, respectively. The treatment needs to take into account the effects of convection on the perturbation in the flux; an additional important fact is the perturbation in the turbulent pressure, whose phase may be such as to damp or drive the oscillation. The computation of such time-dependent effects of convection is highly uncertain; however, using a time-dependent version of a non-local mixing-length description Balmforth (1992a) concluded that the modes observed in the Sun are likely damped.

If the modes are indeed damped, they must be driven by external forcing. A likely source is the near-surface convection which reaches close to sonic speed and hence is an efficient emitter of acoustic noise (Lighthill 1952; Stein 1967). This essentially stochastic source of sound waves couples to the normal modes of the Sun and excites them to the observed amplitudes. A stochastic excitation process is consistent with the observed statistical properties of the amplitude distribution (*e.g.* Chaplin *et al.* 1997; Chang & Gough 1998).

An early analysis of the stochastic excitation of solar modes was carried out by Goldreich & Keeley (1977). A more detailed analysis, although still based on a mixing-length description of convection, was provided by Balmforth (1992b). It was pointed out by Brown (1991) that since the acoustic emission depends on the turbulent velocity to a high power the excitation is probably fairly localized and confined to a small volume near the top of the convection zone. Observational evidence for such localized excitation was obtained by, for example, Goode, Gough & Kosovichev (1992) and Rimmele *et al.* (1995) who found an association between the acoustic exciting events and the dark intergranular lanes associated with downdrafts. A detailed model of the excitation caused by such downdrafts was developed by Rast (1999). Also, hydrodynamical simulations have demonstrated that convection provides the required energy input to the modes, determined observationally from the amplitude and width of the peaks on the observed power spectra[7] (*e.g.* Stein & Nordlund 2001; Stein *et al.* 2004). Since this driving mechanism operates on a small spatial scale and over a range of timescales, it is expected to yield a spectrum of oscillations covering a broad range of degrees and frequencies, as observed.

The properties of the excitation are reflected in the detailed shapes of the observed peaks in the oscillation power spectra, which show distinct asymmetries (Duvall *et al.* 1993), related to the location of the source (see also, for example, Gabriel 1993; Abrams & Kumar 1996; Roxburgh & Vorontsov 1997; Nigam & Kosovichev 1998; Rast & Bogdan 1998; Rosenthal 1998). Analyses of this asymmetry have confirmed that the source of the excitation is indeed confined to the upper part of the convection zone (*e.g.* Kumar & Basu 1999, 2000; Nigam & Kosovichev 1999; Chaplin & Appourchaux 1999). Additional information about the properties of the acoustic sources has been obtained from observations of resonances, or 'pseudo-modes' at frequencies above the acoustical cut-off frequency (*cf.* Eq. 40) (*e.g.* Kumar *et al.* 1990; Kumar & Lu 1991; Kumar 1994; Kumar *et al.* 1994).

The stochastic excitation by convection is expected to operate in any star with vigorous near-surface convection. Very accurate observations over the past few years have in fact found oscillations in a number of stars with outer convection zones, including also subgiants and giants (for a recent brief review, see Bedding & Kjeldsen 2003).

3.3. The Dependence of Frequencies on Solar Properties

From the point of view of investigating solar properties based on the oscillation frequencies the dependence of the frequencies on solar structure is obviously crucial.

---

[7] The spectral width of a damped and stochastically excited oscillator is determined by the damping rate of the undriven oscillator (*e.g.* Batchelor 1956; Christensen-Dalsgaard, Gough & Libbrecht 1989).

This is implicitly given by the coefficients in Eqs (27) – (29). However, these are constrained by the fact that the equilibrium model satisfies the equations of stellar structure. In fact, given the density distribution $\rho(r)$ the mass distribution $m(r)$ can be obtained from Eq. (2); the pressure distribution $p(r)$ then follows by integration of Eq. (1). If in addition $\gamma_1(r)$ is given, it is easy to show that all required coefficients can be determined. Thus, from the point of view of computing adiabatic oscillation frequencies, the model is completely specified by specifying $\rho(r)$ and $\gamma_1(r)$. It follows that analysis of the oscillation frequencies can at most give information about these 'dynamical' variables. In particular, no direct information is obtained about temperature; of course, if an equation of state is assumed and if the composition is given, the temperature can be inferred from p, ρ and the composition. Other pairs of functions equivalent to (ρ, γ_1) can be substituted; since the observed modes are mostly acoustic modes their properties are to a large extent determined by the adiabatic sound speed, suggesting the pair (c^2, ρ) which is often used.

Analysis of the observed frequencies show that present solar models are likely to be relatively close to the true solar structure. In this case changes in the computed frequencies can be obtained by linearizing the oscillation equations around a reference solar model. The analysis is greatly simplified by noting that the adiabatic oscillation frequencies satisfy a variational principle (Chandrasekhar 1964). We let $\delta\omega_{nl} = \omega_{nl}^{(\mathrm{obs})} - \omega_{nl}^{(\mathrm{mod})}$ be the difference between the observed frequencies and the frequencies of the reference model; also, we characterize solar structure by the pair (c^2, ρ) and let $\delta_r c^2 = c^2 - c_{\mathrm{mod}}^2$ and $\delta_r \rho = \rho - \rho_{\mathrm{mod}}$ be the differences between the solar and model values of c^2 and ρ, at fixed r. Then we can express the frequency differences as

$$\frac{\delta\omega_{nl}}{\omega_{nl}} = \int_0^R \left[K_{c^2,\rho}^{nl}(r) \frac{\delta_r c^2}{c^2}(r) + K_{\rho,c^2}^{nl}(r) \frac{\delta_r \rho}{\rho}(r) \right] dr + Q_{nl}^{-1} \mathscr{G}(\omega_{nl}), \qquad (35)$$

(*e.g.* Dziembowski, Pamyatnykh & Sienkiewicz 1990; Gough & Thompson 1991), where the *kernels* $K_{c^2,\rho}^{nl}$ and K_{ρ,c^2}^{nl} are determined by the eigenfunctions computed for the reference model.

In Eq. (35) the last term describes the effects of near-surface errors in the frequency calculation. The expression in terms of the kernels assumes that the frequencies of solar oscillation are correctly described by the assumed adiabatic pulsation equations. In fact, in the near-surface layers, where the thermal timescale is comparable to the pulsation period, the adiabatic approximation certainly breaks down. Also, the frequencies may be affected by other uncertain aspects of the physics of the model and the oscillations, not least the treatment of convection and its interaction with the pulsations. It may be shown that these effects, at least for modes of low or moderate degree, can be described by the form indicated, where $\mathscr{G}(\omega)$ is a function of frequency alone that depends on the unknown physical processes. Q_{nl} is a simple scaling that depends on the inertia of the modes, normalized to be unity for radial modes; it compensates for the fact, discussed below, that higher-degree acoustic modes occupy a smaller fraction of the solar interior and hence are more strongly affected by near-surface errors (*e.g.* Christensen-Dalsgaard & Berthomieu 1991).

It is evident that γ_1 is determined by the equation of state as $\gamma_1 = \gamma_1(p,\rho,Y,Z)$ where, as before, we characterize the composition by the abundances X, Y and Z of hydrogen, helium and heavier elements. Furthermore, using that $c^2 = \gamma_1 u$, where $u = p/\rho$, and utilizing the equations of stellar structure, Eq. (35) can be written as

$$\frac{\delta\omega_{nl}}{\omega_{nl}} = \int_0^R K_{u,Y}^{nl}(r) \frac{\delta_r u}{u}(r) dr + \int_0^R K_{Y,u}^{nl}(r) \delta_r Y(r) dr$$
$$+ \int_0^R K_{c^2,\rho}^{nl}(r) \left(\frac{\delta\gamma_1}{\gamma_1}\right)_{\text{int}} dr + Q_{nl}^{-1} \mathcal{G}(\omega_{nl}) \qquad (36)$$

(Basu & Christensen-Dalsgaard 1997), where for simplicity I neglected the effect of the difference in Z; here $(\delta\gamma_1/\gamma_1)_{\text{int}}$ is the *intrinsic* difference between the solar and the model equation of state, *i.e.*, the difference at fixed p, ρ and composition. Evidently, constraints on $(\delta\gamma_1/\gamma_1)_{\text{int}}$ would be an important contribution to the investigations of the equation of state through helioseismology.

A great deal of insight into the properties of the solar modes can be obtained from a very simple asymptotic analysis. The modes are predominantly standing sound waves which locally satisfy the dispersion relation

$$\omega^2 = c^2 |\mathbf{k}|^2 = c^2 \left[k_r^2 + \frac{l(l+1)}{r^2} \right] ; \qquad (37)$$

here $\mathbf{k}$ is the local wave vector which I have separated into a radial component $k_r \mathbf{a}_r$ and a horizontal component, the length k_h of which satisfies Eq. (22). It follows that k_r is given by

$$k_r = \left[\frac{\omega^2}{c^2} - \frac{l(l+1)}{r^2} \right]^{1/2} , \qquad (38)$$

for $r \geq r_t$, where the *inner turning point* radius r_t is determined by

$$\frac{c(r_t)}{r_t} = \frac{\omega}{\sqrt{l(l+1)}} ; \qquad (39)$$

at $r = r_t$ $k_r = 0$ and the waves travel horizontally, corresponding to a total internal reflection at this point, and below it the wave is evanescent, decaying exponentially.

This simple description does not provide reflection of the waves near the solar surface. This takes place where the wavelength becomes comparable with the density scale height H and where therefore the approximation of plane sound waves in a locally uniform medium no longer holds. It may be shown, from a simple analysis of waves in an isothermal atmosphere or from a more complete asymptotic analysis (*e.g.* Deubner & Gough 1984), that reflection takes place only when the frequency ω is below the *acoustical cut-off frequency* ω_{ac}; in the case of an isothermal atmosphere it is given by

$$\omega_{ac} = \frac{c}{2H} . \qquad (40)$$

In the solar case this corresponds to a cyclic frequency of around 5.3 mHz. Waves with higher frequencies can propagate out through the solar atmosphere, hence loosing energy and being strongly damped. Thus the acoustical cut-off frequency provides an upper limit to the frequencies of standing waves in a star.

The condition for a standing wave must be that the phase change in the radial direction between r_t and the near-surface reflection is an integer times π, with a possible correction for the phase changes at the surface and the inner turning point; thus

$$\int_{r_t}^{R} k_r \, dr = (n + \alpha)\pi \,, \tag{41}$$

where α accounts for the phase change at the surface (and at the inner turning point). Using Eq. (38) it follows that the frequencies satisfy

$$\int_{r_t}^{R} \left(1 - \frac{L^2 c^2}{\omega^2 r^2}\right)^{1/2} \frac{dr}{c} = \frac{[n + \alpha(\omega)]\pi}{\omega} \,, \tag{42}$$

where $L = \sqrt{l(l+1)}$ and I have made explicit that the surface phase change in general depends on frequency. A relation of this form, known as the *Duvall law*, was first obtained for the solar oscillation frequencies on the basis of observational data by Duvall (1982). Although derived here on the basis of highly simplified arguments, it can be obtained from a more rigorous asymptotic analysis of the full oscillation equations (*e.g.*, Gough 1993).

As discussed in connection with Eq. (35) it is useful to analyse the observed frequencies in terms of differences relative to a reference model. Thus we can linearize Eq. (42) in terms of a correction $\delta_r c$ to the sound speed and a correction $\delta \alpha$ to the phase function. The result can be written

$$\mathscr{S}_{nl} \frac{\delta \omega_{nl}}{\omega_{nl}} \simeq \mathscr{H}_1\left(\frac{\omega_{nl}}{L}\right) + \mathscr{H}_2(\omega_{nl}) \,, \tag{43}$$

where

$$\mathscr{S}_{nl} = \int_{r_t}^{R} \left(1 - \frac{L^2 c^2}{r^2 \omega_{nl}^2}\right)^{-1/2} \frac{dr}{c} - \pi \frac{d\alpha}{d\omega} \,, \tag{44}$$

$$\mathscr{H}_1(w) = \int_{r_t}^{R} \left(1 - \frac{c^2}{r^2 w^2}\right)^{-1/2} \frac{\delta_r c}{c} \frac{dr}{c} \,, \tag{45}$$

and

$$\mathscr{H}_2(\omega) = \frac{\pi}{\omega} \delta \alpha(\omega) \tag{46}$$

(Christensen-Dalsgaard, Gough & Pérez Hernández 1988). These expressions are clearly quite similar to Eq. (35) although, in this simple asymptotic description, there is no dependence on density. In particular, the effect of the phase difference, in $\mathscr{H}_2(\omega)$ corresponds to the term $\mathscr{G}(\omega)$ included in the non-asymptotic expression. Observationally, an expression of the functional form in Eq. (43) can be fitted to differences between observed and model frequencies, to determine the functions $\mathscr{H}_1$ and $\mathscr{H}_2$ (Christensen-Dalsgaard, Gough & Thompson 1989).

A given mode is essentially only sensitive to solar conditions outside its inner turning point, *i.e.* for $r \geq r_t$. For the observed range of degrees, from 0 to more than 1000, and frequencies in the vicinity of 3 mHz r_t ranges from the solar centre to very near the solar surface. It is to a large extent this variation in penetration depth and sensitivity which allows the observed frequencies to be combined in such a way as to provide localized information about the solar interior. An explicit illustration of this follows from the Duvall law or its differential form, Eqs (42) or (45). These are essentially in the form of Abel integral equations which may be solved for c, or $\delta_r c/c$, if the right-hand side of Eq. (42) or $\mathcal{H}_1$ have been determined from observations (*e.g.* Gough 1984; Christensen-Dalsgaard *et al.* 1985; Christensen-Dalsgaard *et al.* 1989; Vorontsov & Shibahashi 1991; Marchenkov, Roxburgh & Vorontsov 2000; Vorontsov, this volume).

The information content in the oscillation frequencies goes beyond such simple asymptotic descriptions, however. As discussed by Houdek (this volume) an interesting example is the effect of relatively sharp features, relative to the wavelength of the modes, in the equilibrium structure, which give rise to a signal in the frequencies determined by the phase of the modes at the features. The localized decrease in γ_1 associated with the second ionization of helium is one such feature, and the resulting signal may be used to determine the helium abundance in the convective envelope (*e.g.* Vorontsov *et al.* 1991; Pérez Hernández & Christensen-Dalsgaard 1994; Basu & Antia 1995).

3.4. Helioseismic Inversion

Although asymptotic techniques are powerful, intuitive and efficient, they obviously do not make the fullest possible use of the information contained in the oscillation frequencies. Similarly, the inferences based on sharp features discussed by Houdek (this volume), while very powerful, are also limited to specific aspects of the solar interior. More general inferences that do not depend on the asymptotic properties of the modes can be made through inverse analyses of linearized relations such as Eq. (35). This is discussed in detail by Gough & Thompson (1991) and Gough (1996). Here I summarize some important features. Further detail on inversion procedures and results is provided by Vorontsov (this volume).

The goal of the analysis is to obtain localized information about the corrections to the solar model, taking into account also the errors in the data. To be specific, I consider inversion to determine $\delta_r c^2/c^2$. Most inversion techniques effectively correspond to making linear combinations of the differences $\delta\omega_{nl}/\omega_{nl}$; the result, attempting to infer the solution at some point $r = r_0$, can be written

$$\begin{aligned}\left\langle \frac{\delta_r c^2}{c^2} \right\rangle (r_0) &= \sum_i c_i(r_0) \frac{\delta\omega_i}{\omega_i} \\ &= \int_0^R \mathcal{K}_{c^2,\rho}(r_0,r) \frac{\delta_r c^2}{c^2}(r) \mathrm{d}r + \int_0^R \mathcal{C}_{\rho,c^2}(r_0,r) \frac{\delta_r \rho}{\rho}(r) \mathrm{d}r \\ &\quad + \sum_i c_i(r_0) Q_i^{-1} \mathcal{G}(\omega_i) + \sum_i c_i(r_0) \varepsilon_i \,, \end{aligned} \quad (47)$$

where for simplicity I have used i to label the modes rather than nl. Here ε_i are the errors in the observed $\delta\omega_i/\omega_i$, assumed uncorrelated, with standard deviations σ_i. Also,

$$\mathcal{K}_{c^2,\rho}(r_0,r) = \sum_i c_i(r_0) K^i_{c^2,\rho}(r) \tag{48}$$

is the so-called *averaging kernel*, usually normalized such that $\int \mathcal{K}_{c^2,\rho}(r_0,r)dr = 1$, and

$$\mathcal{C}_{\rho,c^2}(r_0,r) = \sum_i c_i(r_0) K^i_{\rho,c^2}(r) \tag{49}$$

is the *cross-term kernel*. The goal of the inversion procedure is to determine the *inversion coefficients* $c_i(r_0)$, either explicitly or implicitly, such that $\mathcal{K}(r,r_0)$ is localized near $r = r_0$ and the remaining terms on the right-hand side of Eq. (47) are small. In this case the combination of the data provides a localized average of the sound-speed correction $\delta_r c^2/c^2$ near r_0.

The inversion can be carried out through a least-squares fit of parametrized versions of the functions $\delta_r c^2/c^2$, $\delta\rho/\rho$ and $\mathcal{G}$ in Eq. (35), using suitable regularization to reduce the errors in the solution and suppress unphysical variations (*e.g.* Dziembowski *et al.* 1990). From this fit the inversion coefficients can be determined (for the simpler case of rotational inversion, see for example Christensen-Dalsgaard, Schou & Thompson 1990). However, most inversions for solar structure have used the so-called optimally localized averages techniques, where the coefficients $c_i(r_0)$ are explicitly determined. As an example, I consider the Subtractive Optimally Localized Averages (SOLA) technique, originally developed by Pijpers & Thompson (1992, 1994) for rotational inversion. Here the coefficients are determined by matching $\mathcal{K}_{c^2,\rho}(r,r_0)$ to a suitably chosen target function $\mathcal{T}(r,r_0)$, while keeping the other contributions small. Specifically, one minimizes

$$\int_0^R \left[\mathcal{K}_{c^2,\rho}(r_0,r) - \mathcal{T}(r_0,r)\right]^2 dr + \beta \int_0^R \mathcal{C}_{\rho,c^2}(r_0,r)^2 dr + \mu \sum_i \sigma_i^2 c_i(r_0)^2 , \tag{50}$$

subject to

$$\sum_i c_i(r_0) Q_i^{-1} \psi_\lambda(\omega_i) = 0 , \lambda = 0,\ldots,\Lambda , \tag{51}$$

for a suitably chosen set of functions ψ_λ. In the expression (50) the first term serves to ensure that $\mathcal{K}$ is close to $\mathcal{T}$ while the second and third term control the contribution from $\delta\rho/\rho$ and the observational errors to the solution. The relative weight given to these terms is controlled by the *trade-off* parameters β and μ: if they are increased, the contributions from the unwanted terms are reduced, albeit generally at the expense of the match to the target function or more generally the resolution in the inversion. The additional constraints in Eqs (51) serve to suppress the surface contributions (*e.g.* Däppen *et al.* 1991; Kosovichev *et al.* 1992): since $\mathcal{G}$ is generally a slowly varying function of frequency the contribution is suppressed if the ψ_λ, *e.g.* chosen to be low-order polynomials, span such functions. This technique is generally successful in providing localized information about solar structure in most of the solar interior, with the exception of the innermost 5 – 10 per cent of the radius and the region very near the surface. Since the radial wavelength of the eigenfunctions, and hence in some sense the resolution of the kernels, scale as the sound speed c (*e.g.* Thompson 1993), it is common to choose the target functions as Gaussians with a width proportional to c. The determination of the scaling factor, and the other parameters (β and μ) characterizing the inversion is an important part of the procedure (for details, see for example Rabello-Soares, Basu & Christensen-Dalsgaard 1999).

The SOLA technique can be immediately generalized to determine, for example, the intrinsic correction $(\delta\gamma_1/\gamma_1)_{\rm int}$ from Eqs (36) (Basu & Christensen-Dalsgaard 1997). An important limitation for the investigations of the equation of state, however, is the assumption that the surface term is just a function of frequency; this breaks down for the high-degree modes that are particularly interesting for the study of the ionization zones of hydrogen and helium (*e.g.* Antia 1995; Gough & Vorontsov 1995). A technique to include corrections to the surface term in the inversion was developed by Di Mauro *et al.* (2002); with modes of degree as high as 1000 it is possible to obtain high resolution throughout the ionization zones of helium (see also Rabello-Soares *et al.* 2000).

4. OBSERVATION OF SOLAR OSCILLATIONS

Solar oscillations have been observed with high precision using a variety of techniques; a detailed review of observing and data analysis techniques was given by Brown (1996). The most extensive data have been obtained by means of observations of the line-of-sight velocity, using the Doppler shift; with suitable instrumentation it is possible to obtain *Doppler images* of the solar surface, such that each point of the images provides a measure of the surface velocity. Very extensive data have been obtained with the GONG[8] 6-station network of observing stations, suitably placed around the World (Harvey *et al.* 1996), and from instruments on the SOHO[9] satellite, including the Michelson Doppler Imager (MDI) (Scherrer *et al.* 1995). Additional data of very high precision on low-degree modes have been obtained by observing the Sun as a star, in light integrated over the solar disk, with the BiSON (Chaplin *et al.* 1996) and IRIS (Fossat 1991) groundbased networks as well as with the GOLF instrument (Gabriel *et al.* 1997) on SOHO.

Modes are excited in the Sun at degrees ranging from 0 to more than 1000, and over a range of frequencies. Thus the observed signal is a superposition of literally millions of modes, each with a spatial dependence on the solar surface corresponding to the projection of the displacement (Eq. 24) on the line of sight and with different frequencies. To determine the individual frequencies these contributions must be separated. The first step of the analysis is to carry out a spatial projection of the signal with suitable weight functions, aiming to isolate the component corresponding to given spherical-harmonic parameters (l_0, m_0). Had observations been available over the entire solar surface this could in principle have been achieved using a weight function based on $Y_{l_0}^{m_0}$, given the orthogonality of the spherical harmonics. With data for only slightly less than one solar

---

[8] Global Oscillation Network Group
[9] SOlar and Heliospheric Observatory

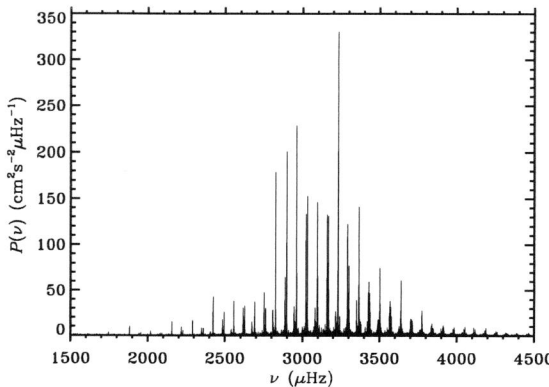

FIGURE 3. Power spectrum of solar oscillations, obtained from Doppler observations in light integrated over the disk of the Sun. The frequency is expressed in terms of cyclic frequency $\nu = \omega/2\pi$. The ordinate is normalized to show velocity power per frequency bin. The data were obtained from six observing stations and span approximately four months. (See Elsworth *et al.* 1995.)

hemisphere such complete separation is not possible and the result of the projection will contain components from other (l,m) in the vicinity of (l_0, m_0). By applying the projection to a series of Doppler images one obtains a timeseries which may then be Fourier analysed in time to isolate the individual modes.

The actual analysis is greatly complicated by the dense spectrum of modes, including the cross-talk between different spherical harmonics, and by the stochastic nature of the excitation and the finite lifetime of the modes. However, efficient and reliable techniques have been developed to determine the mode properties, and comparison of independent datasets and analysis methods have provided some confidence that reasonably consistent results on solar structure can be obtained (*e.g.* Basu *et al.* 2003). Even so, as discussed by Vorontsov (this volume) further work is needed to reduce remaining systematic errors in the results, which are becoming increasingly important as the helioseismic investigations are pushed towards ever finer aspect of the physics of the solar interior.

Some properties of the observed modes are illustrated in Figure 3, based on observations in integrated light with the BiSON network and hence restricted to modes of degree $l \leq 3$. The general distribution of mode power with frequency is common to all low and moderate degrees; the maximum amplitude for an individual mode is around 15 $\mathrm{cm\,s^{-1}}$ and modes have been detected with amplitudes as low as a few $\mathrm{mm\,s^{-1}}$. At low frequencies the natural linewidth of the modes is not resolved in these observations; here the lifetime is several months. At higher frequencies the lifetime is only a few days, and the finite width of the peaks in the spectrum is clearly visible.

Figure 4 show observed multiplet frequencies ν_{nl} obtained with the LOWL instrument (Tomczyk *et al.* 1995). The quality of the frequency determination is illustrated by the fact that the data are shown with 1000σ error bars; the most accurately determined frequencies have a relative error of less than one part in 10^6, making them the most accurately known quantities related to the Sun. Evidently, it is this extremely high

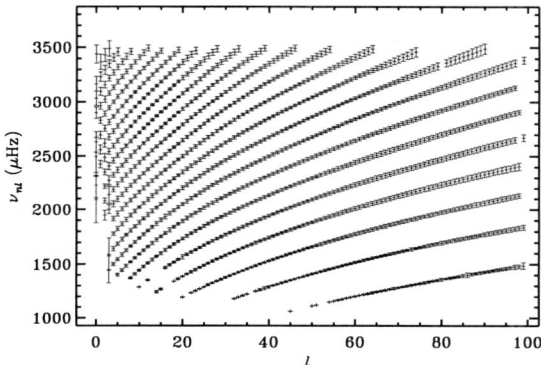

FIGURE 4. Observed solar p-mode frequencies, averaged over the azimuthal order m, as a function of the degree l, from one year of observations. The vertical lines show the 1000σ error bars. Each ridge corresponds to a given value of the radial order n, the lowest ridge having $n = 1$. (See Tomczyk, Schou & Thompson 1996).

accuracy that allows detailed inferences to be made about the properties of the solar interior.

5. SOME RESULTS ON SOLAR STRUCTURE

The simplest analysis of the observed frequencies is evidently to compare them with frequencies of a solar model. Here I use as reference the so-called 'Model S' of Christensen-Dalsgaard *et al.* (1996). This is based on the Livermore equation of state (Rogers, Swenson & Iglesias 1996) and an early generation of the OPAL opacities (Iglesias, Rogers & Wilson 1992). Diffusion and settling of helium and heavier elements were included using the formulation of Michaud & Proffitt (1993). Figure 5a shows relative differences between the observed frequencies and those of Model S. To highlight the effect of near-surface errors, which are certainly present in the calculated frequencies, the differences have been scaled by the inertia ratio Q_{nl}; had the only error in the model been localized near the surface, the resulting scaled differences would have been solely a function of frequency. It is evident that, in fact, the strongest dependence is on frequency but with substantial variation also at given frequency. The nature of this variation is shown in panel (b), where a fitted function of frequency has been subtracted and the residuals plotted against $v/(l+1/2)$ or, as indicated, equivalently against the location r_t of the inner turning point (*cf.* Eq. 39).[10] It should be noted that the residuals are highly systematic, with little scatter; this once more illustrates the extremely small

---

[10] A more careful asymptotic analysis of low-degree modes shows that $\sqrt{l(l+1)}$ must be replaced by $l+1/2$, in Eq. (39) and the following equations.

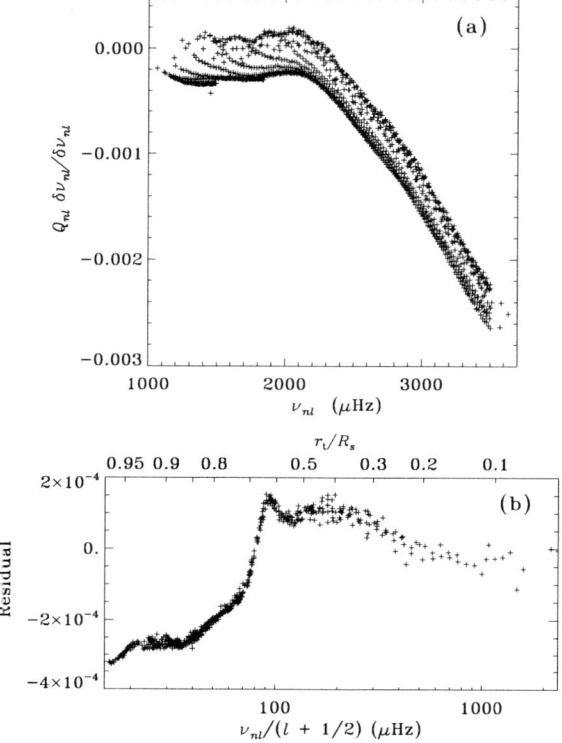

FIGURE 5. (a) Relative frequency differences, in the sense (observation) − (model), scaled by the inertia ratio Q_{nl} (cf. Eq. 35). The observations are a combination of BiSON whole-disk measurements (e.g., Elsworth et al., 1994) and LOWL observations (Tomczyk et al., 1995), as described by Basu et al. (1997), while the computed frequencies are for Model S. (b) Scaled differences after subtraction of a fitted function of frequency, plotted against $\nu_{nl}/(l+1/2)$ which determines the inner turning point r_t, shown as the upper abscissa.

random errors in the observed frequencies.

The separation of the differences is essentially equivalent to the asymptotic separation in Eq. (43) and thus the residuals can be identified with $\mathcal{H}_1$ (cf. Eq. 44), apart from an arbitrary constant shift. It is evident that the contribution for modes trapped within the convection zone, with $r_t \gtrsim 0.7$, is modest, whereas clearly there is a drastic variation just below this point, with further smaller effects contributing to even more deeply penetrating modes.

The detailed cause of this behaviour is revealed by inverse analysis of the frequency differences, based on the linearized relation (35) (see Section 3.4), to determine the relative difference in sound speed between the Sun and the model. The result is shown by

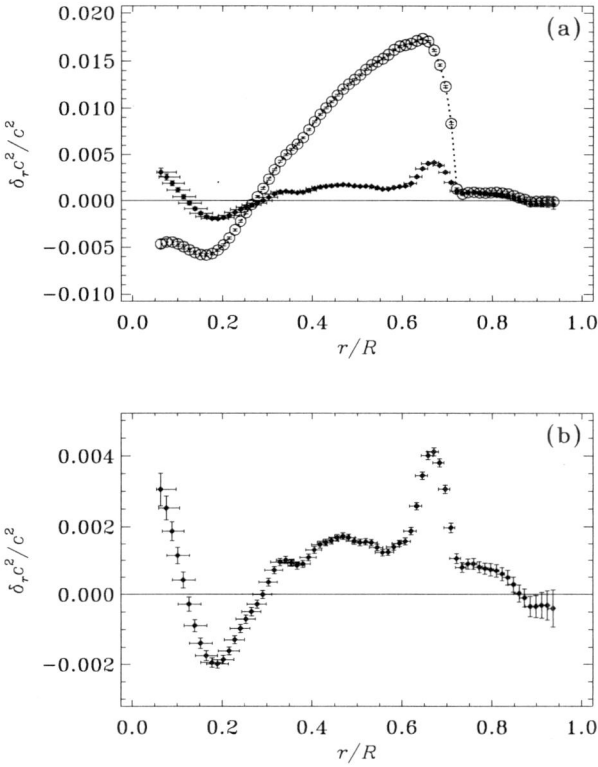

FIGURE 6. Results of sound-speed inversion. (a) Difference in squared sound speed, in the sense (Sun) − (model), inferred from inversion of the differences between the observed BiSON and LOWL frequencies and the frequencies of two solar models: closed circles are for Model S, and open circles for a similar model, but ignoring element diffusion and settling. (b) Results for Model S, on an expanded scale. The vertical error bars are 1-σ errors on the inferred differences, while the horizontal bars provide a measure of the resolution of the inversion. (Adapted from Basu *et al.*, 1997.)

the closed symbols in Fig. 6. It is evident that there is indeed a sharp feature in $\delta_r c^2/c^2$ just below the convection zone, with relatively modest variation elsewhere except in the core. Also, the overall magnitude of the differences is quite small, indicating that the model is a good approximation to the actual solar structure. On the other hand, the differences are evidently highly systematic and, given the inferred errors derived from the quoted standard errors in the observations, highly significant. For comparison, panel (a) also shows results for a model that did not include diffusion and settling, but otherwise with the same physics; clearly the inclusion of diffusion and settling has resulted in major improvements in the agreement with solar structure.

The cause of the remaining differences between Model S and the Sun is obviously

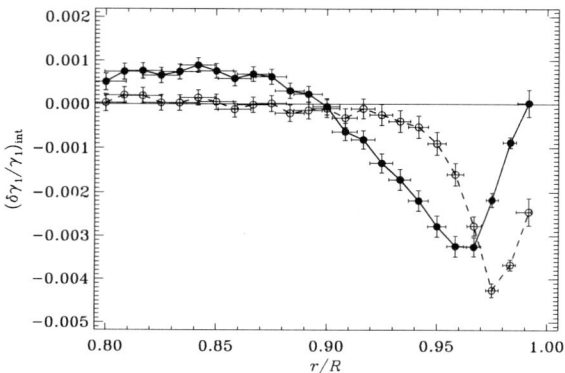

FIGURE 7. Relative difference between γ_1 obtained from an inversion of helioseismic data and γ_1 for two solar models, in the sense "Sun – model". Only the "intrinsic" difference in γ_1 is shown, that is, the part of the difference due to the equation of state (see text). Lines have been drawn through the results to guide the eye. The closed circles connected by a solid line are results obtained with an MHD model, the open circles connected with a dashed line are results with an OPAL model. (Adapted from Basu, Däppen and Nayfonov, 1999.)

of great interest. It is striking that the largest differences occur in regions of strong variations in the hydrogen abundance X: just beneath the convection zone, where a sharp gradient is established by helium settling, and at the edge of the core where the gradient is set up by nuclear burning. In both cases partial mixing would change the composition and hence the sound speed in such a way as to improve the agreement between the model and the Sun. In the case of the bump just beneath the convection zone such mixing could result from overshoot from the convection zone or rotationally induced instabilities (*e.g.* Brun, Turck-Chièze & Zahn 1999; Elliott & Gough 1999; Brun *et al.* 2002). More generally, with suitable modification of the physics of the solar interior it is possible to produce 'seismic solar models' which closely match the structure inferred from helioseismology (*e.g.* Turck-Chièze *et al.* 2001; Couvidat, Turck-Chièze & Kosovichev 2003).

As discussed in connection with Eq. (36) it is possible to formulate the inverse problem for the solar frequencies in such a way as to allow a determination of the intrinsic error in γ_1. Results of such analyses are discussed extensively elsewhere in this volume; as an illustration, Fig. 7 shows the results for models using the MHD[11] and the OPAL equations of state. It is evident that in most of the region shown, the error in γ_1 is smaller for the OPAL than for the MHD formulation, although with a possible reversal of this trend very near the surface. The results, including the formal error bars, are clear indications of the sensitivity of this type of helioseismic analysis to the thermodynamic properties of solar matter.

---

[11] Mihalas, Hummer & Däppen; see, for example, Mihalas *et al.* (1990); Däppen (this volume).

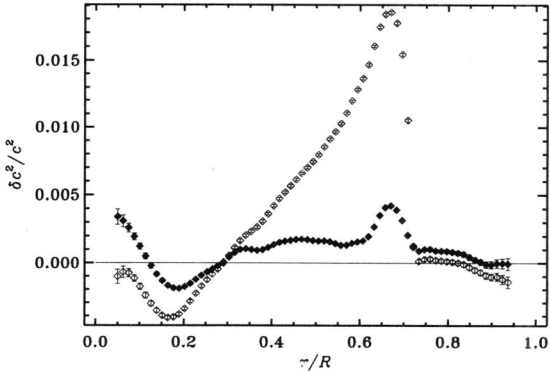

FIGURE 8. Effects of the revision of the inferred solar surface abundance on solar models. The filled symbols show the helioseismically inferred difference between squared sound speed in the Sun and in Model S, as in Fig. 6b, with $(Z/X)_s = 0.0245$. The open symbols show the corresponding results for a model approximately matching the revised abundances of Asplund *et al.* (2004) and consequently with $(Z/X)_s = 0.0183$. (Adapted from Pijpers *et al.* 2004.)

The results shown in Fig. 6 might motivate some complacency, or possibly pride, concerning our ability to model the Sun. After all, given the improvements in the physics of the solar interior up to the computation of Model S, but with no explicit adjustments of parameters, it was possible to match the solar sound speed to within a fraction of a per cent. It appears that such complacency would be misplaced. Recent spectroscopic analyses of the composition of the solar atmosphere, taking into account the three-dimensional dynamic nature of the atmosphere and departures from local thermodynamic equilibrium, have substantially decreased the inferred abundances of carbon, nitrogen and in particular oxygen (*e.g.* Allende Prieto, Lambert & Asplund 2002; Asplund *et al.* 2004). Since oxygen is a dominant source of opacity just below the convection zone, this leads to a corresponding reduction in the opacity and hence a change in the structure of solar models. In particular, the depth of the convection zone is reduced from $0.2885R$ in Model S, very close to the helioseismically inferred value of $0.287R$ (*e.g.* Christensen-Dalsgaard, Gough & Thompson 1991), to around $0.277R$. The effect on the comparison with helioseismic data is illustrated in Fig. 8, based on calculations by Pijpers *et al.* (2004). Owing to the reduction in the depth of the convection zone, the sound speed increases less rapidly with depth in the revised model, leading to the strongly enhanced peak in the sound-speed difference. The effects on the predicted solar neutrino flux were considered by Bahcall & Pinsonneault (2004). It was noted by Basu & Antia (2004) that a significant upward revision of the opacity tables, by more than 10 per cent, would be required to recover the agreement with the helioseismically inferred structure.[12] More detailed analyses of the required opacity changes have been carried out by Bahcall, Serenelli & Pinsonneault (2004) and Bahcall et al. (2004). Interestingly, Seaton & Badnell (2004) have found indications that the recent opacities from the Opacity Project are higher than the OPAL values generally used in current solar modelling, although possibly by an amount inadequate to account for the discrepancy between the Sun and the new models.

6. CONCLUDING REMARKS

It should be obvious that helioseismology has provided very detailed and precise information about the properties of the solar interior, including its thermodynamic state. This motivates its importance for the investigation of the equation of state of matter in near equilibrium under extreme conditions. It is also evident that the information so obtained is incomplete: the adiabatic oscillation frequencies do not yield information about the temperature or the internal energy of solar matter, although working in the nearly adiabatically stratified convection zone yields additional constraints that partly compensate for this. The information about the composition of solar matter is not very accurate; in the case of helium fairly precise constraints can be obtained from the oscillation frequencies, but only subject to uncertainties in the equation of state, making it difficult (and, at some level, impossible) to separate effects of errors in the helium abundance and the equation of state. Also, evidently, information is only available along the single (p, ρ) trace that is represented in the structure of the present Sun. Even so, the information obtained from the Sun clearly provide stringent constraints on the thermodynamical description and, as shown in Fig. 7, even sophisticated implementations of the equation of state fail to satisfy these constraints, to within a large margin.

In many ways the most interesting region, form a thermodynamic point of view, is the region relatively near the surface where ionization of hydrogen and helium takes place (cf. Fig. 1). Somewhat surprisingly, this region has been difficult to probe with helioseismology. To do so is greatly helped by data on high-degree modes, with turning points in the relevant region (e.g. Di Mauro et al. 2002). However, reliable determination of their frequencies is complicated by the fact that the individual modes cannot be separated, owing to cross-talk and the finite lifetime. As a result, different types of fit have to be developed. A detailed analysis of these problems, involving possibly also imperfections in the optics and the detectors, was presented by Korzennik, Rabello-Soares & Schou (2004). Also, the novel analysis techniques under development by Jefferies & Vorontsov (2004), discussed also by Vorontsov in this volume, are extremely promising.

Observations of other stars obviously allow to extend the space of parameters over which the equation of state is investigated. For the foreseeable future such observations

---

[12] Such effects on solar structure of corrections to the opacity were discussed in considerable detail by Elliott (1995), who set up an inverse problem for the opacity errors, and Tripathy & Christensen-Dalsgaard (1998). Tripathy, Basu & Christensen-Dalsgaard (1998) attempted to estimate the opacity corrections required to match the differences between the Sun and Model S.

will be restricted to low-degree modes; however, even for these the observable effects of the sharp features associated with helium ionization may in principle be investigated (*e.g.* Pérez Hernández & Christensen-Dalsgaard 1998; see also Houdek, this volume), although data of the required quality are not imminent. Extreme conditions are found in white dwarfs, for which buoyancy driven oscillations (g modes) have been observed in many cases. Amongst hot white dwarfs cooling due to neutrino emission, certainly reflecting the thermodynamic state of their interiors, may be observable through period changes (*e.g.* O'Brien & Kawaler 2000); also, analysis of observed oscillation frequencies has provided some evidence for a crystallized core in a massive cool white dwarf (Metcalfe, Montgomery & Kanaan 2004). With further observations of the Sun and of this broad variety of stars there seems little doubt that our ability to test observationally the predictions of increasingly sophisticated treatments of the equation of state will expand greatly in the coming years.

ACKNOWLEDGMENTS

I am very grateful to the organizers of this workshop, Vladan Čelebonović, Werner Däppen and Douglas Gough, for bringing together such a broad and exciting programme and such a nice group of people, and to the Lorentz Center for providing the perfect setting for the workshop. I thank Regner Trampedach for providing the results presented in Fig. 1 and Tim Brown for very useful comments on an earlier version of the manuscript. The editors are thanked for their patience and flexibility in connection with the preparation of this paper. The National Center for Atmospheric Research is supported by the National Science Foundation of the United States.

REFERENCES

Abrams, D. & Kumar, P. 1996, *Astrophys. J.* **472**, 882–390.
Allende Prieto, C., Lambert, D. L. & Asplund, M. 2002, *Astrophys. J.* **573**, L137–L140.
Anders, E. & Grevesse, N. 1989, *Geochim. Cosmochim. Acta* **53**, 197–214.
Antia, H. M. 1995, *Mon. Not. R. astr. Soc.* **274**, 499–503.
Asplund, M., Grevesse, N., Sauval, A. J., Allende Prieto, C. & Kiselman, D. 2004, *Astron. Astrophys.* **417**, 751–768.
Bahcall, J. N. & Pinsonneault, M. H. 2004, *Phys. Rev. Lett.* **92**, 121301-(1–4).
Bahcall, J. N., Basu, S., Pinsonneault, M. & Serenelli, A. M. 2004, submitted [astro-ph/0407060v1].
Bahcall, J. N., Serenelli, A. M. & Pinsonneault, M. 2004, *Astrophys. J.*, in the press [astro-ph/0403604v1].
Balmforth, N. J. 1992a, *Mon. Not. R. astr. Soc.* **255**, 603–631.
Balmforth, N. J. 1992b, *Mon. Not. R. astr. Soc.* **255**, 639–649.
Basu, S. & Antia, H. M. 1995, *Mon. Not. R. astr. Soc.* **276**, 1402–1408.
Basu, S. & Antia, H. M. 2004, *Astrophys. J.* **606**, L85–L88.
Basu, S. & Christensen-Dalsgaard, J. 1997, *Astron. Astrophys.* **322**, L5–L8.
Basu, S., Chaplin, W. J., Christensen-Dalsgaard, J., Elsworth, Y., Isaak, G. R., New, R., Schou, J., Thompson, M. J. & Tomczyk, S. 1997, *Mon. Not. R. astr. Soc.* **292**, 243–251.
Basu, S., Christensen-Dalsgaard, J., Howe, R., Schou, J., Thompson, M. J., Hill, F. & Komm, R. 2003, *Astrophys. J.* **591**, 432–445.

Basu, S., Däppen, W. & Nayfonov, A. 1999, *Astrophys. J.* **518**, 985–993.
Batchelor, G. K. 1956, *The theory of homogeneous turbulence*, Cambridge University Press.
Bedding, T. R. & Kjeldsen, H. 2003, *Publ. Astron. Soc. Australia* **20**, 203–212.
Böhm-Vitense, E. 1958, *Z. Astrophys.* **46**, 108–143.
Brown, T. M. 1991, *Astrophys. J.* **371**, 396–401.
Brown, T. M. 1996, in *Proc. VI IAC Winter School "The structure of the Sun"*, edited by T. Roca Cortés & F. Sánchez, Cambridge University Press, pp. 1–45.
Brun, A. S., Antia, H. M., Chitre, S. M. & Zahn, J.-P. 2002, *Astron. Astrophys.* **391**, 725–739.
Brun, A. S., Turck-Chièze, S. & Zahn, J. P. 1999, *Astrophys. J.* **525**, 1032–1041. (Erratum: *Astrophys. J.* **536**, 1005).
Canuto, V. M. & Mazzitelli, I. 1991, *Astrophys. J.* **370**, 295–311.
Chandrasekhar, S. 1964, *Astrophys. J.* **139**, 664–674.
Chang, H.-Y. & Gough, D. O. 1998, *Solar Phys.* **181**, 251–263.
Chaplin, W. J. & Appourchaux, T. 1999, *Mon. Not. R. astr. Soc.* **309**, 761–768.
Chaplin, W. J., Elsworth, Y., Howe, R., Isaak, G. R., McLeod, C. P., Miller, B. A., van der Raay, H. B., Wheeler, S. J. & New, R. 1996, *Solar Phys.* **168**, 1–18.
Chaplin, W. J., Elsworth, Y., Howe, R., Isaak, G. R., McLeod, C. P., Miller, B. A. & New, R. 1997, *Mon. Not. R. astr. Soc.* **287**, 51–56.
Christensen-Dalsgaard, J. 2002, *Rev. Mod. Phys.* **74**, 1073–1129.
Christensen-Dalsgaard, J. & Berthomieu, G. 1991, in *Solar interior and atmosphere*, edited by A. N. Cox, W. C. Livingston & M. Matthews, Space Science Series, University of Arizona Press, pp. 401–478.
Christensen-Dalsgaard, J., Däppen, W., Ajukov, S. V., Anderson, E. R., Antia, H. M., Basu, S., Baturin, V. A., Berthomieu, G., Chaboyer, B., Chitre, S. M., Cox, A. N., Demarque, P., Donatowicz, J., Dziembowski, W. A., Gabriel, M., Gough, D. O., Guenther, D. B., Guzik, J. A., Harvey, J. W., Hill, F., Houdek, G., Iglesias, C. A., Kosovichev, A. G., Leibacher, J. W., Morel, P., Proffitt, C. R., Provost, J., Reiter, J., Rhodes Jr., E. J., Rogers, F. J., Roxburgh, I. W., Thompson, M. J., Ulrich, R. K. 1996, *Science* **272**, 1286–1292.
Christensen-Dalsgaard, J., Duvall, T. L., Gough, D. O., Harvey, J. W. & Rhodes Jr, E. J. 1985, *Nature* **315**, 378–382.
Christensen-Dalsgaard, J., Gough, D. O. & Libbrecht, K. G. 1989, *Astrophys. J.* **341**, L103–L106.
Christensen-Dalsgaard, J., Gough, D. O. & Thompson, M. J. 1989, *Mon. Not. R. astr. Soc.* **238**, 481–502.
Christensen-Dalsgaard, J., Gough, D. O. & Thompson, M. J. 1991, *Astrophys. J.* **378**, 413–437.
Christensen-Dalsgaard, J., Gough, D. O. & Pérez Hernández, F. 1988, *Mon. Not. R. astr. Soc.* **235**, 875–880.
Christensen-Dalsgaard, J., Schou, J. & Thompson, M. J. 1990, *Mon. Not. R. astr. Soc.* **242**, 353–369.
Couvidat, S., Turck-Chièze, S. & Kosovichev, A. G. 2003, *Astrophys. J.* **599**, 1434–1448.
Cowling, T. G. 1941, *Mon. Not. R. astr. Soc.* **101**, 367–375.
Däppen, W., Gough, D. O., Kosovichev, A. G. & Thompson, M. J. 1991, in *Challenges to theories of the structure of moderate-mass stars*, Lecture Notes in Physics vol. **388**, edited by D. O. Gough & J. Toomre, Springer, Heidelberg, pp. 111–120.
Deubner, F.-L. & Gough, D. O. 1984, *Annu. Rev. Astron. Astrophys.* **22**, 593–619.
Di Mauro, M. P., Christensen-Dalsgaard, J., Rabello-Soares, M. C. & Basu, S. 2002, *Astron. Astrophys.* **384**, 666–677.
Duvall, T. L. 1982, *Nature* **300**, 242–243.
Duvall, T. L., Jefferies, S. M., Harvey, J. W., Osaki, Y. & Pomerantz, M. A. 1993, *Astrophys. J.* **410**, 829–836.
Dziembowski, W. A., Pamyatnykh, A. A. & Sienkiewicz, R. 1990, *Mon. Not. R. astr. Soc.* **244**, 542–550.
Elliott, J. R. 1995, *Mon. Not. R. astr. Soc.* **277**, 1567–1579.
Elliott, J. R. & Gough, D. O. 1999, *Astrophys. J.* **516**, 475–481.
Elsworth, Y., Howe, R., Isaak, G. R., McLeod, C. P., Miller, B. A., van der Raay, H. B. & Wheeler, S. J. 1995, in *Proc. GONG'94: Helio- and Astero-seismology from Earth and Space*, edited by R. K. Ulrich, E. Rhodes Jr & W. Däppen, Astronomical Society of the Pacific Conference Series vol. **76**, San Francisco, pp. 392–397.
Elsworth, Y., Howe, R., Isaak, G. R., McLeod, C. P., Miller, B. A., New, R., Speake, C. C. & Wheeler, S. J. 1994, *Astrophys. J.* **434**, 801–806.
Fossat, E. 1991, *Solar Phys.* **133**, 1–12.

Gabriel, A. H., Charra, J., Grec, G., Robillot, J.-M., Roca Cortés, T., Turck-Chièze, S., Ulrich, R., Basu, S., Baudin, F., Bertello, L., Boumier, P., Charra, M., Christensen-Dalsgaard, J., Decaudin, M., Dzitko, H., Foglizzo, T., Fossat, E., García, R. A., Herreros, J. M., Lazrek, M., Pallé, P. L., Pétrou, N., Renaud, C. & Régulo, C. 1997, *Solar Phys.* **175**, 207–226.
Gabriel, M. 1993, *Astron. Astrophys.* **274**, 935–939.
Goldreich, P. & Keeley, D. A. 1977, *Astrophys. J.* **212**, 243–251.
Goode, P. R., Gough, D. O. & Kosovichev, A. 1992, *Astrophys. J.* **387**, 707–711.
Gough, D. O. 1977. in *Problems of stellar convection, IAU Colloq. No. 38, Lecture Notes in Physics* vol. **71**, edited by E. A. Spiegel & J.-P. Zahn, Springer-Verlag, Berlin, pp. 15–56.
Gough, D. O. 1984, *Phil. Trans. R. Soc. London, Ser. A* **313**, 27–38.
Gough, D. O. 1993, in *Astrophysical fluid dynamics, Les Houches Session XLVII*, edited by J.-P. Zahn & J. Zinn-Justin, Elsevier, Amsterdam, pp. 399–560.
Gough, D. 1996, in *Proc. VI IAC Winter School "The structure of the Sun"*, edited by T. Roca Cortés & F. Sánchez, Cambridge University Press, pp. 141–228.
Gough, D. O. & Thompson, M. J. 1991, in *Solar interior and atmosphere*, edited by A. N. Cox, W. C. Livingston & M. Matthews, Space Science Series, University of Arizona Press, pp. 519–561.
Gough, D. O. & Vorontsov, S. V. 1995, *Mon. Not. R. astr. Soc.* **273**, 573–582.
Gough, D. O. & Weiss, N. O. 1976, *Mon. Not. R. astr. Soc.* **176**, 589–607.
Harvey, J. W., Hill, F., Hubbard, R. P., Kennedy, J. R., Leibacher, J. W., Pintar, J. A., Gilman, P. A., Noyes, R. W., Title, A. M., Toomre, J., Ulrich, R. K., Bhatnagar, A., Kennewell, J. A., Marquette, W., Partrón, J., Saá, O. & Yasukawa, E. 1996, *Science* **272**, 1284–1286.
Iglesias, C. A., Rogers, F. J. & Wilson, B. G. 1992, *Astrophys. J.* **397**, 717–728.
Jefferies, S. M. & Vorontsov, S. 2004, *Solar Phys.* **220**, 347–359.
Kippenhahn, R. & Weigert, A. 1990, *Stellar structure and evolution*, Springer-Verlag, Berlin.
Korzennik, S. G., Rabello-Soares, M. C. & Schou, J. 2004, *Astrophys. J.* **602**, 481–515.
Kosovichev, A. G., Christensen-Dalsgaard, J., Däppen, W., Dziembowski, W. A., Gough, D. O. & Thompson, M. J. 1992, *Mon. Not. R. astr. Soc.* **259**, 536–558.
Kumar, P. 1994, *Astrophys. J.* **428**, 827–836.
Kumar, P. & Basu, S. 1999, *Astrophys. J.* **519**, 396–399.
Kumar, P. & Basu, S. 2000, *Astrophys. J.* **545**, L65–L68.
Kumar, P. & Lu, E. 1991, *Astrophys. J.* **375**, L35–L39.
Kumar, P., Duvall, T. L., Harvey, J. W., Jefferies, S. M., Pomerantz, M. A. & Thompson, M. J. 1990, in *Progress of seismology of the sun and stars, Lecture Notes in Physics* vol. **367**, edited by Y. Osaki & H. Shibahashi, Springer, Berlin, pp. 87–92.
Kumar, P., Fardal, M. A., Jefferies, S. M., Duvall, T. L., Harvey, J. W. & Pomerantz, M. A. 1994, *Astrophys. J.* **422**, L29–L32.
Lighthill, M. J. 1952, *Proc. Roy. Soc. London* **A211**, 564–587.
Marchenkov, K., Roxburgh, I. & Vorontsov, S. 2000, *Mon. Not. R. astr. Soc.* **312**, 39–50.
Metcalfe, T. S., Montgomery, M. H. & Kawaler, S. D. 2003, *Mon. Not. R. astr. Soc.* **344**, L88–L92.
Michaud, G. & Proffitt, C. R. 1993, in *Proc. IAU Colloq. 137: Inside the stars*, edited by A. Baglin & W. W. Weiss, Astronomical Society of the Pacific Conference Series vol. **40**, San Francisco, pp. 246–259.
Mihalas, D., Hummer, D. G., Mihalas, B. W. & Däppen, W. 1990, *Astrophys. J.* **350**, 300–308.
Nigam, R. & Kosovichev, A. G. 1998, *Astrophys. J.* **505**, L51–L54.
Nigam, R. & Kosovichev, A. G. 1999, *Astrophys. J.* **514**, L53–L56.
O'Brien, M. S. & Kawaler, S. D. 2000, *Astrophys. J.* **539**, 372–378.
Pérez Hernández, F. & Christensen-Dalsgaard, J. 1994, *Mon. Not. R. astr. Soc.* **269**, 475–492.
Pérez Hernández, F. & Christensen-Dalsgaard, J. 1998, *Mon. Not. R. astr. Soc.* **295**, 344–352.
Pijpers, F. P. & Thompson, M. J. 1992, *Astron. Astrophys.* **262**, L33–L36.
Pijpers, F. P. & Thompson, M. J. 1994, *Astron. Astrophys.* **281**, 231–240.
Pijpers, F. P., Asplund, M., Christensen-Dalsgaard, J. & Houdek, G. 2004, in preparation.
Rabello-Soares, M. C., Basu, S. & Christensen-Dalsgaard, J. 1999, *Mon. Not. R. astr. Soc.* **309**, 35–47.
Rabello-Soares, M. C., Basu, S., Christensen-Dalsgaard, J. & Di Mauro, M. P. 2000, *Solar Phys.* **193**, 345–356.
Rast, M. P. 1999, *Astrophys. J.* **524**, 462–468.
Rast, M. P. & Bogdan, T. J. 1998, *Astrophys. J.* **496**, 527–537.
Rimmele, T. R., Goode, P. R., Harold, E. & Stebbins, R. T. 1995, *Astrophys. J.* **444**, L119–L122.

Robinson, F. J., Demarque, P., Li, L. H., Kim, Y.-C., Chan, K. L. & Guenther, D. B. 2003, *Mon. Not. R. astr. Soc.* **340**, 923–936.

Rogers, F. J., Swenson, F. J. & Iglesias, C. A. 1996, *Astrophys. J.* **456**, 902–908.

Rosenthal, C. S. 1998, *Astrophys. J.* **508**, 864–875.

Roxburgh, I. W. & Vorontsov, S. V. 1997, *Mon. Not. R. astr. Soc.* **292**, L33–L36.

Sackmann, I.-Juliana & Boothroyd, A. I. 2003, *Astrophys. J.* **583**, 1024–1039.

Scherrer, P. H., Bogart, R. S., Bush, R. I., Hoeksema, J. T., Kosovichev, A. G., Schou, J., Rosenberg, W., Springer, L., Tarbell, T. D., Title, A., Wolfson, C. J., Zayer, I., and the MDI engineering team 1995, *Solar Phys.* **162**, 129–188.

Seaton, M. J. & Badnell, N. R. 2004, *Mon. Not. R. astr. Soc.*, submitted [astro-ph/0404437].

Spiegel, E. A. 1963, *Astrophys. J.* **138**, 216–225.

Stein, R. F. 1967, *Solar Phys.* **2**, 385–432.

Stein, R. F. & Nordlund, Å. 1989, *Astrophys. J.* **342**, L95–L98.

Stein, R. F. & Nordlund, Å. 1998, *Astrophys. J.* **499**, 914–933.

Stein, R. F. & Nordlund, Å. 2001, *Astrophys. J.* **546**, 585–603.

Stein, R., Georgobiani, D., Trampedach, R., Ludwig, H.-G. & Nordlund, Å. 2004, *Solar Phys.* **220**, 229–242.

Thompson, M. J. 1993, in *Proc. GONG 1992: Seismic investigation of the Sun and stars*, edited by T. M. Brown, Astronomical Society of the Pacific Conference Series vol. **42**, San Francisco, pp. 141–154.

Thompson, M. J., Christensen-Dalsgaard, J., Miesch, M. S. & Toomre, J. 2003, *Annu. Rev. Astron. Astrophys.* **41**, 599–643.

Tomczyk, S., Schou, J. & Thompson, M. J. 1996, *Bull. Astron. Soc. India* **24**, 245–250.

Tomczyk, S., Streander, K., Card, G., Elmore, D., Hull, H. & Cacciani, A. 1995, *Solar Phys.* **159**, 1–21.

Tripathy, S. C. & Christensen-Dalsgaard, J. 1998, *Astron. Astrophys.* **337**, 579–590.

Tripathy, S. C., Basu, S. & Christensen-Dalsgaard, J. 1998, in *Poster Volume; Proc. IAU Symposium No 181: Sounding Solar and Stellar Interiors, Nice, Sept. 30 – Oct. 3, 1996*, edited by J. Provost & F. X. Schmider, Université de Nice, pp. 129–130.

Turck-Chièze, S., Couvidat, S., Kosovichev, A. G., Gabriel, A. H., Berthomieu, G., Brun, A. S., Christensen-Dalsgaard, J., García, R. A., Gough, D. O., Provost, J., Roca-Cortes, T., Roxburgh, I. W. & Ulrich, R. K. 2001, *Astrophys. J.* **555**, L69–L73.

Unno, W., Osaki, Y., Ando, H., Saio, H. & Shibahashi, H. 1989, *Nonradial Oscillations of Stars, 2nd Edition*, University of Tokyo Press.

Vorontsov, S. V. & Shibahashi, H. 1991, *Publ. Astron. Soc. Japan* **43**, 739–753.

Vorontsov, S. V., Baturin, V. A. & Pamyatnykh, A. A. 1991, *Nature* **349**, 49–51.

Helioseismic Inversions and the Equation of State

Sergei V. Vorontsov

*Astronomy Unit, Queen Mary, University of London, Mile End Road, London E1 4NS, UK;
Institute of Physics of the Earth, B. Gruzinskaya 10, Moscow 123810, Russia*

Abstract. Techniques of helioseismic inversions targeted at the diagnostics of the equation of state are reviewed, with particular emphasis on the uniqueness of the solutions. Measurement of the adiabatic exponent deep in the solar convective envelope with p-mode frequencies currently available from SOHO MDI data is presented. Further efforts in the field of global helioseismic inversions are briefly discussed.

INTRODUCTION

A vast amount of high-quality measurements of solar acoustic oscillations is now available from the ground-based (GONG, BiSON, TON) and space projects (SOHO MDI, GOLF). With a variety of efficient tools of helioseismic inversions, developed along two decades, helioseismic measurements provide a unique source of precise information about solar internal structure and dynamics. For a review of basic properties of solar p modes, the reader is referred to Christensen-Dalsgaard (these procedings), and for a recent general review on solar seismology—to Christensen-Dalsgaard [11].

The precise helioseismic data make the Sun a unique laboratory for the diagnostics of the equation of state. This diagnostic is based on the possibility of addressing the adiabatic exponent $\Gamma_1 = (\partial \ln p / \partial \ln \rho)_S$, which influences the oscillation frequencies—predominantly, through the direct effects in the adiabatic sound speed $c = \sqrt{\Gamma_1 p / \rho}$ in the solar interior.

However, despite of the significant improvement of the amount and quality of helioseismic data in recent years, measuring the adiabatic exponent in solar interior remains a difficult task, and the results which can be found in the current literature are sometimes in variance which each other.

There are two main problems behind these uncertainties. First, the relevant effects in the p-mode frequencies are rather small. Below the H and He ionization regions (limited by the outer 3 percent of solar radius), the adiabatic exponent deviates from its simple monoatomic ideal gas value of 5/3 by only about $2 \cdot 10^{-3}$. One part in 10^3 in the sound speed is quite a measurable quantity; but there is a second problem. The oscillation frequencies are specified by the hydrostatic "seismic model" which, in turn, needs two independent radial profiles to be specified (Christensen-Dalsgaard, these proceedings). The measurable difference $\delta c^2(r)$ between the Sun and a reference model can in general be attributed to either an inconsistency in the adiabatic exponent, or in the density profile (or both). In the inversions, we are facing two options:

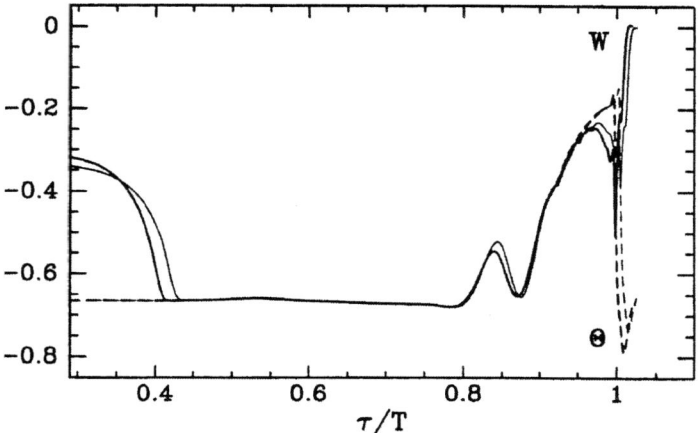

FIGURE 1. Functions W and Θ, defined by Eq. (1), plotted against acoustic radius (the base of the convection zone is at about 0.4, $\tau/T = 1$ is the photospheric level). Two solar models are shown, which differ in helium abundance. The peak near $\tau/T = 0.83$ results from HeII ionization. (Adapted from Gough [16].)

(i) Simultaneous inversion for the two independent functions of radius, e.g. $\delta\rho(r)$ and $\delta\Gamma_1(r)$;

(ii) Constraining the problem with using some *a'priory* information which would relate the two unknown functions with each other, and thus allow to target the inversion on measuring one function of radius only.

Option (ii) was suggested 20 years ago by Gough [15], with using, as the required supplementary constraint, a well-established property that in the bulk of the solar convection zone, where convective energy transport is far more efficient than the radiative one, the convection zone is quite accurately adiabatically stratified. Interestingly, in the adiabatically-stratified domain we then have [15]

$$W(r) \equiv \frac{r^2}{Gm}\frac{dc^2}{dr} = 1 - \Gamma_1\left[1 + \left(\frac{\partial \ln \Gamma_1}{\partial \ln p}\right)_S\right] \equiv \Theta(r): \qquad (1)$$

function W on the left, governed by the sound-speed gradient, gives a direct measure of function Θ on the right, which is a pure thermodynamic quantity (Fig.1)

The sound speed is a principal quantity accessible by heliosesmic measurements. Solar p modes are trapped acoustic waves, modified only slightly by the effects of buoyancy and gravity. When working with observational frequencies contaminated by errors, the sound speed is thus the first quantity which profile can be reliably measured. Separation of the two functions which contribute to $\delta c^2(r)$ appeals to the small effects of buoyancy and gravity (more exactly, to the relatively small difference between two small quantities—in the Sun and in the reference model), and requires data of far better quality.

As will be argued in this paper, the required accuracy in the p-mode frequency measurements is not yet achieved (or only marginally achieved). The robust inferences on the equation of state in solar interior are still largely limited by option (ii).

THE SOLAR ACOUSTIC CAVITY

To discuss the diagnostic properties of solar p modes, it is convenient to employ the so-called Cowling approximation, which discards small effects of perturbations in gravity. These effects are taken into account in all the recent inversions; but they change nothing in the qualitative discussion. In this approximation, the differential equations of small adiabatic oscillations (Christensen-Dalsgaard, these proceedings) are reduced to the second order and can be written in a form of stationary Schrödinger equation as (e.g. [7])

$$\frac{d^2}{d\tau^2}\zeta + [\omega^2 - V(\tau)]\zeta = 0, \tag{2}$$

with

$$\tau = \text{sgn}(s^2) \int_r^R |s|\,dr, \qquad s^2 = \frac{1}{c^2} - \frac{\tilde{w}^2}{r^2}, \tag{3}$$

$$\tilde{w} = \frac{\ell + 1/2}{\omega}, \qquad \zeta = s^{-1/2} r \xi_r, \tag{4}$$

where ξ_r is radial displacement function. The acoustic potential V is shown in Fig. 2, for three different values of parameter $\tilde{w}$, which specifies the position of the inner turning point (located where $s^2(r) = 0$).

Closer to the surface, the acoustic potential becomes independent of $\tilde{w}$; this property comes from the decreasing values of the sound speed towards the solar surface. The solutions to the wave equations in the outer layers thus become independent on the degree ℓ, when the degree is not too high. This property is of crucial importance for all the helioseismic studies. Current understanding of the turbulent surface layers and of the physics of oscillations there is still largely uncertain. In the inversions with p modes of low and intermediate degree, targeted at sounding deep solar interior, these uncertain effects are filtered away using the property that they have to depend on frequency only, but not on the degree.

Mathematically, this property leads to the expansion of the wave solutions in the outer solar layers in powers of $\tilde{w}$ (only even powers of $\tilde{w}$ enter this asymptotic expansion). The leading-order term, which depends on frequency only, provides an adequate accuracy when the degree ℓ is not too high; at higher degree, the accuracy is improved by taking into account the next term.

There is another type of asymptotic expansion, also suggested by Fig. 2. In the acoustic cavity far from the turning points, the acoustic potential V is small compared with ω^2. This property suggests an expansion of the solutions in powers of $1/\omega^2$—the high-frequency asymptotic expansion. In the leading order, the solution describes a pure acoustic wave; modification of this trapped acoustic wave by the effects of buoyancy and

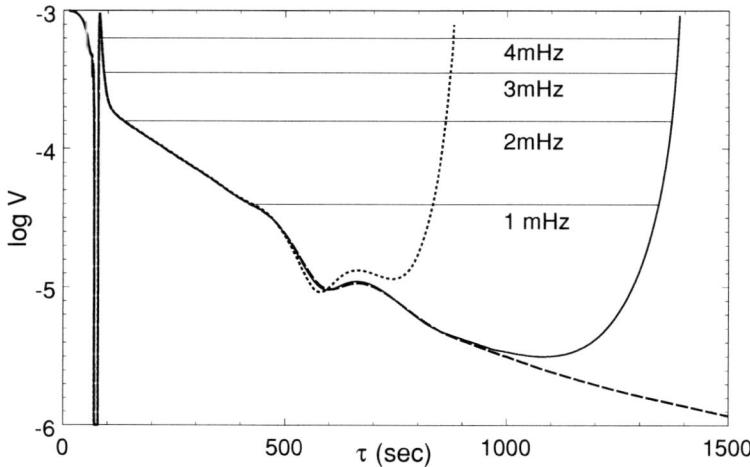

FIGURE 2. Acoustic potential $V(\tau)$ in a solar model envelope. The solid line is for $\tilde{w} = 4000$ s (lower turning point $r_1 \simeq 0.85R$), the short-dashed line for $\tilde{w} = 8000$ s ($r_1 \simeq 0.94R$), and the long-dashed line is for radial modes ($\tilde{w} = 0$). Surface radius R (where $\tau = 0$) is taken at the temperature minimum. Horizontal bars indicate schematically the trapping region for p modes with $\tilde{w} = 4000$ s and with the frequencies indicated. Rapid variation of the potential near $\tau = 600$ s results from HeII ionization. (From Gough & Vorontsov [17].)

gravity (including gravity perturbations, discarded by Cowling approximation) enter the high-frequency expansion starting from its second term [34].

High-Frequency Asymptotic Inversions

When both the two asymptotic expansions (high-frequency expansion for the inner solution, $\tilde{w}$-expansion for the outer solution) are matched together, the resulted second-order eigenfrequency equation is [7]

$$F(\tilde{w}) + \frac{1}{\omega^2}\Phi(\tilde{w}) \simeq \pi \frac{n + \alpha_0(\omega) + \tilde{w}^2 \alpha_2(\omega)}{\omega}, \qquad (5)$$

where n is radial order, and the leading-order term on the left is

$$F(\tilde{w}) = \int_{r_1}^{R} \left(\frac{1}{c^2} - \frac{\tilde{w}^2}{r^2}\right)^{1/2} dr. \qquad (6)$$

Matching the eigenfrequency equation with observational frequencies [e.g. 17] allows to measure different terms in this equation separately, and each of them has a diagnostic value. The sound-speed profile is recovered from $F(\tilde{w})$ using Abel integral transform. Modification of the asymptotic inversion can be developed by linearizing the eigenfrequency equation in the vicinity of a reference solar model ([8]; Christensen-Dalsgaard, these proceedings).

A region of primary interest for helioseismic inversions is the region of HeII ionization (Figs 1, 2); in particular, it provides a possibility of measuring the solar He abundance. Modes which are most sensitive to HeII ionization are p-modes with inner turning points in and around this region. Unfortunately, at high degree ℓ the solar frequency measurements still suffer from big uncertainties [19]. The high-precision measurements, required by helioseismic inversions, are still limited by modes with turning points well below the HeII ionization region. For these modes, the acoustic ray paths in the HeII ionization region are nearly vertical, and the signal of HeII ionization is detectable in the "surface" phase shift $\alpha_0(\omega)$ (and in $\alpha_2(\omega)$, which gives an additional information of a sort of "stereoscopic" effect [17]). The uncertain near-surface effects operate near and above the upper turning points, and only affect $\alpha(\omega)$ with a contribution which varies slowly with frequency ω. The rapid variation in $V(\tau)$, caused by the HeII ionization, is localized deep in the acoustic cavity, and produces a rapidly-varying quasi-periodic component in $\alpha(\omega)$. Using a proper spectral filtration [36, 26] allows to separate these two components (Fig. 3).

The amplitude of the fast component in the "surface" phase shift gives a measure of the He abundance. First measurements of He abundance, based on this approach but realized with different techniques [35, 9, 26], brought the same result of $Y \simeq 0.25$. The HeII signal in the data is sensitive not only to He abundance, but also to the value of the specific entropy in the adiabatic part of the convection zone (two-parametric grids of envelope models are used in He abundance measurements to account for this dependence), and to the equation of state. Uncertainties in the equation of state, in turn, affect the accuracy of helioseismic He abundance measurements [20, 1, 29, 2].

The signal in the acoustic phase shifts is defined by the HeII hump in the acoustic potential (Fig. 2), and a direct approach would consist in the inversion for the acoustic potential itself [22]. This inversion (a typical example of an inverse problem of potential scattering) suffers from the limited frequency range of observational data, which makes the inverse problem extremely ill-conditioned. It was found more productive to calibrate the grids of envelope models, calculated with different values of He abundance, specific entropy, and different versions of the equation of state, directly against the measurable phase-shift data.

These calibrations have shown unambiguously [25, 36] that models computed with simplest possible version of the equation of state, provided by the Saha equation for the ionization equilibrium of the ideal gas, are completely ruled out by the observations. Inclusion of electrostatic corrections in the Debye-Hückel approximation improves the results dramatically [36]; fits of comparable quality (though still in variance with observations) were obtained with Mihalas, Hammer and Däppen (MHD) equation of state [24, 12, and Däppen, these proceedings]. We will get back to the calibration of the equation of state in the HeII ionization region when discussing "differential-response" technique later in this section.

Below the HeII ionization, helioseismic data allow quite accurate inversion for the sound-speed profile; asymptotic inversion of $c^2(r)$ in the deep layers of the solar convective envelope also rules out the models computed without electrostatic corrections, and permits us to detect the influence of terms in the equation of state beyond the Debye-

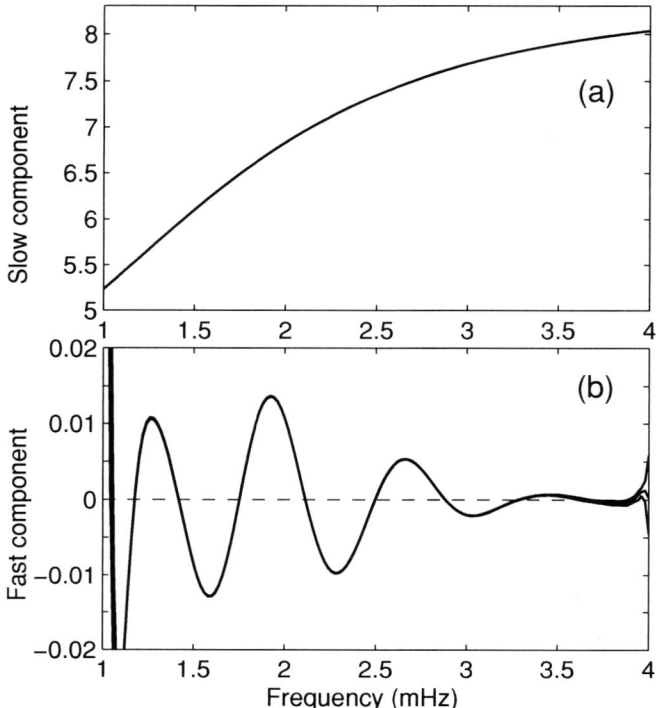

FIGURE 3. Slow (a) and fast (b) components in the surface phase shift $\alpha_0(\omega)/\omega$, in normalized units. The spectral filtration was performed using orthogonal polynomials. Three curves, distinguishable in the lover panel, show the solution and its envelope obtained when the input frequencies were added with white noise of amplitudes corresponding to the reported observational errors. The solar p-mode frequencies are from the "low-resolution" set of Libbrecht, Woodard & Kaufman [21]. (Adapted from Vorontsov, Baturin & Pamiatnykh [36].)

Hückel correction (Baturin et al. [5]). It also allows to probe the composition of the mixture of heavy elements; Fig. 4 illustrates a comparison with observations of the two envelope models constructed with two forms of the MHD equation of state, one with heavy elements represented by C, N, O and Fe (MHD-5), and another with oxygen only (MHD-2). All the models tested were unable to fit observations with adequate accuracy, including models computed with OPAL [30, and Däppen, these proceedings] equation of state, which allows to improve the agreement somewhat; interestingly, it was found possible to essentially eliminate the mismatch by employing a very simple parametric model of pressure ionization [5].

Results of a most recent non-asymptotic structural inversion deep in the convective envelope, expressed directly in terms of the adiabatic exponent, will be described later in this paper.

The high-frequency asymptotic approximation provides rather high accuracy: even when it is truncated to the leading order, the frequency errors are well within 1 percent

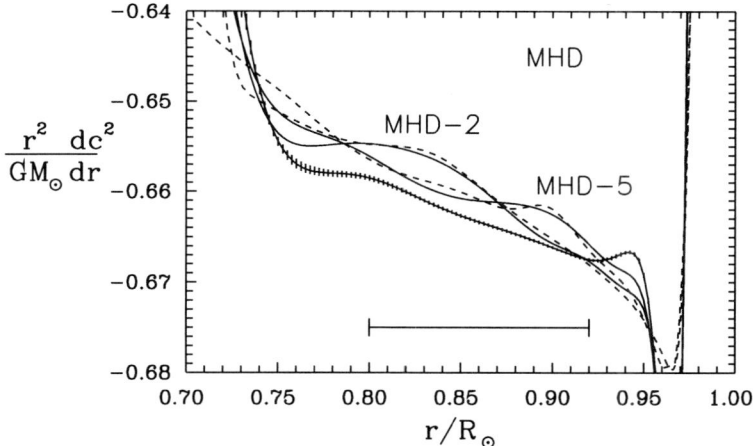

FIGURE 4. Function $W(r)$ defined by Eq. 1 (but with $m(r)$ replaced by total solar mass $M_\odot$). Line with error bars show the result of the asymptotic inversion for the sound-speed profile; the error bars indicate the response of the solution to the expected random errors in the data. Horizontal bar indicates the region where the inversion is considered to be reliable. Two other solid curves show the similar result but when the inversion is applied to theoretical p-mode frequencies of two solar models (see text). The dashed curves show the actual sound-speed gradient in the models. (From Baturin et al. [5].)

level (except of low-frequency modes of low degree ℓ, affected strongly by gravity perturbations in the high-density solar core, where the accuracy degrades to some few percent [34]).

This accuracy, however, can not meet the demands of the precise helioseismic measurements. Even when taken to the higher order, the high-frequency approximation suffers from inaccuracies related with localized regions of non-asymptotic behaviour (rapid variations at the base of the convection zone, strong effects of gravity perturbations in the high-density core). The accuracy can be improved by supplementing the high-frequency asymptotic analysis with distorted-wave Born approximation [31]. This approach allows to construct an efficient algorithm of non-linear iterative inversion of high resolution [23], but requires some further developments.

Non-Asymptotic Inversions

These are techniques which do not employ the high-frequency asymptotic approximation. Asymptotic expansion in powers of $\tilde{w}$, targeted at eliminating the near-surface uncertainties, will stay with us until better understanding of the turbulent convection and its interaction with p-modes will be achieved. These uncertainties are accounted for in any structural inversion, by allowing an arbitrary smooth function of frequency, corresponding to $\alpha_0(\omega)$ in Eq. 5, or two functions of frequency, corresponding to the two-term expansion $\alpha_0(\omega) + \tilde{w}^2 \alpha_2(\omega)$.

The high-frequency approximation is replaced by linearizing the relation between the seismic model and its oscillation frequencies in the vicinity of a reference model. This linearization employs the variational principle, which allows to express the small variations in the eigenfrequencies in terms of integrals of parameter variations, with kernels specified by the undistorted eigenfunctions (Christensen-Dalsgaard, these proceedings). After being linearized, the problem can be dealt with using a variety of techniques of linear inversion.

Classical Tikhonov regularization is based on the idea of replacing a naive minimization of the quadratic functional, which defines a (squared) norm of the residuals, with minimizing its linear combination with another (regularizing) functional, which penalizes the solution for having unphysical (e.g. non-smooth) behaviour. In recent literature, Tikhonov regularization is sometimes referred to as "regularized least squares". Another approach is based on the localized averaging of Backus and Gilbert. The idea behind is to consider a linear combination of the variational expressions for frequency corrections, taken such as to make the linear combination of kernels in the right-hand side close to Dirac's δ-function: the same linear combination of frequency corrections in the left-hand side will then give a localized measure of the required variation of the relevant seismic parameter. A convenient extension of this approach, developed by Pipers and Thompson [27, 28], is known as SOLA technique (for Subtractive Optimally Localized Averaging). For a short (but comprehensive) overview of these techniques, the reader is referred to Christensen-Dalsgaard [11]. When applied to helioseismic structural inversion, these techniques are targeted at separate measurement of the two functions which define the seismic model.

Recent results obtained with SOHO MDI data indicate that, although the OPAL equation of state fits observations better in the convective envelope for $r < 0.97 R_\odot$ when compared with the MHD equation of state, the situation is reversed in the outer 2%–3% of the solar radius (Basu, Däppen & Nayfonov [2], Di Mauro et al. [13]). Another impressive finding (Elliott & Kosovichev [14]) is that relativistic effect of electrons is well measurable, which only affects the adiabatic exponent in the central solar core by 0.1%–0.2%.

These results were obtained in inversions free of any additional constraints on solar hydrostatic stratification, and shall be taken with caution before they are confirmed by further detailed investigations. One worrying sign is that in the extensive study reported by Basu, Pinsonneault & Bahcall [3], it was found not possible to make satisfactory inversion for two separate functions of radius, when the SOHO MDI data set was replaced by GONG data of nearly the same quality.

Differential-Response Technique

This is an alternative approach of structural diagnostics [33], a framework in which both asymptotic and non-asymptotic inversions can be developed. The idea is to truncate the solar model somewhere near or below the upper turning points of the acoustic modes (0.999 $R_\odot$ was used in results which follow), and to eliminate the rest of the model, together with uncertainties in the near-surface layers, from the analysis at all, by transferring the outer boundary conditions to the truncation level. The boundary

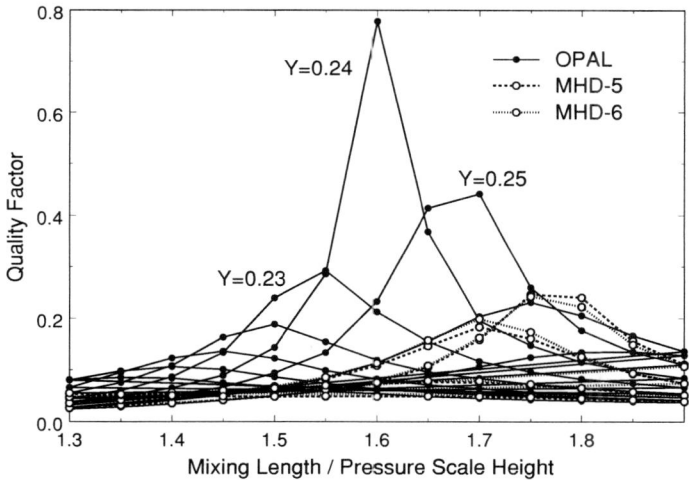

FIGURE 5. Calibration of grids of solar envelope models computed with OPAL and with MHD equation of state. Parameter along the horizontal axis determines the specific entropy in the adiabatic part of the convection zone. The calibration was targeted on matching the signal in the p-mode frequencies coming from the HeII ionization region. The solar p-mode frequencies are from the "low-resolution" set of Libbrecht, Woodard & Kaufman [21]. (From Baturin & Vorontsov [4].)

conditions at the edge of the truncated model are not known explicitly. But since the solutions above this level do not depend on the degree ℓ (or more accurately, can be represented by two functions of frequency in their two-term asymptotic expansion in powers of $\tilde{w}^2$), the boundary conditions, transferred to this level, must obey exactly the same behaviour.

When solar model is tested against observations, there is no need to calculate its p-mode frequencies. Solutions to the oscillation equations are calculated by shooting technique from below towards the surface with using the *observational* frequencies. If the model fits the data, phases of all the wave solutions at the boundary must follow a slowly-varying function of frequency only (or two-term expansion in powers of $\tilde{w}^2$); any deviation from this behaviour signals a misfit which needs to be eliminated. This is a sensitive device, working similar to differential bridge used in electric engineering.

Direct calibration of solar envelope models with signal coming from the HeII ionization zone is illustrated by Fig. 5. The calibration consisted in suppressing the fast component in frequency dependence of phases of wave solutions at the truncation boundary. In this calibration, the OPAL equation of state allows to obtain an almost perfect fit; models constructed with the MHD equation of state appear to be less consistent with observations. It is most interesting to note that the reference solar model S ([10], and Christensen-Dalsgaard, these proceedings) passes this test almost perfectly. More elaborated calibrations with OPAL equation of state and with using more recent SOHO MDI data, were reported in [6].

Linearized inversions for the internal structure [37, 38] employ an extension to the standard variational principle, which allows variation of boundary conditions. The seismic model is described by the profile of running mean density $m(r)/r^3$ (cubic B-splines on a dense mesh) and by the profile of the adiabatic exponent. New algorithm of adaptive gradient regularization [32] was implemented in the inversions as an alternative to Tikhonov regularization.

Adiabatic Exponent Deep in the Convective Envelope

We here apply the differential-response technique to measuring $\Gamma_1(r)$ deep in the convective envelope, using data now available from eight years of SOHO MDI measurements. The solar p-mode frequencies inferred from individual 72d data sets were corrected for variations with solar activity, to bring them to the same epoch before averaging, as discussed in [38]. Correction for solar-activity effects is possible because they appear to operate in the near-surface layers only.

The initial residuals obtained with reference solar model S are shown in Fig. 6, and on a bigger scale—in the upper panel of Fig. 7. The residuals are plotted in units of Phase Propagation Time, which is time needed for the acoustic wave to travel between the truncation boundary and the upper turning point; 1 sec mismatch in the PPT scale corresponds roughly to 1 μHz mismatch in frequency. The input data was limited by p modes with $\tilde{w} < 10000$ s (inner turning points $r_1 < 0.97 R_\odot$) to ensure that the two-term expansion of the outer boundary condition imposed at the truncation boundary provides enough accuracy. The boundary condition was then fixed using modes with turning points above 0.88 $R_\odot$. These modes fit the reference model almost perfectly—another astonishing success of the model S. To improve this accuracy somewhat, the model was rescaled to slightly bigger radius (using homology rescaling), by $1.5 \cdot 10^{-4}$ in relative units, which is about 100 km.

The initial residuals collapse to a single curve, which is nothing else but the manifestation of the high-degree asymptotic properties of solar p modes. Particularly pronounced is an almost abrupt change of slope of this curve at about 0.87–0.88 $R_\odot$, to very small residuals above this level. It indicates that starting almost abruptly from this level, the sound speed deviates from the model prediction, and the deviation growth with depth.

The inversion of the difference $\delta c^2(r)$ between the sun and the model in the convection zone (but not too close to its bottom) can be done by any technique, asymptotic or not. To make the inversion as transparent as possible, here I am choosing to do it by hand: increase the sound speed in the model slightly, just below the level of rapid variation in the residuals, get new model and new residuals, and move further down. The procedure works similar to solving a linear algebraic system with triangular matrix; after slight polishing of the solution, the residuals are reduced to an average level of some few nHz (lower panel of Fig. 7). The sound-speed difference allows to get the difference in the adiabatic exponent, by using the additional constraint of keeping stratification adiabatic. The result is shown in Fig. 8.

Using the constraint of adiabatic stratification is crucial in this inversion. In the final solution, about a half of the difference in the sound speed comes from the difference in the adiabatic exponent, another half—from the difference in density stratification.

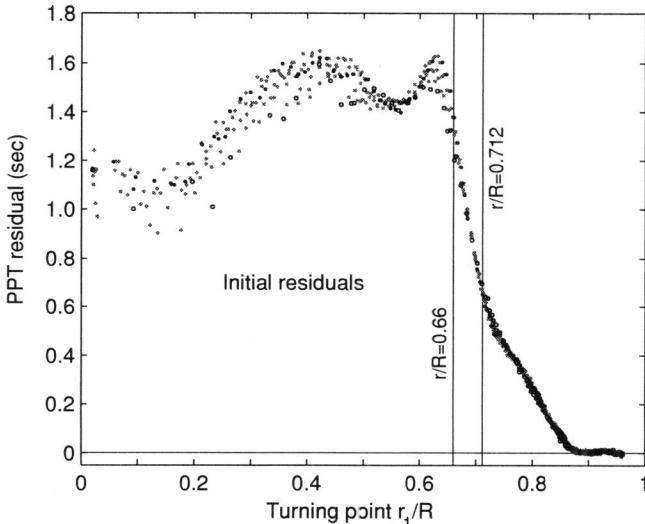

FIGURE 6. The initial mismatch between the reference solar model S and the input data, versus the position of the inner turning point. Gray circles indicate modes with frequencies between 2 and 3 mHz, black—with frequencies higher than 3 mHz, and open circles—with frequencies below 2 mHz.

Modes with turning points well above the base of the convection zone, which were used in this inversion, are essentially pure acoustic waves. They know only about the sound speed, and can not distinguish whether the sound-speed difference was caused by the difference in the adiabatic exponent, or in the density (mass distribution) profile.

Uniqueness of the Structural Inversions

To address all the family of possible solutions which satisfy the input data is an enormously difficult task. There is a way, however, which allows to at least get some feeling about the possible uncertainty in the result. It consists in constructing solutions which are "transparent" to the acoustic waves and which are not detectable by the measurement; these solutions can be added to the result of linear inversion without destroying its agreement with observations.

In Fig. 9, the reference solar model was distorted in its density profile such as to produce 10^{-3} variation in $c^2(r)$ at $r = 0.5R_\odot$, if $\Gamma_1(r)$ stays the same. The variations induced in the p-mode frequencies are shown in gray in the lower panel. We can supplement this variation in density profile with variation in $\Gamma_1(r)$, taken such as to make the sound-speed variation zero. Variations in frequencies are now shown in black. This signal in the solar p-mode frequencies is not detectable (or only marginally detectable). It means that the variation in the model with $\delta\rho$ shown in gray, $\delta\Gamma_1$ shown in black and with $\delta c^2 = 0$ can be added to solution without destroying its agreement with data.

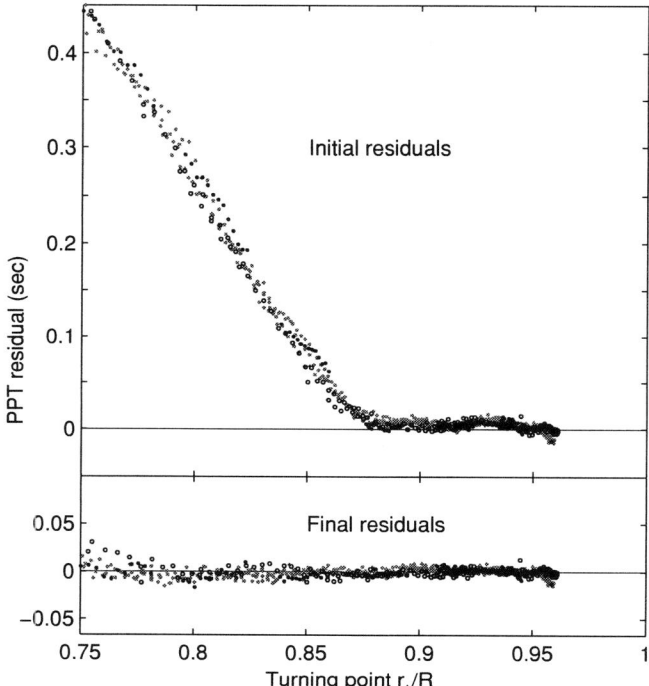

FIGURE 7. Upper panel—same as Fig. 6, but on enlarged scale. Lower panel shows the residuals which remain after the solar model was improved by the inversion.

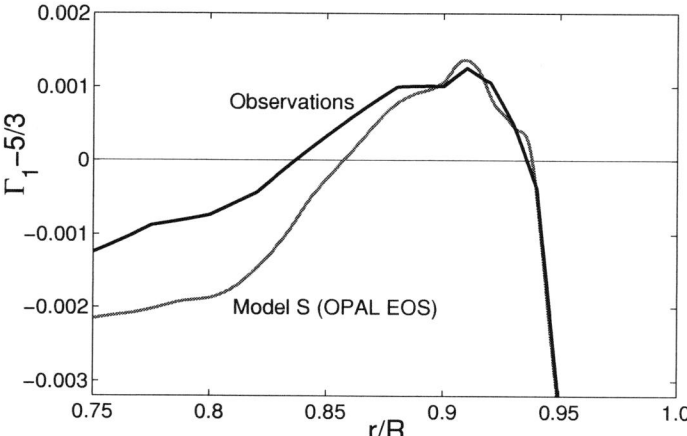

FIGURE 8. Profile of the adiabatic exponent resulted from the inversion (black line), and the initial profile in the reference model (gray line).

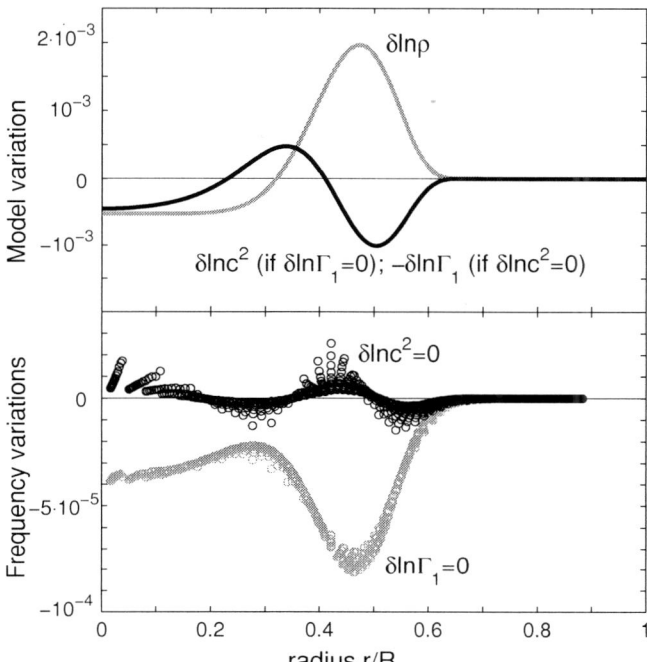

FIGURE 9. Artificial variations in the seismic solar model (upper panel), and the resulted variations in the p-mode frequencies (lower panel). The p-mode frequency range is 1.5–3.5 mHz.

Similar exercise, but with variation localised near the solar center is illustrated by Fig. 10. The frequency variations shown in black are not detectable at all (the accuracy of p-mode frequency measurements drops significantly when going to low degree ℓ).

These examples illustrate quantitatively the difficulties in measuring the adiabatic exponent (and density), if no additional constraints are imposed on the hydrostatic stratification. Possibility of separating the effects of $\delta\rho(r)$ and $\delta\Gamma_1(r)$ in the p-mode frequencies depends not only on the magnitude, but also on the spatial scale of the variation in the model compared with radial wavelength. The rapid variation near the base of the convection zone (Fig. 6) induces significant non-asymptotic effects in frequency variations. These effects can not be reproduced by e.g. simple changes in the adiabatic exponent only, and require mass redistribution in the model.

FURTHER PROSPECTS

The limiting factor in recent helioseismic inversions is the accuracy of frequency measurements. Using longer observations, we expect the precision of frequency measurements to improve as a square root of the observing time. Unfortunately, this is not being

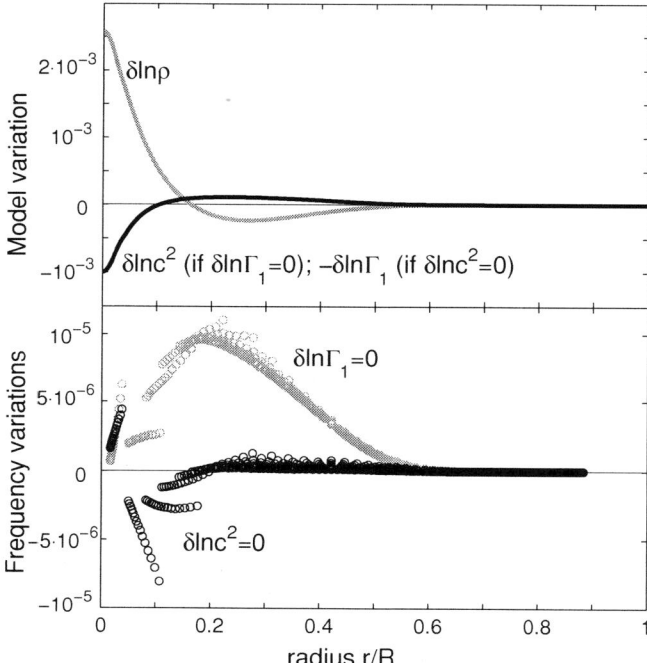

FIGURE 10. Same as Fig. 9, but with variation in the hydrostatic model localized towards the solar center.

realized. This is illustrated in Fig. 11, where the residuals of global inversion for the interior structure using eight years of SOHO MDI measurements do not show the expected reduction compared with inversion based on two years of data. Instead, we see an obvious correlation between the residuals, and their tendency to have an opposite sign at lower and higher frequencies. The reason behind is systematic errors in the measured p-mode frequencies.

The source of the systematic errors lies in the procedures of raw data analysis, when the p-mode frequencies are measured from the observational power spectra. Asymmetry of the resonant line profiles, cross-talks between spectra corresponding to different spherical harmonics, uncertainties in the background noise level make these measurements vulnerable to systematic errors.

New approach of streamlined helioseismic inversion is now under development, targeted at eliminating the frequency measurements from the data analysis, with replacing the inversion of the p-mode frequencies by the direct inversion of the observational power spectra [18]. This approach requires the precise and accurate global modelling of the observational power spectra over a wide range of frequency and degree. Recent progress in this direction is illustrated by Fig. 12, where two small pieces of the observational and synthetic power spectra are compared with each other, showing essentially

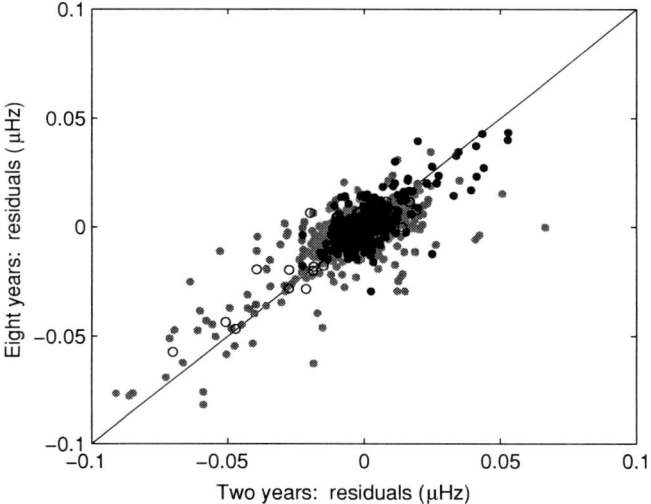

FIGURE 11. Residuals of global structural inversion with p-mode frequencies coming from two and from eight years of SOHO MDI measurements. Gray circles indicate frequencies between 2 and 3 mHz, black circles—above 3 mHz, and open circles—below 2 mHz.

a perfect agreement.

Using power spectra directly in the seismic inversion will require quite significant efforts. But as a byproduct of these studies, better spectral modelling which is now in hand will provide better measurement of the p-mode frequencies. Higher accuracy of the p-mode frequencies, expected in the near future, will improve the potential of solar seismology in the diagnostics of the equation of state.

ACKNOWLEDGMENTS

Author thanks the workshop organizers for their efforts resulted in the excellent, very productive and stimulating meeting. The data obtained from SOHO MDI measurements was provided by J. Schou. This work was supported in part by the UK PPARC under grant PPA/G/S/2003/00074. SOHO is a project of international cooperation between ESA and NASA.

REFERENCES

1. Antia, H. M. & Basu, S. 1994, Astrophys. J., 426, 801
2. Basu, S., Däppen, W. & Nayfonov, A. 1999, Astrophys. J., 518, 985
3. Basu S., Pinsonneault, M. H. & Bahcall, J. N. 2000, Astrophys. J., 529, 1084
4. Baturin, V.A. & Vorontsov, S.V. 1998, in: Sounding Solar and Stellar Interiors - Poster Volume, eds J. Provost and F.-X. Schmider, Nice: Obs. Cote d'Azur, p.67

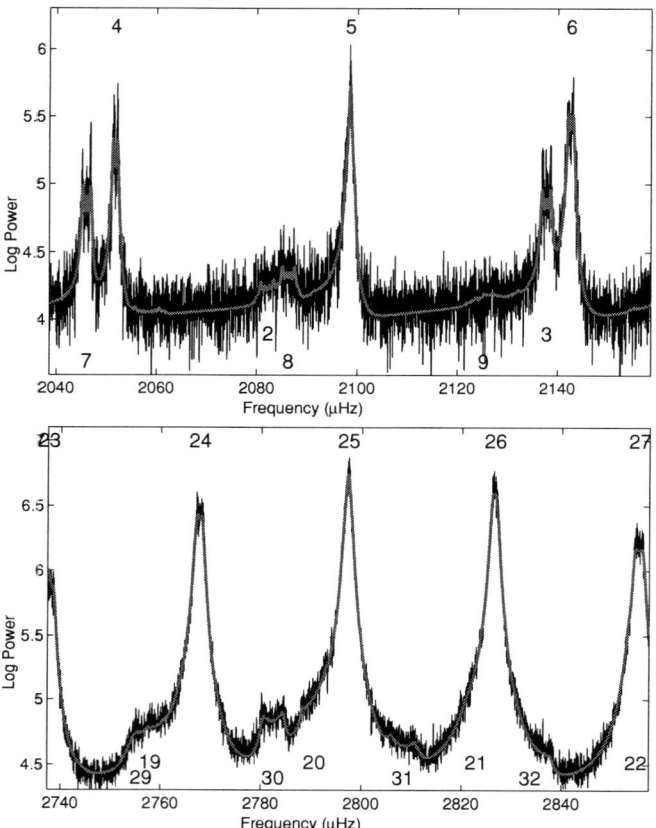

FIGURE 12. One-year, m-averaged SOHO MDI power spectrum (black line) and synthetic power spectrum (gray line), at target degree $\ell = 5$ (top) and $\ell = 25$ (bottom). Numbers at the top of each plot indicate degree ℓ of spectral leaks from modes of the same $n = 12$ ridge, position of the target peak is centered horizontally. Numbers at the bottom indicate degree ℓ of leaks coming from the neighboring $n = 13$ ridge (upper row) and $n = 11$ ridge (lower row).

5. Baturin, V. A., Däppen, W., Gough, D. O. & Vorontsov, S. V. 2000, Mon. Not. R. astr. Soc., 316, 71
6. Baturin, V. A. & Vorontsov, S. V. 2001, in: Helio- and Asteroseismology at the Dawn of the Millenium, ed. A. Wilson, ESA SP-464, p.615
7. Brodsky, M. A. & Vorontsov, S. V. 1993, Astrophys. J., 409, 455
8. Christensen-Dalsgaard, J., Thompson, M. J. & Gough, D. O. 1989, Mon. Not. R. asrt. Soc., 238, 481
9. Christensen-Dalsgaard, J. & Pérez Hernández, F. 1991, in: Challenges to Theories of the Structure of Moderate-Mass Stars, eds D. Gough & J. Toomre (Heidelberg: Springer), p.43
10. Christensen-Dalsgaard, J., Däppen, W., Ajukov, S. V. et al. 1996, Science, 272, 1286
11. Christensen-Dalsgaard, J. 2002, Rev. Mod. Phys., 74, 1073
12. Däppen, W., Mihalas, D., Hummer, D. G. & Mihalas, B. W. 1988, Astrophys. J., 332, 261
13. Di Mauro, M. P., Christensen-Dalsgaard, J., Rabello-Soares, M. C. & Basu, S. 2002, Astron. Astrophys., 384, 666
14. Elliott, J. R. & Kosovichev, A. G. 1998, Astrophys. J., 500, L199

15. Gough, D. O. 1984, Mem. Soc. Astron. Ital., 55, 13
16. Gough, D. O. 1991, Ann. Rev. Astron. Astrophys., 29, 627
17. Gough, D. O. & Vorontsov, S. V. 1995, Mon. Not. R. astr. Soc., 273, 573
18. Jefferies, S. M. & Vorontsov, S. V. 2004, Solar Phys., 220, 347
19. Korzennik, S. G., Rabello-Soares, M. C. & Schou, J. 2004, Astrophys. J., 602, 481
20. Kosovichev, A. G., Christensen-Dalsgaard, J., Däppen, W., Dziembowski, W. A., Gough, D. O. & Thompson, M. J. 1992, Mon. Not. R. astr. Soc., 259, 536
21. Libbrecht, K.G., Woodard, M. F. & Kaufman, J. M. 1990, Astrophys. J. Suppl., 74, 1129
22. Marchenkov, K. I. & Vorontsov, S. V. 1991, Solar Phys. 133, 149
23. Marchenkov, K. I., Roxburgh, I. W. & Vorontsov, S. V. 2000, Mon. Not. R. astr. Soc., 312, 39
24. Mihalas, D., Däppen, W. & Hummer, D. G. 1988, Astrophys. J., 331, 815
25. Pamiatnykh, A. A., Vorontsov, S. V. & Däppen, W. 1991, Astron. Astrophys., 248, 263
26. Pérez Hernández, F. & Christensen-Dalsgaard, J. 1994, Mon. Not. R. astr. Soc., 269, 475
27. Pijpers, F. P. & Thompson, M. J. 1992, Astron. Astrophys., 262, L33
28. Pijpers, F. P. & Thompson, M. J. 1994, Astron. Astrophys., 281, 231
29. Richard, O., Dziembowski, W. A., Sienkiewicz, R. & Goode, P. R. 1998, Astron. Astrophys. 338, 756
30. Rogers, F. J., Swenson, F. J. & Iglesias, C. A. 1996, Astrophys. J., 456, 902
31. Roxburgh, I. W. & Vorontsov, S. V. 1996, Mon. Not. R. astr. Soc., 278, 940
32. Strakhov, V. N. & Vorontsov, S. V. 2001, in: Helio- and Asteroseismology at the Dawn of the Millenium, ed. A. Wilson, ESA SP-464, p.539
33. Vorontsov, S.V. 1998, in: Sounding Solar and Stellar Interiors - Poster Volume, eds J. Provost and F.-X. Schmider, Nice: Obs. Cote d'Azur, p.135
34. Vorontsov, S. V. 1991, Astr. Zh., 68, 808 (English translation: Soviet Astr. 35, 400)
35. Vorontsov, S. V., Baturin, V. A. & Pamiatnykh, A. A. 1991, Nature, 349, 49
36. Vorontsov, S. V., Baturin, V. A. & Pamiatnykh, A. A. 1992, Mon. Not. R. astr. Soc., 257, 32
37. Vorontsov, S. V. 2001, in: Helio- and Asteroseismology at the Dawn of the Millenium, ed. A. Wilson, ESA SP-464, p.563
38. Vorontsov, S. V. 2002, in: From Solar Min to Max: Half a Solar Cycle with SOHO, ed. A. Wilson, ESA SP-508, p.107

PART II

PLASMAS OF STELLAR INTERIORS:
Basic Principles

The limit on mean field theories and nuclear screening in stellar plasmas

Giora Shaviv

Department of Physics, Israel Institute of Technology, Haifa, 32,000 Israel

Abstract. We discuss the physical conditions in the core of the Sun and Main Sequence stars and show that the basic requirement for the validity of the Debye mean field theory is not satisfied. We apply Molecular Dynamics to a gedanken numerical experiment and investigate the implications to the screening problem. We show how the thermodynamic limit is approached and under what conditions. Examples of actual screening enhancement calculated under these conditions are brought.

INTRODUCTION

Schatzman (1948) and Salpeter (1954) suggested that the environment affects the rate of nuclear reactions. The detailed theory in the weak screening limit was developed by Salpeter (1954) and extensions to the strong screening were developed later. The classical Salpeter theory is based on statistical mechanics and the assumption of mean field, which in the case of weak screening is the Debye potential.

We first show that the conditions in the Sun and Main Sequence stars are such that the number of particles in the screening cloud is ~ 1 and hence the mean field assumption is strictly speaking not valid. The invalidity of standard mean field approach requires special methods to explore the physics of systems with a small number of particle in the neutralizing (screening) cloud. We developed a special Molecular Dynamics method to handle this problem from first principles. The method allows us to trace the energy exchange between a scattering pair and the environment. We then explore the implications to the predicted enhancement of the rate of nuclear reactions in stellar plasmas.

The paper is structured as follows: We discuss the theory of nuclear reactions in stars and the scpecial conditions prevailing in stellar cores in general and in the Sun in particular.

The nature of the problem calls for re examining the foundations of the statistical theory and the mean field theory used to evaluate the screening enhancement. Next we investigate the physical problem using Molecular Dynamics with particular boundary conditions. We then discuss a numerical gedanken experiment which allows us to assess the role and influence of the potential parameters. Further, we show the connection between our results and dynamic friction and the Langevin equation.

Some results for the pp reaction in the solar core are presented and we conclude with conclusions. In general, the application of the mean field under such conditions over estimates the screening by large factors depending on the particular reaction.

NUCLEAR REACTIONS IN STARS AND SCREENING

The reaction rate in stellar plasma is given by:

$$R = n_1 n_2 \int f_2(\mathbf{x}_1, \mathbf{x}_2, \mathbf{v}_1, \mathbf{v}_2) \sigma(E_{kin-rel}) |\mathbf{v}_1 - \mathbf{v}_1| d\mathbf{v}_{12} \tag{1}$$

where $f_2(\mathbf{x}_1, \mathbf{x}_2, \mathbf{v}_1, \mathbf{v}_2)$ is the two body correlation function, n_i are the corresponding number densities of the reacting species, σ is the nuclear cross section and the integration is carried over all relative velocities. The standard BBGKY expansion starts with the representation:

$$f_2(\mathbf{x}_1, \mathbf{x}_2, \mathbf{v}_1, \mathbf{v}_2) = f_1(\mathbf{v}_1) f_1(\mathbf{v}_2)(1 + g(\mathbf{x}_1, \mathbf{x}_2, \mathbf{v}_1, \mathbf{v}_2)). \tag{2}$$

If the interaction between any two particles does not depend on momentum, like the simple Coulomb interaction, then in the thermodynamic limit we get that

$$g(\mathbf{x}_1, \mathbf{x}_2, \mathbf{v}_1, \mathbf{v}_2) \to g(\mathbf{x}_1, \mathbf{x}_2), \tag{3}$$

namely, we find that in the thermodynamic limit the pair correlation function becomes a product of a function which depends only on momentum and a function which depends only on space. In this limit we define a mean potential field as:

$$1 + g(\mathbf{x}_1, \mathbf{x}_2) = 1 + g(|\mathbf{x}_1, \mathbf{x}_2|) = e^{-\phi(x)/kT}, \tag{4}$$

where $x = |\mathbf{x}_1 - \mathbf{x}_2|$. Let the classical turning point (at the relevant Gamow peak) be r_{tp}, and let the scale for the changes in ϕ be R_ϕ, then if

$$R_\phi \gg r_{tp} \tag{5}$$

then we can approximate the rate by:

$$R = n_1 n_2 e^{-\phi(0)/kT} \int f_1(|\mathbf{v}_2 - \mathbf{v}_1|) \sigma(E_{kin-rel}) |\mathbf{v}_1 - \mathbf{v}_1| d\mathbf{v}_{12} \tag{6}$$

which is known as the Salpeter weak screening approximation. Salpeter assumed for mean field $\phi(r)$ the Debye potential, namely

$$\phi \to \phi_{Debye} = \frac{e^{-r/r_D}}{r}; \quad r_D = \sqrt{\frac{kT}{8\pi e^2 n}}, \tag{7}$$

where e is the electron charge and n the number density (when there is a mixture of species one has to sum over all components including the electrons). Define now the plasma parameter as:

$$\Gamma = \frac{Z_1 Z_2 e^2}{\langle r \rangle kT}, \quad \langle r \rangle = n^{-1/3}, \tag{8}$$

so that the classical turning point becomes:

$$\frac{r_{tp}}{\langle r \rangle} = \frac{Z_1 Z_2 e^2}{\langle r \rangle \eta kT} = \frac{\Gamma}{\eta}, \tag{9}$$

where $\eta = E_{Gamow}/kT$, namely the peak Gamow energy for the reacting particles given in units of kT. In the case of the Sun, for $Z_1 = Z_2 = 1$, $\Gamma = 0.06$ and $\eta_{pp} = 5$, $\eta_{Be+p} = 18$. On the other hand for the solar core $r_D/\langle r \rangle = 0.88$. Thus for the pp reaction we $r_{tp} \sim 0.01$ and the condition for weak screening is satisfied. Note however, that the width of the Gamow peak for the pp reaction is $\sim 5kT$ and hence a significant part of the pp reaction takes place at a larger r_{tp} where the approximation in marginal. The situation improves as the charge of the reacting species increases.

THE CONDITIONS IN THE SUN

For the Debye theory to be valid, the number of particles in a Debye sphere N_D, must be much larger then unity. The calculation of N_D in the case of the solar core ($T_6 = 15.5, \rho = 155$ and $X = 0.35$ and $\rho_{crit} = 1451 g/cc$) yields $N_D \sim 3 - 4$ and hence the mean field theory is not expected to be strictly valid. We need a new way which is not based on the statistics of a large number of particles, to solve the problem.

WHEN MEAN FIELD IS NO GOOD

A good way to approach the problem when the number of particles in the Debye sphere does not justify a statistical approach is Molecular Dynamics (MD). In Molecular Dynamics we start from first principles and should get the statistical limit results when appropriate.

To this goal we developed a Molecular Dynamic program which treats the solar plasma from first principles, for example a frequently used approximation in plasma simulations is the Debye potential. This approximation simplifies the calculations significantly. We did not use this approximation as it already implies some correlation in the plasma. The potential used is the Kelbg potential which takes only quantum effects (which under these specific conditions are small anyway).

The canonical MD method assumes periodic boundary conditions. The advantage is the small number of particles in the cell and the ability to use the Ewald sums. However, since we are interested in the interaction between the environment and the scattering protons, the energy exchange between the two systems depends on the recoil of the particles in the environment (not the recoil of the scattering pair) with which our specific pair interacts. The imaging method, the basis of the Ewald sum, ignores this effect (all images have the same recoil). For this reason, we calculated the direct sum over all interacting particles going up to several Debye lengths and including up to 1200 particles (in contrast to the $3 \sim 4$ in the Debye sphere). Thus, the recoil and its associated energy, are properly taken into account.

We define the screening energy as follows: The screening energy is the relative kinetic energy of the scattering particles at the classical turning point minus the relative kinetic energy when far apart, all evaluated in the Center of Mass (CM) system.

Recall, the pair interacts with the plasma and may lose/gain energy to the plasma. Here we calculate this energy exchange by comparing the dynamic quantities when the

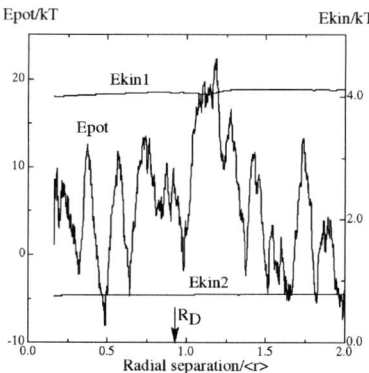

FIGURE 1. The time dependence of the potential energy as a function of distance for a typical a pair of scattering protons in an actual simulation of the Solar plasma. Also shown are the kinetic energies which show a significantly smaller change. The arrow marks the Debye radius.

particles are at the point of closest approach and when far away. First, a close encounter between two particles is identified and then the pair is then followed through the point of closest approach to the separation of few Debye radii. The comparison of dynamic quantities is only for a given pair. The approaching (and separating) pair interacts with the the environment in a non adiabatic way.

In fig.(1) we show the dependence of the potential and kinetic energy of two scattering protons under the conditions prevailing in the solar core. Also shown is the location of the Debye radius. In this particular example, which is calculated for the conditions in the solar core, we see (a) the potential field fluctuates and a smooth mean field potential does not exists. (b) the kinetic energies do not respond to the changes in the potential energies because (1) there is no conservation of energy for the pair of scattering particles only, and (2) the fluctuations in the potential energy are due mostly to the surrounding electrons and very little energy is transfered form the light electron to the massive protons.

In Fig.(2) we see the same results as a function of time on a short and an expanded scales. We see that the fluctuations in both particles take place even when the two particles are close to r_{tp}. The fluctuations are not identical in the two particles (they are not inside the same fluctuating potential). The discontinuity observed in the potential energy near $t = 1.5$ is due to the fact that the present pair separated and a new pair is formed. The right panel shows the behaviour of one particle on a longer time scale. Again, the lack of correlation between the changes in the potential and kinetic energies changes is clear. Also observed is the gradual drift in the mean kinetic energy.

The conclusions so far are:

- Fluctuations control the p-p scattering.
- The protons scatter from a (time dependent) stochastic potential created by the fluctuations (caused by random passage of few particles).
- The expectation that mean field approximation should not be a good description for p-p scattering under such conditions is fully verified.

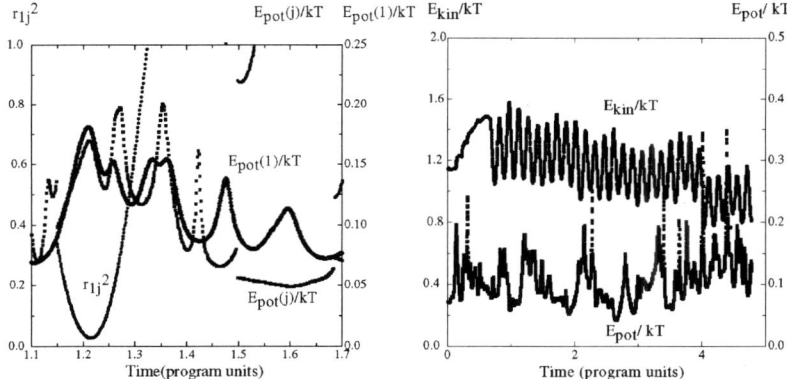

FIGURE 2. The same as figure 1 but as a function of time. The left panel provides the information for the two scattering particles on a short time scale while the right panel described the variation of the energies of a single particle on a longer time scale. The unit of time is the time needed for a particle with energy kT to cross a distance $\langle r \rangle$.

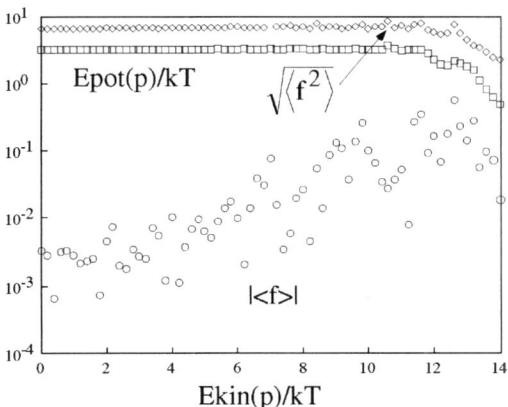

FIGURE 3. The average potential energy, the mean value of the absolute force $\sqrt{\langle f^2 \rangle}$ and the mean force $|\langle f \rangle|$ as a function of absolute kinetic energy. All averages are over thermodynamics time scale, ensemble averages.

THE DIFFERENCE BETWEEN THERMODYNAMIC LONG RANGE AVERAGING AND SHORT TIME AVERAGING

To appreciate the discussion here we feel that it is necessary to clarify the difference between the long time thermodynamic averaging and the short collision time scale averaging.

In fig.(3) we show the long time average of the force and potential energy as a function of the kinetic energy in the laboratory. The figure was obtained in the following way:

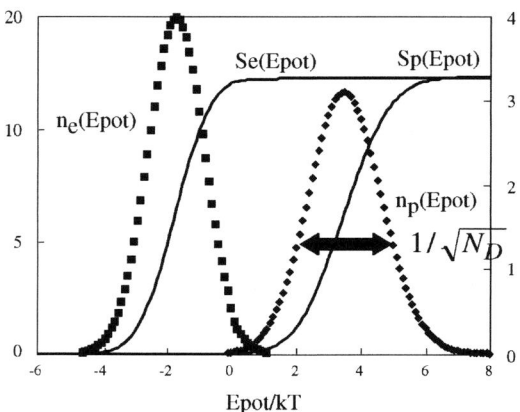

FIGURE 4. The potential energy distribution of protons and electrons for pure Hydrogen plasma at $n = 10^{26}$ and $T = 1.5 \times 10^7$K. As the plasma is neutral the integral $S_e = S_p$. The width is proportional to $1/\sqrt{N_D}$.

we checked all particles and found inside what energy bin each of them is. Then the potential energy was stored for later averaging. The bins are bins of kinetic energy in the laboratory. Contrary to previous claims (Carraro et al 1988) about the existence of a dynamic screening in the sense that the potential energy of the particle is a function of the kinetic energy, we confirm that the long time thermodynamic average of the potential energy is not a function of time.

In fig.(4) we show a snap shot of the potential energy distribution among the particles. We find that the potential energies has a Gaussian distribution the width of which is $1/\sqrt{N_D}$, and since in the present case $N_D \sim 1$ the width is significant. Thus, contrary to the standard screening theory which is appropriate in the limit $N_D \gg 1$ when the distributions are very narrow, the colliding particles in the case of the Sun come with a spread of energies. This fact alone has an effect on the effective enhancement but it also demonstrates once more that the conditions are far from the applicability of the mean field. The energy per particle is the difference between the two Gaussian distributions. As the difference is between two large numbers, expect large fluctuations.

THE GEDANKEN NUMERICAL EXPERIMENT

The fundamental interaction in the plasma is a long range Coulomb interaction. However, in the plasma the interaction between positive and negative charges leads to a finite effective range. The existence of positive and negative charges dictates right away an effective finite interaction radius. Many of the conceptual difficulties in appreciating the effect of the finite radius and small number of particles in the Debye sphere result from the so dictated *effective* interaction range which leads to the result of a small number of particles in a screening cloud. Why there are under such conditions deviations from the statistical mechanics result? To understand the role of a finite effective interaction with a range of the order of the inter-particle distance we decided to carry out a gedanken numerical experiment in which the gedanken potential has a controllable range and strength which are not dictated by the density and temperature in an unchangeable manner. The basic idea therefore is: conduct a numerical experiment with a hypothetical potential in such a way that the parameters of the potential can be varied at will keeping the thermodynamic conditions fixed. In this way we hope to learn how the thermodynamic limit is reached and what happens before this limit becomes a good approximation. Such a potential is for example:

$$\begin{aligned} V(r > R_n) &= 0 \\ V(r < R_n) &= C_f(1/r - 1/R_n). \end{aligned} \qquad (10)$$

The strength of the interaction C_f and the range R_n are free parameters.

Results

In fig.(5) we show the distribution of potential energy for $R_n > 1$. The distribution approaches a guassian with a decreasing width. Thus, as N_D increases, the distribution narrows. On the right panel we see the potential energy distribution measured when the particles are at the distance of closest approach and far away. In this particular case we see that when the particles are close, the mean as well as the distribution of the potential energies are higher. In any case, measuring the potential energy with a bias leads to results which deviate from the ensemble distribution. Note that the c distribution is significantly more asymmetric than the f distribution. The arrow marks the ensemble average. It corresponds to the peak of the f distribution.

In fig.(6) we see the distribution of the potential energy for $R_n < 1$. We find that the distribution is not narrow nor guassian. The distribution vanishes for $E_{pot} = 0$ then rises very quickly to a maximum value from which it declines slowly. Also shown is the difference between the distribution of potential energy when close and when far away for one particular case. In the ensemble mean, the cases of close particles are averaged over and have no effect. Here these cases are singled out (a bias) and hence the difference.

The ensemble average of the potential energy per particle should be summed over all positions of particles. However, if we bias the averaging, say average over particles only when they are close to each other, a different result may be obtained. This bias is seen in the right panels of figs. (5) and (6).

The resulting screening potential for $R_n = 0.4 \langle r \rangle$ and $R_n = 10 \langle r \rangle$ is shown in fig.(7). The dependence on the relative kinetic energy is clear. For high R_n most of the change takes place for $R_n \sim 1$ while for $R_n < 1$ it is spread over a wide range in relative kinetic energy. In the thermodynamic limits ($N_D \gg 1$) we should get a constant function, namely complete independence on relative kinetic energy. A summary of several cases is shown in fig.(8), where the distributions of the screening potential (as a function of relative kinetic energy) for several ranges are given on the same scale. It is evident that as R_n increases the dependence on energy decreases. As one approaches the thermodynamic limit ($N_D >> 1$) the energy dependence disappears, as expected. On the right panel of

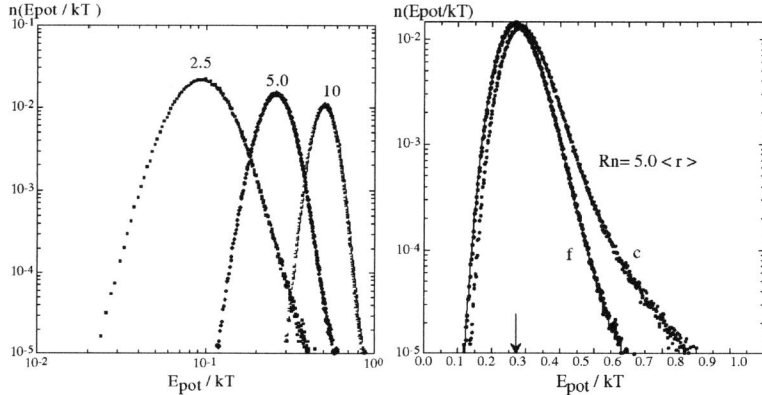

FIGURE 5. The left panel shows the potential energy distribution for potential ranges larger than unity. (The numbers are $R_n/\langle r \rangle$). The right panel compares the potential energy distribution when the particles are at the distance of closest approach with the case when they are far apart. The arrow points to the ensemble mean potential energy.

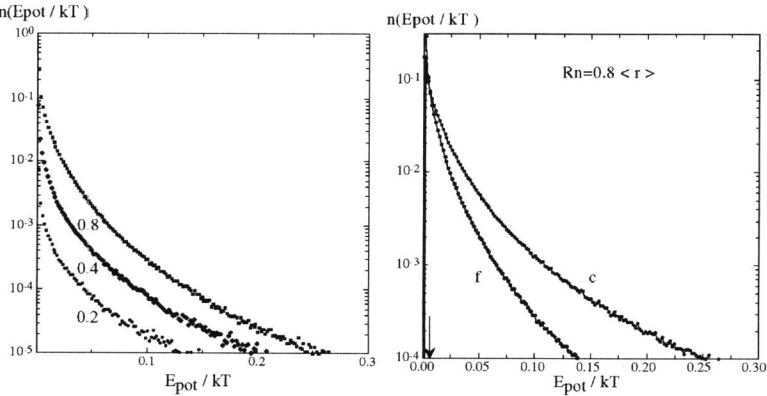

FIGURE 6. The distribution of potential energies for a potential range smaller than unity. (The numbers are $R_n/\langle r \rangle$). The left panel shows the very asymmetric distribution in this case. The right panel shows the comparison between the distribution when the particles are close and far away. The arrow marks the ensemble average.

fig.(8) we show the standard deviations. We see that the relative fluctuations decay with increasing R_n.

The non linearity of the effect is shown in fig.(9) where we compare the results for two coupling constants, ($C_f = 1$ and $C_f = 10$). The ratio of the screening is significantly larger than the ratio of the potential strengths for large relative kinetic energies and vice versa.

The tendency of the ensemble average towards the potential energy (per particle) at the distance of closest approach for particles in the Gamow peak is shown on the right panel of fig.(10) along with the ratio of the present screening calculation to the classical Salpeter value. For $R_n \gg 1$ the Salpeter thermodynamic value is essentially recovered, while for $R_n \leq 1$, where the mean field Debye theory is not expected to be valid, large deviations are found. The actual screening is smaller than the thermodynamic value. The very large fluctuations do not average to the thermodynamic ensemble average in this range. Probably because the amplitude of the fluctuations is much greater than the mean value ($\sigma(E) \gg \langle E \rangle$).

When $N_D \sim 1$ the energy change by the center of mass due to the non spherically symmetric interaction (as well as fluctuating potential) is not negligible. The average absolute energy change of the center of mass is shown in fig.(11). Again, as $N_D \to \gg 1$ the energy change of the CM diminishes and vice versa, as $N_D \sim 1$ the energy change increases along with the increase in the fluctuations. The energy gained/lost by the CM system is an expression for the deviation of the effective interaction between the scattering particles from a simple central force interaction. On the other hand, the energy change of the center of mass is a critical part of the energy exchange between the scattering pair and the environment.

As is well known, $g(r,E) \to g(r)$ in the thermodynamic limit. In fig.(12) we show that for $R_n = 5\langle r \rangle$ this is almost the case while for $R_n = 0.4\langle r \rangle$ the energy dependence of the correlation function is still obvious.

We conclude that:

- The range of the potential in units of the mean inter-particle distance has profound effect on the validity of the mean field approximation. For the latter to be valid and for the results from statistical mechanics to have a good accuracy, the range must be sufficiently large to include a high number of particles in the screening cloud.
- The effect is seen in the dependence of the screening potential as well as the dependence of the correlation function, on relative kinetic energy.

THE FOKKER-PLANCK EQUATION

The gedanken experiment has shown how N_D affects the applicability of the statistical limit. General results can be examined from a wider point of view of energy exchange in particle collision in plasmas. What do we expect in steady state? Let $P(E - \Delta E, \Delta E)$ be the probability that a particle with energy E changes its energy by ΔE during collision time τ. The particle distribution must satisfy

$$f(E,t) = \int f(E - \Delta E, t - \tau) P(E - \Delta E, \Delta E) d\Delta E. \tag{11}$$

Expanding the integral yields:

$$f(E - \Delta E, t - \tau) P(E - \Delta E, \Delta E) = f(E,t) P(E, \Delta E) - \tau \frac{\partial f(E,t)}{\partial t} P(E, \Delta E)$$

$$- \Delta E \frac{\partial}{\partial E} [f(E,t) P(E, \Delta E)] + \frac{(\Delta E)^2}{2} \frac{\partial^2}{\partial E^2} [f(E,t) P(E, \Delta E)] \tag{12}$$

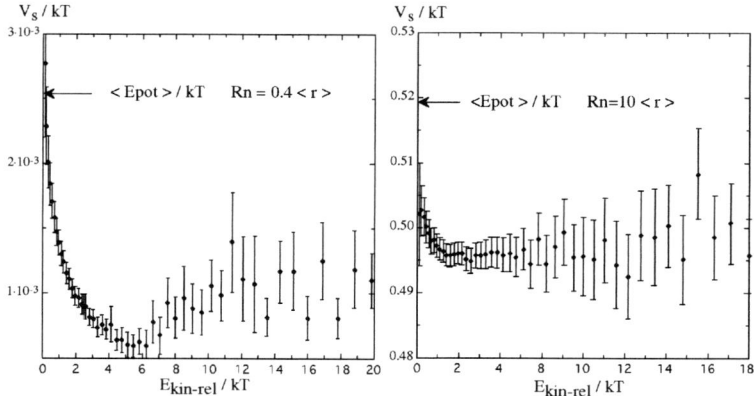

FIGURE 7. The screening potential as a function of relative kinetic energy for two cases: on the left panel $Rn = 0.4\langle r\rangle$ and on the right panel $Rn = 10.\langle r\rangle$. The arrow marks the ensemble mean. The error bars are due to statistics only.

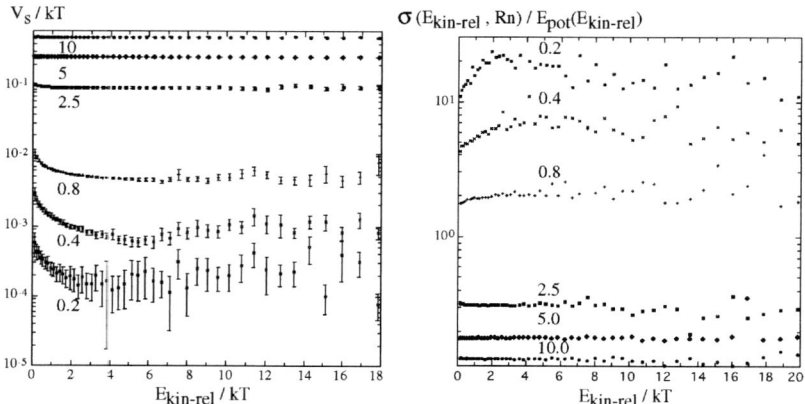

FIGURE 8. The screening potential (left panel) and the standard deviation (right panel) for several potential ranges. Note that σ is relative.

Clearly, in steady state:

$$\int f(E,t)P(E,\Delta E)d\Delta E = 0. \tag{13}$$

Define:

$$\langle \Delta E(E)\rangle = \int_{-\infty}^{\infty} P(E,\Delta E)\Delta E\, d\Delta E, \quad \langle \Delta E^2(E)\rangle = \int_{-\infty}^{\infty} P(E,\Delta E)(\Delta E)^2 d\Delta E. \tag{14}$$

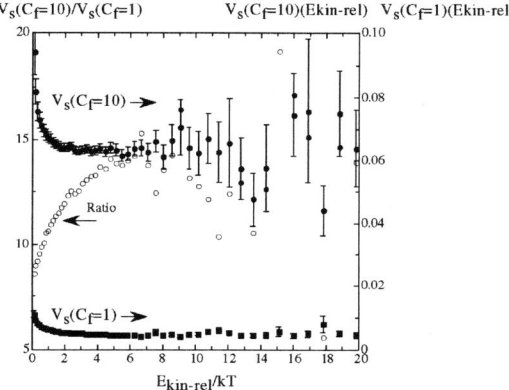

FIGURE 9. A comparison between two potential strengths. The open circles are the ratio between the two results (full circles $C_f = 10$) and full squares $C_f = 1$) and as can be seen the ratio (open circles) in the screening is not constant and is not equal to the ratio between the two strengths. The error bars on the results are small for low energies only in the case $C_f = 10$ because of a shorter run and hence poorer statistics. This is not the case for $C_f = 1$.

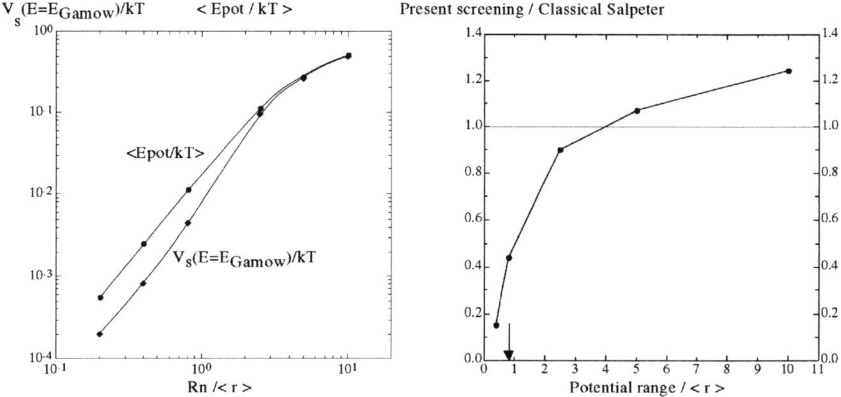

FIGURE 10. The ratio of the ensemble average screening potential to the mean screening potential at the Gamow peak as a function of potential range (left panel). The ratio of the present screening (in the gedanken numerical experiment) to the Salpeter thermodynamic value as a function of potential range.

These are the coefficients of the Fokker-Plank equations. Consider a steady state, then

$$\frac{\partial}{\partial E}[f(E,t)\langle \Delta E(E)\rangle] = \frac{1}{2}\frac{\partial^2}{\partial E^2}[f(E,t)\langle \Delta E^2(E)\rangle] \quad (15)$$

which integrates to

$$f(E,t)\langle \Delta E(E)\rangle = \frac{\partial}{\partial E}[f(E,t)\langle \Delta E^2(E)\rangle]. \quad (16)$$

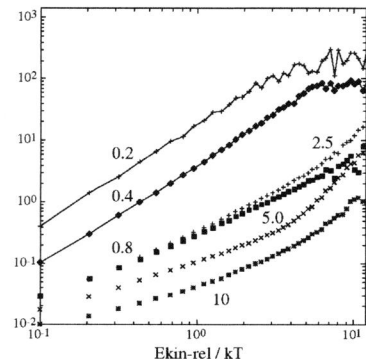

FIGURE 11. The change in the absolute value of the energy of the CM (relative to the laboratory) for different potential ranges.

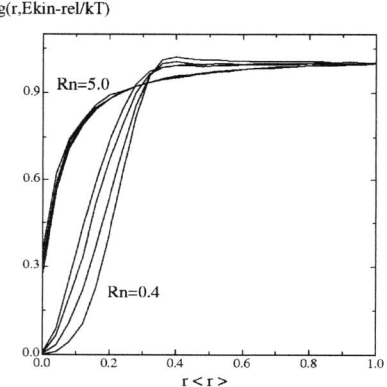

FIGURE 12. The pair correlation function for two values of the potential range. The different curves are for different relative kinetic energies ($E = 0 - 2\text{kT}, 2 - 4\text{kT}, 4 - 6\text{kT}, > 6\text{kT}$).

Integrate once more to get:

$$\int_0^\infty f(E)\langle \Delta E(E)\rangle dE = \int_0^\infty \frac{\partial}{\partial E}\left[f(E,t)\langle \Delta E^2(E)\rangle\right] dE = \left|f(E,t)\langle \Delta E^2(E)\rangle\right|_0^\infty = 0 \quad (17)$$

As $f(E)$ is a positive function, it means that $\langle \Delta E(E)\rangle$ either vanishes identically or changes sign. Hence, for some energies there is an energy gain while for others there must be energy loss. In the Salpeter theory $\langle \Delta E(E)\rangle \equiv 0$. A particle does not gain nor lose energy in a collision. The potential is rigid as if it had an infinite mass.

In reality, $\langle \Delta E(E)\rangle \neq 0$ irrespective if the plasma is in steady state or not. What is the connection to the screening? While here we discuss the entire energy change in a collision, the screening problem discusses just the approach or the separation. Define $P_{in}(E_{in}, \Delta E)$ as the probability that a particle with energy E_{in} will gain energy ΔE_{in} as it comes from infinity, scatters from particles that compose the environment and reaches the distance of closest approach. Similarly, define $P_{out}(E_{out}, \Delta E_{out})$ as the probability that a particle with energy E_{out} (at the point of closest approach) will gain energy ΔE_{out} as it moves out and separates. The probability for the particle to come close and then separate and gain energy $\Delta E = \Delta E_{in} + \Delta E_{out}$ is:

$$P(E_{in}, \Delta E) = \int_0^\infty P_{in}(E_{in}, \Delta E_{in}) P_{out}((E_{in} + \Delta E_{in}, \Delta E - \Delta E_{in}) d\Delta E_{in} \quad (18)$$

where $\Delta E = \Delta E_{in} + \Delta E_{out}$. The classical Salpeter approximation is:

$$\begin{aligned} P_{in}(E, \Delta E_{in}) &= \delta(U - \Delta E_{in}) \\ P_{out}(E_{close}, \Delta E_{out}) &= \delta(U - \Delta E_{out}) = \delta(U - \Delta E + \Delta E_{in}) \end{aligned} \quad (19)$$

where U is the mean electrostatic energy of the particle, or $P(E, \Delta E) = 0$ for $\Delta E \neq 0$. Clearly, it is a special approximation. We calculated the mean energy gain/loss per collision. The results are shown in fig.(13) for the conditions of the solar core. The figure provides the probabilities of particles with given energy to gain energy (marked by 'up') and lose energy (marked by 'down'). The difference between these two probabilities creates the result for $\langle \Delta E(E) \rangle$. We see how the mean energy change per collision changes sign at about 1kT. Hence, the energy gained by a pair of particles approaching each other to the distance of closest approach is not constant.

The traditional FP is expressed in terms of changes in velocities so that $\langle \Delta E(E) \rangle \to \langle \Delta \mathbf{v} \rangle$ and $\langle \Delta E^2(E) \rangle \to \langle \Delta \mathbf{v} \Delta \mathbf{v} \rangle$ and the FP eq. is then:

$$\tau \frac{\partial f}{\partial t} = -\frac{\partial}{\partial \mathbf{v}} (f \langle \Delta \mathbf{v} \rangle) + \frac{1}{2} \frac{\partial^2}{\partial \mathbf{v} \partial \mathbf{v}} (f \langle \Delta \mathbf{v} \mathbf{v} \rangle). \quad (20)$$

In the case of pure and uniform Hydrogen plasma

$$\langle \Delta \mathbf{v}_2 \rangle = \frac{ne^4}{4\pi\varepsilon_0^2 \mu m} \ln \Gamma \int f(\mathbf{v}_1) \frac{\mathbf{v}_2 - \mathbf{v}_1}{|\mathbf{v}_2 - \mathbf{v}|^3} d\mathbf{v}_1 \quad (21)$$

Define now a generalised potential H_{12} so that

$$\langle \Delta \mathbf{v}_2 \rangle = C \frac{\partial H_{12}}{\partial \mathbf{v}_2}; \quad C = \left(\frac{e^2}{2\varepsilon_0 m}\right)^2 \frac{\ln \Lambda}{\pi}. \quad (22)$$

In the case of Hydrogen plasma and ignoring the electrons one can show that

$$H_{12}(v) = n \frac{m}{\mu} \int \frac{f(v_1)}{|v - v_1|} dv_1 \quad (23)$$

where n is the number density, m the mass of the particle and μ is the reduced mass. H_{12} is the first Rosenbluth potential. The second Rosenbluth potential yields the second coefficient of the FP eq. We show in fig.(14) the first Rosenbluth potential. It leads to energy gain for low energy particles and energy loss for high energy particles, just as expected.

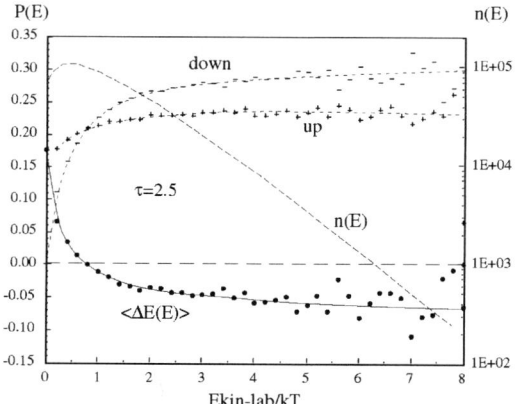

FIGURE 13. The mean energy gain/loss per collision for a plasma in steady state. The calculation is for a collision time of 2.5 in units of the thermal velocity.

THE LANGEVIN EQUATION FOR THE SCATTERING OF TWO PROTONS IN THE PLASMA

Consider two protons in the plasma. The governing equation in the center of mass is:

$$\mu \frac{d^2 \mathbf{r}_{12}}{dt^2} = \frac{e^2 \mathbf{r}_{1,2}}{r_{12}^3} + \sum_{j \neq 12} \frac{e^2 \mathbf{r}_{jCM}}{r_{jCM}^3}. \qquad (24)$$

The first term is the interaction between the scattering protons. The second term, where r_{CM} is the center of mass, is due to the interaction between each of the scattering pair and the environment. In a mean field theory, the second term is replaced by a mean field (which is central and hence yields no change in the energy of the center of mass system). In the classical Langevin equation theory we have:

$$\langle \sum_{j \neq i} \frac{e^2 \mathbf{r}_{ij}}{r_{ij}^3} \rangle \rightarrow m\mathbf{F}_d(\mathbf{v}) + m\mathbf{Q}(v) \cdot \mathbf{S}(t). \qquad (25)$$

Here $\mathbf{S}(t)$ is a random vector and $\mathbf{Q}$ is a tensor from which the Langevin force is derived. The dynamic friction is given by the Rosenbluth potential. The detailed solution is described elsewhere. Here we show the result, namely the energy loss/gain by a pair of scattering protons in the plasma. The solution of the Langevin equation assuming the Rosenbluth potential is shown in fig.(14).

RESULTS FOR THE SOLAR PLASMA

A typical result for the screening enhancement under solar conditions in shown in fig. (15). We see a reduction of about 75% in the enhancement factor for the *pp* reaction.

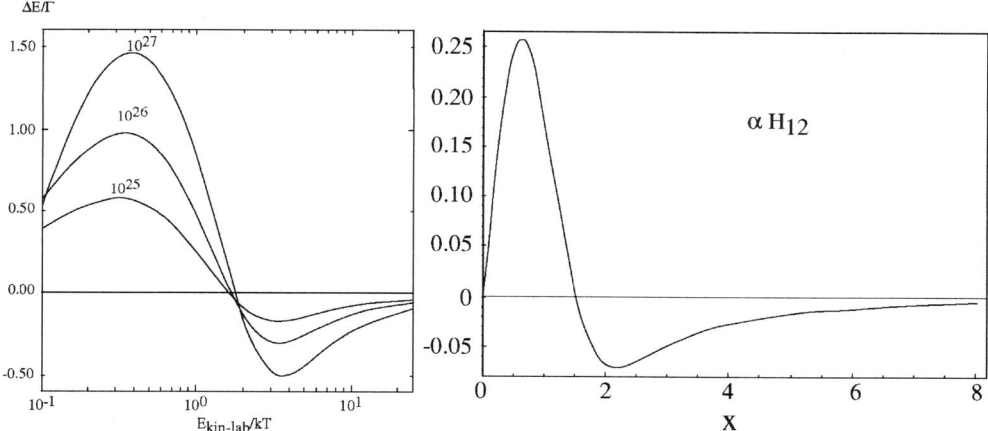

FIGURE 14. The energy gain/lost by a pair of colliding particles as a function of the kinetic energy in the laboratory for different densities, base on the Langevan equation and Rosenbluth potential.

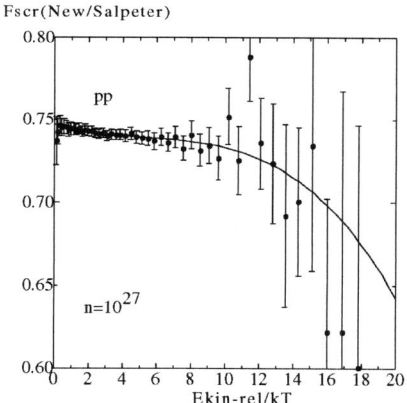

FIGURE 15. The ratio of the new screening factor to the classical Salpeter value as a function of relative kinetic energy. The result should be averaged over the Gamow peak to obtain the figure required for stellar rates calculation.

The amount of decrease in enhancement increases with the charges of the ions.

CONCLUSION

Due to the particular thermodynamic conditions in the Sun and Main Sequence stars, the classical Debye mean field theory is not strictly valid as $N_D \sim 1$. Several results affecting the screening potential and the pair correlation function were discussed. When this basic fact is ignored (cf. Bahcall et al 2001) the effect of the screening is exaggerated by a factor of two and more.

The implication for nuclear reactions under stellar conditions is a reduced enhancement by screening relative to the predictions of the classical Salpeter theory which is based on the assumption that the number of particle in the screening cloud is very large.

ACKNOWLEDGMENTS

I would like to thank the Asher Space Research Fund for partial support of this research.

REFERENCES

Bahcall, N.J., Brown, L.S., Gruzinov, A. & Sawyer, R.F. 2001, A & A, 383, 291
Carraro,C.; Schafer,A.; Koonin,S.E., 1988, ApJ, 331, 565C
Salpeter, E.E. 1954, Austranlian J. Physics, 7, 373
Schatzman, E. 1948, J. Phys. Rad. 9,46
Shaviv, N.J. & Shaviv, G., 1997, ApJ SSLib 214,3
Shaviv G. & Shaviv ,N.J. 2000, ApJ. 529, 473
Shaviv, G., 2000 astroph-0010152

Corrected EOS of Weakly-Nonideal Hydrogen Plasmas without Mysteries

Andrey N. Starostin[1] and Vitali C. Roerich

Troitsk Institute for Innovation and Fusion Research, Troitsk, Moscow region, 142190 Russia

Abstract. The detailed derivation of hydrogen weakly-nonideal plasmas EOS is presented. For the model hydrogen plasmas the values of computed sound velocity and adiabatic exponent along the Sun trajectory are presented. In our computations the relativistic corrections, degeneracy of electrons, radiation pressure in plasma, Coulomb interaction in Debye-Hückel approximation together with diffraction and exchange corrections, and the contribution of bound and scattering states are taken into account.

1. INTRODUCTION

Helioseismology opens a unique possibility to check within accuracy better than 10^{-4} the equation of state (EOS) of weakly-nonideal hydrogen plasmas due to the inversion of local sound velocity from optical observations [1–3]. The comparison of different theoretical models with experiment permits to check the existing ways to count bound and scattering states contributions, presented in physical literature for the second virial coefficient (SVC) [4–6].

It is established, at least in physical literature, that following from the papers [7–12], also reviewed in monographs [13, 14] (where numerous references are done), that the question of EOS for weakly-nonideal hydrogen is solved in principle up to the Γ_D^2, where $\Gamma_D = (e^2 \varkappa_D)/T$ — is the Debye nonideality parameter, e — is the electron electrical charge, $\varkappa_D$ — is the inverse Debye radius, T — is plasma temperature in energy units. In particular, it is commonly agreed, that a contribution from bound states (atomic partition functions) into plasma pressure is presented by convergent Planck-Brillouin-Larkin expression, or simply Planck-Larkin (P-L) [6–11, 13, 14].

At the same time, in astrophysical literature the atomic partition function is used in the form, presented in [4], and P-L formula arouses critics [15], due to the fact of P-L formula gives convergence at principal quantum numbers $n \lesssim n_{max} \sim \sqrt{Ry/T}$ ($Ry = \frac{\mu e^4}{2\hbar^2} \approx 13.598$ eV — is the hydrogen ionization potential, $\mu = \frac{m_e m_p}{m_e + m_p}$ — is reduced mass for electron-proton system, m_e, m_p — are masses of electron and proton, $\hbar$ — is the Planck constant), while in the Sun photosphere the radiation from Balmer series with $n \approx 17 > 6 \gtrsim n_{max}$ is observed [16].

---

[1] Corresponding author e-mail: A.Starostin@relcom.ru

An expression, different from P-L formula for the bound states contribution was published in the paper [17]. This formula was later confirmed in the works [18, 19]. In this paper we draw attention to internally contradicting way of obtaining convergent results for existing in literature SVC formulas, (see for example [13, 14]), as for their derivation Beth-Uhlenbeck formula [20, 21], Levinson theorem [22] and the regularization methods for divergent expressions [7–9, 11] are used, which applicability for long-range Coulomb interaction is not obvious.

2. THERMODYNAMIC PERTURBATION THEORY

It is possible to use Matsubara technique [23, 24], or nonequilibrium Kadanoff-Baym-Keldysh technique [25–27] to compute the thermodynamic functions of weakly-nonideal hydrogen plasmas. In some cases we also use the second way, as it permits to have the generalization for nonequilibrium situation and to take into account the effects of broadening of atomic states, observed in spectroscopy, on thermodynamic functions.

According to [25] we can compute plasmas pressure P, using corrections to the pressure P_0 of ideal gas, consisting from electrons and protons [28]:

$$P = P_0 + P_H + P_{exch} + P_{D-H} + \delta P, \qquad (1)$$

where we include the following corrections, discussed later in this chapter: P_H — Hartree correction, P_{exch} — electron-electron exchange interaction, P_{D-H} — plasma Coulomb interaction in Debye-Hückel approximation, and higher order correction δP, which takes into account contribution from ladder diagrams.

We shall consider protons as non-degenerate particles, their ideal gas concentration, called like activity ζ_p, is connected in grand canonical ensemble with their chemical potential μ_p and temperature ($\beta = T^{-1}$):

$$\zeta_p = 2\lambda_p^{-3} \exp(\beta \mu_p), \qquad (2)$$

$\lambda_p = \sqrt{\frac{2\pi\hbar^2}{m_p T}}$ — is thermal de Broglie wavelength. The pressure of ideal gas of protons

$$P_{0p} = T\zeta_p. \qquad (3)$$

Electrons may be degenerate (e.g. at the Sun center a parameter $n_e \lambda_e^3 \approx 0.6$), so we shall express their activity ζ_e [28] via ideal gas concentration n_e^0, for general case of degenerate particles, taking into account relativistic correction.

$$n_e^0(\mu_e) \equiv \zeta_e = \frac{2}{\sqrt{\pi}} \frac{2}{\lambda_e^3} \int_0^\infty \frac{x^{1/2}\left(1 + \frac{5}{4}\frac{T}{mc^2}x\right)}{\exp(x-y)+1} dx \qquad (4)$$

Here $x = \varepsilon/T$, $y = \mu_e/T$. For electron gas pressure we have

$$P_{0e} = \frac{4T}{3\sqrt{\pi}} \frac{2}{\lambda_e^3} \int_0^\infty \frac{x^{3/2}\left(1 + \frac{3}{4}\frac{T}{mc^2}x\right)}{\exp(x-y)+1}. \qquad (5)$$

At this approximation the electroneutrality condition is written in the following form

$$\zeta_p = \zeta_e. \tag{6}$$

The next, after ideal gas approximation $P_0 = P_{0p} + P_{0e}$, Hartree correction has the following presentation [12, 13, 28] for Helmholtz thermodynamic potential $\Omega = -PV$ (V — is the system volume)

$$\delta\Omega_H V^{-1} = \tilde{V}(0) \cdot (\zeta_e - \zeta_p)^2. \tag{7}$$

Here $\tilde{V}(0)$ — is the Fourier-transform of the Coulomb potential at zero transferred momentum. Using the regularization of the integral with the help of $\exp(-\varkappa r)$, $\varkappa \to 0$ — infinitesimal parameter, we shall obtain ($\mathbf{r} = (r_1, r_2, r_3)$, $|\mathbf{r}| = r$, $d\mathbf{r} = dr_1 dr_2 dr_3$)

$$\tilde{V}(\mathbf{q}) = \lim_{\varkappa \to 0} \int \frac{e^2}{r} \exp(i\mathbf{q}\mathbf{r} - \varkappa r) \, d\mathbf{r} = \frac{4\pi e^2}{q^2 + \varkappa^2}, \qquad \tilde{V}(0) = 4\pi e^2 \varkappa^{-2}. \tag{8}$$

This way of regularization Fourier-component of the Coulomb potential, gives zero result for the expression (7), due to neutrality condition (6). For general case of multicomponent plasma, the neutrality condition (6) has the form (z_k — is a particle of k-kind charge in units of electron charge e)

$$\zeta_e = \sum_k z_k \zeta_k, \tag{9}$$

and expression (7) generalized in an obvious way

$$\frac{\delta\Omega_H}{V} = \tilde{V}(0) \cdot \left(\zeta_e - \sum_k z_k \zeta_k\right)^2. \tag{10}$$

In the next, Hartree-Fock approximation, we obtain known convergent result [25, 28, 29] for electron-electron exchange interaction. For non-degenerate electrons, for instance we have [28] in the first Born approximation

$$\frac{\delta\Omega_{exch}}{V} = -\frac{2\pi e^2 \hbar^2}{m_e T} \zeta_e^2. \tag{11}$$

Here for ζ_e it is needed to use non-degenerate limit ($\exp(-y) \gg 1$) in the expression (4): $\zeta_e = 2\lambda_e^{-3} \exp(\mu_e/T)$.

Next order terms in interaction potential, named ring diagrams [30, 31], correspond to Debye-Hückel contribution (see e.g. [7–14, 23])

$$\frac{\delta\Omega_{D\text{-}H}}{V} = -T \frac{\varkappa_D^3}{12\pi}. \tag{12}$$

Here $\varkappa_D$ — is the inverse Debye radius [23, 32]

$$\varkappa_D^2 = 4\pi e^2 \sum_k z_k^2 \left(\frac{\partial n_k}{\partial \mu_k}\right)_T. \tag{13}$$

Expression (13) was derived strictly by Fradkin [32], who showed, that in the right hand side of eq. (13) the derivatives of physical concentrations over chemical potentials must be taken. In perturbation theory we can be restricted by the activities values $n_k^0(\mu_k) = \zeta_k$.

We must remind, that the physical concentrations are connected with chemical potentials by the relation [28]

$$n_k = -\left(\frac{\partial(\Omega/V)}{\partial \mu_k}\right)_T. \tag{14}$$

For physical concentrations the true electrical neutrality condition takes place:

$$n_e = \sum_k z_k n_k. \tag{15}$$

To adjust conditions (9) and (15), the first one needed for canceling divergence of Hartree term (10), when parameter $\varkappa$ goes to zero, we shall use the following method. We define physical concentrations from the condition (14), taking into account bounds, going from (9), when calculating derivatives of thermodynamic potential over chemical potentials. Let us find value of n_e from (15), and ζ_e from (9).

$$n_e + \sum_k n_k = \sum_k (z_k + 1) n_k = -\beta \sum_k \left(\frac{\partial(\Omega/V)}{\partial \zeta_k}\right)_T \zeta_k. \tag{16}$$

In the expression (16) all derivations and summation take place only over ionic activities, and ζ_e in potential Ω is expressed from (9). Within the framework of the model, defined by equation $\Omega = \Omega_0 + \delta\Omega_{\text{D-H}}$, from the condition (16) follows

$$\zeta_k = \frac{n_k}{1 + \frac{\Gamma_D}{2} z_k}, \quad \zeta_e = \sum_k \frac{z_k n_k}{1 + \frac{\Gamma_D}{2} z_k}. \tag{17}$$

Debye-Hückel approximation, presented in static limit by the contribution of ring diagrams, contains some corrections [23], which can be obtained with the help of diagram technique [25, 26] (spins of the particles for simplicity are assigned to $\frac{1}{2}$):

$$\frac{\Delta\Omega}{V} = -4 \sum_{i,j} \int_0^1 \frac{d\lambda}{2\lambda} \lambda^2 \int \frac{d\mathbf{P}\, d\mathbf{q}\, d\mathbf{k}}{(2\pi)^9} \cdot \frac{\exp\left(-\beta(\varepsilon_k^{ij} - \varepsilon_q^{ij})\right) - 1}{\varepsilon_q^{ij} - \varepsilon_k^{ij}} \times$$

$$\times n_i\left(\frac{m_i}{M}\mathbf{P}+\mathbf{q}\right)\left(1 - n_i\left(\frac{m_i}{M}\mathbf{P}+\mathbf{k}\right)\right) n_j\left(\frac{m_j}{M}\mathbf{P}-\mathbf{q}\right)\left(1 - n_j\left(\frac{m_j}{M}\mathbf{P}-\mathbf{k}\right)\right) \times \tag{18}$$

$$\times \frac{16\pi^2 e^4 z_i^2 z_j^2}{(\mathbf{q}-\mathbf{k})^2 \left[(\mathbf{q}-\mathbf{k})^2 + 4\pi e^2 \lambda \Pi^R(\mathbf{q}-\mathbf{k}, \mathbf{V}\cdot(\mathbf{q}-\mathbf{k}))\right]},$$

where the integration over parameter λ corresponds to charge integration $e^2 \mapsto e^2\lambda$; $\mathbf{P}$ — is the total momentum of particles i and j; $\mathbf{q}, \mathbf{k}$ — wave vectors of relative motion with regard to mass center before and after interaction; $n_i(\mathbf{q})$ — population numbers for particle of i-kind; m_i, z_i — are mass and charge (in elementary units) for particle of

i-kind; $M = m_i + m_j$ — total mass; $\varepsilon_k^{ij} = \frac{\hbar^2 k^2}{2\mu}$, $\mu = \frac{m_i m_j}{m_i + m_j}$ — is the reduced mass of i and j particles; $\mathbf{V} = \mathbf{P}/M$, $\hbar|\mathbf{V}|q \sim \hbar|\mathbf{V}|\varkappa_{\text{D}} < T$.

For non-degenerate case ($n_i \ll 1$) from (18) in the static limit: $\max\{T, \varepsilon_{Fi}\} \gg \hbar\omega_{pi}$ (here ε_{Fi} — is the Fermi energy for particle of i-kind, ω_{pi} — their plasma frequency [23, 28]), follows

$$\frac{\Delta\Omega}{V} = -\frac{T\varkappa_{\text{D}}^4}{4\pi^2} \int_0^1 \lambda\, d\lambda \int_0^\infty \frac{dq}{q^2 + 4\pi e^2 \lambda \Pi^R(q)}. \tag{19}$$

Here $\varkappa_{\text{D}}^2 = 4\pi e^2 \Pi^R(0)$ — square of inverse Debye radius, $\Pi^R(q) \equiv \Pi^R(|\mathbf{q}|) \equiv \Pi^R(\mathbf{q}, 0)$ — retarded polarization operator (see e.g. [32]); for non-degenerate case $\varkappa_{\text{D}}^2 = 4\pi e^2 \beta \sum_i z_i^2 n_i$. In commonly used approximation, corresponding to the substitution of $\Pi^R(q)$ by its value $\Pi^R(0)$ from (19), (13) follows expression (12).

The ratio of thermal de Broglie wavelength of electrons to the Debye radius characterizes quantum effects during Debye screening, and correspondent corrections are named like diffraction corrections [8].

For non-degenerate case we can obtain in the first approximation over $\lambda\varkappa$ parameter from (18) (see [8]; $\lambda_{ee} = \frac{\hbar}{\sqrt{m_e T}}$; $\lambda_{ek} = \frac{\hbar}{\sqrt{2\mu_{ek}T}}$; $\lambda_{kj} = \frac{\hbar}{\sqrt{2\mu_{kj}T}}$)

$$\frac{\Delta\Omega_{\text{diff}}}{V} = \frac{\pi^{3/2}}{4} T \left(\frac{e^2}{T}\right)^2 \left\{ \lambda_{ee}\zeta_e^2 + 2\zeta_e \sum_k \zeta_k z_k^2 \lambda_{ek} + \sum_{kj}' \zeta_k \zeta_j z_k^2 z_j^2 \lambda_{kj} \right\}. \tag{20}$$

There is one more class of corrections to Debye-Hückel approximation, obtained for the first time in the work [11] (see also [8, 10]). Physically this corrections corresponds to the interaction of particles, taking part in screening effect, with the surrounding medium ("dressed" particles).

Taking this effect into account we obtain "classical" correction to Debye-Hückel approximation (subscript "cl" means, that this result has classical nature)

$$\frac{\delta\Omega_{\text{cl}}}{V} = -\frac{\pi}{3} T \left(\frac{e^2}{T}\right)^3 \left(\sum_i \zeta_i z_i^4\right) \left(\sum_j \zeta_j z_j^2\right). \tag{21}$$

In (21) the summation takes place over all particles.

3. LADDER APPROXIMATION FOR SVC CALCULATION

Let us consider the contribution δP present in the formula (1). From Matsubara technique [23, 24] for $\delta\Omega/V = -\delta P$ we have (integration over parameter λ corresponds to charge integration $e^2 \mapsto e^2\lambda$)

$$\frac{\delta\Omega_L}{V} = \frac{2}{\beta} \sum_{i,\omega} \int_0^1 \frac{d\lambda}{2\lambda} \int \frac{d\mathbf{p}}{(2\pi)^3} G_i(\mathbf{p}, \omega) \Sigma_i(\mathbf{p}, \omega). \tag{22}$$

Here we have the summation over frequencies ω (or p_4); for fermions $\omega = \pi T(2n+1)$, the subscript L stands for "ladder", $G_i(\mathbf{p}, \omega)$ — Green's function of particle of kind i with momentum $\mathbf{p}$ and frequency ω in Matsubara technique [23], and the self-energy operator $\Sigma_i(\mathbf{p}, \omega)$ can be expressed via two-particle vertex part Γ_{ij}, obtained in the ladder approximation [29].

$$\Sigma_i(\mathfrak{p}) = \frac{2}{\beta} \sum_{j,k_4} \int \frac{d\mathbf{k}}{(2\pi)^3} G_j(\mathfrak{k}) \Gamma_{ij}\left(\frac{m_j \mathfrak{p} - m_i \mathfrak{k}}{m_i + m_j}; \frac{m_j \mathfrak{p} - m_i \mathfrak{k}}{m_i + m_j}; \mathfrak{p} + \mathfrak{k}\right). \quad (23)$$

E.g. for electron-proton interaction $m_i = m_e$, $m_j = m_p$, $\mathfrak{p} = (\mathbf{p}, p_4) \equiv (\mathbf{p}, \omega)$ — is 4-vector for electron, $\mathfrak{k} = (\mathbf{k}, k_4)$ — correspondingly for proton. The quantity $\Gamma_{ij}(\mathbf{q}, \mathbf{q}'; \mathfrak{P})$ ($\mathbf{q}, \mathbf{q}'$ — are relative motion momentum before and after scattering, $\mathfrak{P} = \mathfrak{p} + \mathfrak{k} = (\mathbf{P}, P_4)$ — is the 4-vector for total momentum) we have in the ladder approximation [12, 29] for small population numbers ($n \ll 1$):

$$\frac{\Gamma_{ep}(\mathbf{q}, \mathbf{q}'; \mathfrak{P})}{(2\pi)^3} = \sum_n \left(iP_4 - \frac{\hbar^2 P^2}{2M} - \frac{\hbar^2 q^2}{2\mu} + \mu_e + \mu_p\right) \frac{\widetilde{\Psi}_n(\mathbf{q}) \widetilde{\Psi}_n^*(\mathbf{q}') \left(E_n - \frac{\hbar^2 q'^2}{2\mu}\right)}{iP_4 - \frac{\hbar^2 P^2}{2M} - E_n + \mu_e + \mu_p}. \quad (24)$$

Here $M = m_e + m_p$ — is total mass, $\mu = \frac{m_e m_p}{M}$ — is the reduced mass; μ_e, μ_p — are the chemical potentials; E_n — is the binding energy for a state with principal quantum number n; the summation over n takes part over all discrete quantum numbers $\{n\} = (n, l, m)$, characterizing wave functions of electron-proton relative motion (bound states of electrons in the proton field) $\widetilde{\Psi}_n(\mathbf{q})$, as well as over continuum states. The sign $\sim$ over wave function means the Fourier-transform in momentum space. $P_4 = 2\pi nT$ — is the fourth component of total momentum. For the continuum (scattering) states we have instead of summation over discrete states $\{n\}$ integration over wave vectors $\mathbf{k}$, characterizing wave function at the infinity (for short-range interaction-plane wave). Using Schroedinger equation in momentum space and the theorem of completeness for the wave functions ($\widetilde{V}(\mathbf{q})$ — is the Coulomb interaction potential in momentum representation)

$$\left(E_n - \frac{\hbar^2 q^2}{2\mu}\right) \widetilde{\Psi}_n(\mathbf{q}) = \int \widetilde{V}(\mathbf{q} - \mathbf{q}') \widetilde{\Psi}_n(\mathbf{q}') \frac{d\mathbf{q}'}{(2\pi)^3}, \quad (25)$$

it is possible to transform scattering amplitude into the following form:

$$\Gamma_{ep}(\mathbf{q}, \mathbf{q}'; \mathfrak{P}) = \widetilde{V}_{ep}(\mathbf{q} - \mathbf{q}') + (2\pi)^3 \sum_n \frac{\widetilde{\Psi}_n(\mathbf{q}) \widetilde{\Psi}_n^*(\mathbf{q}') \left(E_n - \frac{\hbar^2 q^2}{2\mu}\right)\left(E_n - \frac{\hbar^2 q'^2}{2\mu}\right)}{iP_4 - \frac{\hbar^2 P^2}{2M} - E_n + \mu_e + \mu_p}. \quad (26)$$

With the help of equations (22) and (23) we can write $\delta\Omega$ in the form:

$$\frac{\delta\Omega}{V} = \sum_{i,j} \frac{4T^2}{(2\pi)^6} \sum_{q_4, P_4} \int_0^1 \frac{d\lambda}{2\lambda} \int d\mathbf{q} d\mathbf{P} \, G_i\left(\frac{m_i}{M}\mathfrak{P} + \mathfrak{q}\right) G_j\left(\frac{m_j}{M}\mathfrak{P} - \mathfrak{q}\right) \Gamma_{ij}(\mathbf{q}, \mathbf{q}; \mathfrak{P}). \quad (27)$$

Making summation over frequencies q_4 and P_4 and integration over $d\mathbf{P}$ in the non-degenerate case from (27) and (26) we shall obtain (compare with [18])

$$\frac{\Delta\Omega_L}{V} = \sum_{i,j}\zeta_i\zeta_j\lambda_{ij}^3 \int_0^1 \frac{d\lambda}{2\lambda} \int \frac{d\mathbf{q}}{(2\pi)^3} \sum_n \left(E_n - \frac{\hbar^2 q^2}{2\mu}\right)|\tilde{\Psi}_n(\mathbf{q})|^2 \left(e^{-\beta E_n} - e^{-\beta\varepsilon_q}\right). \quad (28)$$

In (28) for e–p interaction it is necessary to sum over discrete spectrum (bound states) as well as to integrate over scattering states, described by index $\mathbf{k}$. For e–e and p–p interaction only the last action is not unreasonable.

4. CONTRIBUTION TO THE SVC FROM BOUND STATES.

To calculate the contribution of bound states to the SVC we use the exact Fock [33] result for wave functions of non-relativistic hydrogen atom in momentum representation

$$\frac{1}{(2\pi)^3}\sum_{l,m}|\tilde{\Psi}_{n,l,m}(\mathbf{q})|^2 = \frac{8}{\pi^2 a_0^5 n^3 (q^2 + p_n^2)^4}. \quad (29)$$

Here $p_n = (a_0 n)^{-1}$; $a_0 = \frac{\hbar^2}{\mu e^2 \lambda}$ — is the Bohr "radius" with current charge $e^2\lambda$. Taking into account that $E_n \equiv -\frac{\hbar^2 p_n^2}{2\mu}$ and using (29) the part of expression (28) corresponding to bound states may be written in the form:

$$\frac{\delta\Omega^{BS}}{V} = -4\zeta_e\zeta_p \frac{\mu^4 \lambda_{ep}^3 e^{10}}{\pi^2 \hbar^8} \sum_{n=1}^\infty \frac{1}{n^3} \int_0^1 d\lambda\, \lambda^4 \left[\exp\left(\frac{\lambda^2 X}{n^2}\right) \int \frac{d\mathbf{q}}{(q^2+p_n^2)^3} \right.$$
$$\left. - \int \frac{d\mathbf{q}}{(q^2+p_n^2)^3} \exp\left(-\beta\frac{\hbar^2 q^2}{2\mu}\right)\right]. \quad (30)$$

Here $X = \beta \text{Ry}$. From (30) we obtain $\left(\beta\frac{\hbar^2 q^2}{2\mu} = \lambda_{ep}^2 q^2;\ \lambda_{ep} = \frac{\hbar}{\sqrt{2\mu T}}\right)$

$$\int \frac{d\mathbf{q}}{(q^2+p_n^2)^3} = \frac{\pi^2}{4p_n^3}, \quad \int \frac{d\mathbf{q}}{(q^2+p_n^2)^3}\exp\left(-\beta\frac{\hbar^2 q^2}{2\mu}\right) = c_n(X). \quad (31)$$

From (30) we get $\left(R_n = \exp\left(\frac{\lambda^2 X}{n^2}\right)\Gamma\left(\frac{1}{2},\frac{\lambda^2 X}{n^2}\right)\right)$

$$c_n(X) = \pi^{3/2}\left[\frac{1}{4}a_0^3 n^3 R_n - \frac{a_0\beta\hbar^2}{2\mu}nR_n + \frac{\hbar a_0^2 n^2}{2}\sqrt{\frac{\beta}{2\mu}} + \left(\frac{\beta\hbar^2}{2\mu}\right)^{3/2} - \frac{\beta^2\hbar^4}{4\mu^2 a_0 n}R_n\right]. \quad (32)$$

Using (30), (31) and (32) after integration over λ we obtain ($I = \text{Ry}$):

$$\frac{\Delta\Omega^{BS}}{V} = -T\zeta_e\zeta_p\lambda_{ep}^3 \sum_{n=1}^\infty n^2 \exp\left(\frac{I}{n^2 T}\right) F_n(\beta). \quad (33)$$

Here $F_n(\beta)$ has the form ($I_n = 1/n^2$, $\chi = \beta I_n$):

$$F_n(\beta) = F(\chi) = 1 - e^{-\chi}\left[4 - \frac{6}{\sqrt{\pi}}\chi^{1/2} + \frac{4}{\sqrt{\pi}}\chi^{3/2}\right] + \frac{\Gamma(\frac{1}{2},\chi)}{\sqrt{\pi}}\left[3 - 4\chi + 4\chi^2\right]. \quad (34)$$

Asymptotically, for $n \gg 1$ from (34) it follows that

$$F(\chi) \sim 2\chi^2. \quad (35)$$

This expression is four times greater than the similar expression in Planck-Larkin formula

$$F^{P\text{-}L}(\chi) = 1 - e^{-\chi} - \chi e^{-\chi} \to \frac{1}{2}\chi^2. \quad (36)$$

Using (32) the expression (30) can be represented in the form:

$$\frac{\delta\Omega^{BS}}{V} = -\zeta_e\zeta_p T \lambda_{ep}^3 \sum_{n=1}^{\infty} n^2 \left(\exp\left(\frac{X}{n^2}\right) - 1 - \frac{X}{n^2} + \frac{3}{2}\frac{X^2}{n^4} - \frac{64}{15}\frac{X^{5/2}}{\sqrt{\pi}n^5} + \ldots \right). \quad (37)$$

First three terms in summation (37) correspond to the Planck-Larkin formula. The rest of exponent expansion series (see (34)) $\frac{X^2}{2n^4}$ added to the fourth term of the sum (37) produces asymptotic (35).

Bound states contribution may be written as

$$\delta\Omega^{BS}V^{-1} = -\zeta_e\zeta_p T\lambda_{ep}^3\Sigma^{BS}. \quad (38)$$

For expression (34) using X-expansion it can be obtained that:

$$\Sigma_{SRM}^{BS} = \sum_{k=4}^{\infty} \zeta(k-2)\frac{(-\sqrt{X})^k}{\Gamma(\frac{k}{2}+1)}(k-2)^2 + \sum_{k=1}^{\infty} \zeta(2k+1)\frac{X^{k+3/2}}{\Gamma(k+\frac{5}{2})}. \quad (39)$$

Notice that expressions like (28) are obtained in Keldysh' technique [25–27]:

$$\delta P = \sum_a (2S_a+1) \int_0^1 \frac{d\lambda}{2\lambda} \int \frac{d\mathbf{p}}{(2\pi)^3} \int \frac{d\omega d\omega'}{(2\pi)^2} \frac{\Sigma_a^>(\mathbf{p},\omega) G_a^<(\mathbf{p},\omega')}{\omega - \omega'}\left(1 - e^{-\beta(\omega-\omega')}\right). \quad (40)$$

Symbolic expression of $\Sigma^>$ using the imaginary part of amplitude of scattering Γ_{ep} [25]

$$\Sigma_e^> \sim (2\pi)^{-4} \int \text{Im}\Gamma_{ep}(\omega+\Omega)\cdot G_i^<(\mathbf{p}',\Omega)d\mathbf{p}'\,d\Omega, \quad (41)$$

results for Γ_{ep} in (see (26)):

$$\text{Im}\,\Gamma_{ep} \sim \sum_n |\tilde{\Psi}_n(\mathbf{q})|^2 (E_n - \varepsilon_q)^2 \delta_\gamma\left(\omega + \Omega - E_n - \frac{p^2}{2M}\right), \quad (42)$$

$\delta_\gamma(x)$ — Lorentzian shape that tends to δ-function in the limit of zero width $\gamma \to 0$.

Atomic partition function convergence in the lowest order of interaction (up to the SVC) for principal quantum number greater than $n_{max} \sim \sqrt{\beta Ry}$ means that bound states contribution to pressure can be represented by expression like Σ_{SRM} (or $\Sigma_{P\text{-}L}$). While for other effects, e.g. spontaneous radiation of equilibrium plasma, observable contribution of discrete states may occur (corresponding sum converges) for $n > n_{max}$ up to states with comparable Stark' width and ionization energy.

5. CONTRIBUTION TO THE SVC FROM SCATTERING STATES.

To calculate the contribution of continual spectrum states to expressions like (28) one need to evaluate Fourier components of wave functions describing mutual scattering of charged particles. It is convenient to use the system of Coulomb wave functions, represented by sum over orbital moments [34]. For example, for attraction field

$$\Psi_k(\mathbf{r}) = (2\pi)^{-3/2} \exp\left(\pi\tilde{\xi}/2\right) \Gamma\left(1 - i\tilde{\xi}\right) \exp(i\mathbf{k}\mathbf{r}) F\left(i\tilde{\xi}, 1, i(kr + \mathbf{k}\mathbf{r})\right). \tag{43}$$

Here $\tilde{\xi} = (a_0 k)^{-1} = \frac{\mu e^2 \lambda}{\hbar^2 k}$; $F(\alpha, \beta, z)$ — degenerated hypergeometric function. To find the Fourier transform of function (43) a linear regularization should be used

$$\tilde{\Psi}_k(\mathbf{q}) = \int d\mathbf{r} \exp(-i\mathbf{q}\mathbf{r} - \varkappa_1 r) \Psi_k(\mathbf{r}). \tag{44}$$

The $\varkappa_1$ parameter is infinitesimal. From (43), (44) we obtain [35]

$$\tilde{\Psi}_k(q) = (2\pi)^{-3/2} \exp\left(\pi\tilde{\xi}/2\right) \Gamma\left(1 - i\tilde{\xi}\right) J_{\varkappa_1}. \tag{45}$$

$$J_{\varkappa_1} = \left\{ \frac{2\pi\left(1 - i\tilde{\xi}\right)\varkappa_1}{\left[\frac{(\mathbf{q}-\mathbf{k})^2}{2} + \frac{\varkappa_1^2}{2}\right]^2} + \frac{2\pi\tilde{\xi}(k + i\varkappa_1)}{\left[\frac{(\mathbf{q}-\mathbf{k})^2}{2} + \frac{\varkappa_1^2}{2}\right]\left[\frac{q^2 - k^2 + \varkappa_1^2}{2} - ik\varkappa_1\right]} \right\} e^{i\tilde{\xi} \ln \frac{(\mathbf{q}-\mathbf{k})^2 + \varkappa_1^2}{q^2 - k^2 + \varkappa_1^2 - 2ik\varkappa_1}}. \tag{46}$$

The first term in (46) is similar to regularized 3-d δ-function. Notice that for repulsion field for p–p interaction following equality is valid [34, Appendix §f]:

$$\lambda e^2 \int \frac{\exp(-\varkappa r)}{r} |\Psi_k(\mathbf{r})|^2 d\mathbf{r} = \frac{4\pi e^2 \lambda}{(2\pi)^3 \varkappa^2} e^{-\lambda z_p} \frac{\pi\tilde{\xi}}{\operatorname{sh} \pi\tilde{\xi}} F\left(i\tilde{\xi}, -i\tilde{\xi}; 1; \frac{y}{y + \eta_p^2/4}\right). \tag{47}$$

Here $\tilde{\xi} = \frac{\lambda \alpha_p}{2\sqrt{y}}$, $\alpha_p = \sqrt{\frac{m_p e^4}{\hbar^2 T}}$, $z_p = \frac{\alpha_p}{\sqrt{y}} \operatorname{argtg}\left(\frac{\eta_p}{2\sqrt{y}}\right)$, $y = \frac{\hbar^2 k^2}{m_p T}$, $\eta_p^2 = \frac{\hbar^2 \varkappa^2}{m_p T}$, $F(\alpha, \beta; \gamma; z)$ — hypergeometric function. Similar expressions can be obtained also for e–e interaction changing $m_p \to m_e$.

Taking into account subtraction in the factor $(\exp(-\beta\varepsilon_k) - \exp(-\beta\varepsilon_q))$ in formula (28) we can transform expression (28) for continual spectrum (of scattering states - SS) to (e.g. for e–p interaction $\tilde{\xi}_{ep} = \frac{\lambda \alpha_{ep}}{\sqrt{y}}$, $\alpha_{ep} = \sqrt{Ry/T}$)

$$\frac{\Delta \Omega^{ss}_{ep}}{V} = -\frac{\lambda_{ep}}{\pi} \zeta_e \zeta_p T \left(\frac{e^2}{T}\right)^2 \int_0^1 d\lambda\, \lambda \int_0^\infty \int_0^\infty dx\, dy\, \frac{\pi \tilde{\xi}_{ep}}{\operatorname{sh} \pi \tilde{\xi}_{ep}} \exp\left(\pi \tilde{\xi}_{ep}\right) \frac{e^{-y} - e^{-x}}{x - y} \times \tag{48}$$

$$\times \left[\frac{1}{(\sqrt{x}-\sqrt{y})^2 + \tilde{\eta}^2_{ep}} - \frac{1}{(\sqrt{x}+\sqrt{y})^2 + \tilde{\eta}^2_{ep}} \right] e^{-2\tilde{\xi}_{ep} \operatorname{Im} \ln\left(x - y + \tilde{\eta}^2_{ep} + i 2\sqrt{y}\tilde{\eta}^2_{ep}\right)}.$$

3-D numerical integration results in following asymptotic as $\varkappa \to 0$ $(a = e, p)$

$$\frac{\Delta \Omega^{ss}_{aa}}{V} - \frac{\Delta \Omega^{(2)}_{aa}}{V} \to -\zeta_a^2 \left(\frac{e^6}{T^2} \left[\frac{\pi}{3} \ln(\varkappa \lambda_{aa}) - \frac{\pi}{6}(1 - C) \right] + \frac{\lambda^3_{aa} T}{2} \Sigma_{\mathrm{Q}}\left(-\frac{\alpha_a}{2}\right) \right), \tag{49}$$

$$\frac{\Delta \Omega^{ss}_{ep}}{V} - \frac{\Delta \Omega^{(2)}_{ep}}{V} \to 2\zeta_e \zeta_p \left(\frac{e^6}{T^2} \left[\frac{\pi}{3} \ln(\varkappa \lambda_{ep}) - \frac{\pi}{6}(1 - C) \right] + \frac{\lambda^3_{ep} T}{2} \left[\Sigma^{\mathrm{BS}}_{\mathrm{SRM}} - \Sigma_{\mathrm{Q}}(\alpha_{ep}) \right] \right) \tag{50}$$

where $\Sigma_{\mathrm{Q}}(\alpha) = \frac{1}{2} \sum_{n=4}^\infty \frac{\zeta(n-2)}{\Gamma\left(\frac{n}{2}+1\right)} \alpha^n = -\left(\ln|2\alpha| + \frac{3C}{2} - \frac{4}{3} \right) \frac{2\alpha^3}{3\sqrt{\pi}} + o(|\alpha|^3)$ as $\alpha \to -\infty$.

For repulsion field we will obtain

$$\delta \Omega^L_{pp} = \delta \Omega^{\mathrm{d}}_{pp} + \delta \Omega^q_{pp}. \tag{51}$$

The first term correspond to "classical" limit, represented by logarithmic contribution to the SVC (here $\varkappa \to 0$, $\lambda_p = \frac{\hbar}{\sqrt{m_p T}}$).

$$\frac{\delta \Omega^{\mathrm{d}}_{pp}}{V} = -\zeta_p^2 T \left(\frac{e^2}{T}\right)^3 \left\{ \frac{\pi}{3} \ln \varkappa \lambda_p - \frac{\pi}{6}(1 - C) \right\}. \tag{52}$$

In [13] in the second item an excess term with $\ln 3$ in parentheses $(-C - 2\ln 3 + 1)$ is present.

For second quantum term in (51) we obtain series on Born' parameter $\sim e^2/(\hbar v_T)$, in this case on parameter $\alpha_p/2$ that is exactly equal to expression given in [13]:

$$\frac{\delta \Omega^q_{pp}}{V} = -2\pi \lambda_p^3 \zeta_p^2 T \sum_{n=4}^\infty \frac{\zeta(n-2)}{\Gamma\left(\frac{n}{2}+1\right)} \left(-\frac{\alpha_p}{2}\right)^n. \tag{53}$$

Notice that this result is confirmed also by detail comparison to numerical integration taking into account subtraction of $\Delta \Omega^{(1)}_{pp}$ and $\Delta \Omega^{(2)}_{pp}$. For e–e interaction all results (52), (53) coincide up to change $m_p \to m_e$.

Results of numerical integration shows that 3-D integral is approximated by an analytic expression discussed in the next section: $\delta \Omega^L_{ep} = \delta \Omega^{\mathrm{d}}_{ep} + \delta \Omega^q_{ep}$. Here $\delta \Omega^{\mathrm{d}}_{ep}$ represents logarithmic contribution (compare with (52), $\lambda_{ep} = \frac{\hbar}{\sqrt{2\mu T}}$)

$$\frac{\delta \Omega^{\mathrm{d}}_{ep}}{V} = \zeta_e \zeta_p T \left(\frac{e^2}{T}\right)^3 \cdot 2 \left\{ \frac{\pi}{3} \ln \varkappa \lambda_{ep} - \frac{\pi}{6}(1 - C) \right\}. \tag{54}$$

Summation of (52)-like expressions for p–p and e–e interactions together with (54) taking into account that $\zeta_e = \zeta_p$ will give result independent of $\varkappa$:

$$\frac{\delta\Omega^{\text{cl}}}{V} = \zeta_e^2 T \left(\frac{e^2}{T}\right)^3 \cdot \frac{\pi}{6} \ln \frac{m_p}{4m_e}. \tag{55}$$

For $\delta\Omega_{ep}^q$ this result may be represented in the form ($\alpha_{ep} = \sqrt{\beta}\text{Ry}$)

$$\delta\Omega_{ep}^q V^{-1} = -\zeta_e \zeta_p T \lambda_{ep}^3 \Sigma^{\text{ss}}, \tag{56}$$

$$\Sigma^{\text{ss}} = \sum_{n=4}^{\infty} \frac{\zeta(n-2)}{\Gamma\left(\frac{n}{2}+1\right)} \left(\frac{1}{2} - (-1)^n (n-2)^2\right) \alpha_{ep}^n - \sum_{k=1}^{\infty} \frac{\zeta(2k+1)}{\Gamma\left(k+\frac{5}{2}\right)} \alpha_{ep}^{2k+3}. \tag{57}$$

For J_{exch} value we will obtain taking into consideration expressions like (43) for repulsion field (here $\tilde{\xi}_{ee} = \frac{\lambda}{\sqrt{2y}}\sqrt{\frac{m_e e^4}{2\hbar^2 T}}$, $y = \frac{\hbar^2 k^2}{m_e T}$)

$$J_{\text{exch}} = \int d\mathbf{r}\, \Psi_k^*(\mathbf{r}) V(\mathbf{r}) \Psi_k(-\mathbf{r}) = \frac{1}{(2\pi)^3} \frac{\pi \tilde{\xi}_{ee}}{\operatorname{sh} \pi \tilde{\xi}_{ee}} \lambda e^2 \pi. \tag{58}$$

As a result, for exchange contribution $\delta\Omega_{ee}^{\text{exch}}$ we obtain:

$$\frac{\delta\Omega_{ee}^{\text{exch}}}{V} = -\frac{\zeta_e^2}{16\hbar^2} \lambda_{ee}^3 m_e e^4 \int_0^1 \lambda\, d\lambda \int_0^{\infty} \frac{dk}{k} \frac{\exp\left(-\frac{\hbar^2 k^2}{m_e T}\right)}{\operatorname{sh}\frac{\pi \lambda m_e e^2}{2\hbar^2 k}}. \tag{59}$$

From (59) we obtain $\left(\lambda_{ee} = \sqrt{\frac{4\pi\hbar^2}{m_e T}}\right)$ a convergent expression

$$\delta\Omega_{ee}^{\text{exch}} V^{-1} = \zeta_e^2 T \pi^{-1/2} \lambda_{ee}^3 E(\alpha_e). \tag{60}$$

For $E(\alpha_e)$ one can get explicit expression $\left(\alpha_e = \sqrt{\frac{m_e e^4}{\hbar^2 T}}, \alpha_{ee} = -\frac{\alpha_e}{2}\right)$

$$E(\alpha_e) = \frac{\alpha_{ee}}{2} + \frac{\sqrt{\pi}}{4} \ln 2 \cdot \alpha_{ee}^2 + \frac{\pi^2}{72} \alpha_{ee}^3 + \sum_{n=4}^{\infty} \frac{\sqrt{\pi}(1 - 2^{2-n})}{\Gamma\left(\frac{n}{2}+1\right)} \zeta(n-1) \alpha_{ee}^n. \tag{61}$$

Expression (61) is just the same as in [13] for exchange contribution and is confirmed by numerical integration. The first term of the Born series in (61) gives contribution expressed earlier by (11).

6. HYDROGEN WEAKLY-NONIDEAL PLASMAS EOS

Consider the total contribution of expressions (3), (5), (12), (20), (21), (33), (38), (51), (52), (55), (53), (56), (57), (60), (61) to describe the EOS of hydrogen weakly-nonideal plasmas in application to helioseismology problems.

Let us elucidate once more the meaning of these expressions: (3) — ideal gas pressure of protons, (5) — the same for electrons with account of degeneration, (12) — contribution to pressure from Coulomb interaction ($\delta P_{\text{D-H}} = -\delta \Omega_{\text{D-H}}/V$) in the Debye-Hückel (D-H) approximation in grand canonical ensemble, (20) — diffractional (quantum) corrections to D-H approximation; (21) — corrections, taking into account that Debye screening is performed by medium-interacting particles, not but by free ones; (33), (38) — convergent contribution of bound states to pressure (contribution of "atoms" to pressure); (51), (52) — logarithmic contribution $\sim (e^2/T)^3$ of proton-proton interaction to the SVC; (55) — total logarithmic contribution of all interactions $\sim (e^2/T)^3 \ln(m_p/m_e)$ in plasma of protons and electrons; (53), (56), (57) — contributions of scattering states to plasma thermodynamics; (60), (61) — contribution of electron-electron exchange.

Notice, that obtained analytic expressions for convergent contribution of bound states (38) together with finite expression for contribution of scattering states (56), (57) add up to (using expansion (39))

$$\delta\Omega^q_{ep} = \delta\Omega^{\text{BS}} + \delta\Omega^{\text{ss}}_{ep} = -V\zeta_e\zeta_p T \lambda^3_{ep} (\Sigma^{\text{BS}} + \Sigma^{\text{ss}}) = -V\zeta_e\zeta_p T \lambda^3_{ep} \Sigma^{\text{tot}}. \tag{62}$$

Here Σ^{tot} — corresponding sum of series on $\alpha_{ep} = \sqrt{\beta \text{Ry}}$, exactly reproducing expression given in [13, 14] and obtained from formulas like Beth-Uhlenbeck ones, derived for lump contribution of attraction states without subdivision on bound and scattering states

$$\Sigma^{\text{tot}} = \frac{1}{2} \sum_{n=4}^{\infty} \frac{\zeta(n-2)}{\Gamma\left(\frac{n}{2}+1\right)} \alpha^n_{ep}. \tag{63}$$

Subtracting from (63) the contribution of bound states, represented by Planck-Larkin formula (here I — ionization potential of hydrogen $I = \text{Ry}$),

$$\Sigma_{\text{P-L}} = \sum_{n=1}^{\infty} n^2 \left(\exp\left(\frac{I}{n^2 T}\right) - 1 - \frac{I}{n^2 T} \right), \tag{64}$$

we obtain an analogue of scattering states contribution (compare with (56), (57)),

$$(\Sigma^{\text{ss}})' = \Sigma^{\text{tot}} - \Sigma^{\text{BS}}_{\text{P-L}} = -\frac{1}{2} \sum_{n=4}^{\infty} \frac{(-1)^n \zeta(n-2)}{\Gamma\left(\frac{n}{2}+1\right)} \alpha^n_{ep}. \tag{65}$$

Total contribution of bound and scattering states (63) may be represented in two ways

$$\Sigma^{\text{tot}} = \Sigma^{\text{BS}}_{\text{SRM}} + \Sigma^{\text{ss}} = \Sigma^{\text{BS}}_{\text{P-L}} + (\Sigma^{\text{ss}})'. \tag{66}$$

In this sense the Planck-Larkin formula may be considered as "correct" (representing not only bound states, but partly scattering states also). The term $\Sigma^{\text{BS}}_{\text{SRM}}$ describes the bound states contribution only.

To plasma EOS (value of pressure $P(\rho,T)$ and other thermodynamic functions) a contribution of equilibrium thermal radiation in plasma should be added.

Taking into account the relation $k = n\omega/c$, where $n = \sqrt{1 - \omega_p^2/\omega^2}$ — refraction index, the radiation energy in optically thin dense plasma [36] may be calculated as:

$$E_R = \frac{\hbar V}{\pi^2 c^3} \int_{\omega_p}^{\infty} \frac{(\omega^2 - \omega_p^2)^{1/2} \omega^2}{e^{\hbar\omega/T} - 1} d\omega. \qquad (67)$$

Here V — system volume ($V \to \infty$), ω_p — electron plasma frequency, $\omega_p^2 = 4\pi e^2 n_e/m_e$. In the expression (67) it is considered, that radiation in the range $\omega < \omega_p$ does not transfer as free one. For free energy of radiation we obtain

$$F_R = \frac{TV}{\pi^2 c^3} \int_{\omega_p}^{\infty} \omega \sqrt{\omega^2 - \omega_p^2} \ln\left(1 - e^{-\hbar\omega/T}\right) d\omega = -\frac{\hbar V}{3\pi^2 c^3} \int_{\omega_p}^{\infty} \frac{(\omega^2 - \omega_p^2)^{3/2}}{e^{\hbar\omega/T} - 1} d\omega. \qquad (68)$$

From (68) the radiation pressure can be derived $P_R = -\left(\frac{\partial F_R}{\partial V}\right)_T$.

To calculate sound velocity in hydrogen plasmas following formula may be used [37]

$$c_S^2 = \left(\frac{\partial P}{\partial \rho}\right)_T \left[1 + \frac{T\left(\frac{\partial P}{\partial T}\right)_\rho^2}{c_V \rho \left(\frac{\partial P}{\partial \rho}\right)_T}\right]. \qquad (69)$$

Here pressure and heat capacity c_V are represented as sum of matter contribution P_M (c_V^M) and a radiation one P_R (c_V^R); ρ — is plasma density. One more important parameter, used in helioseismology problems is the adiabatic exponent Γ_1, defined as:

$$\Gamma_1 = c_S^2 \rho / P. \qquad (70)$$

7. HYDROGEN EOS ALONG THE SUN TRAJECTORY.

Using relations (2), (3), (6), (4), (5), (12), (20), (21), (33), (38), (52), (55), (53), (56), (57), (60), (61), (67), (68) we calculated hydrogen weakly-nonideal plasmas EOS, that is dependence of total pressure $P(T)$ along the S-model' Sun interior distribution [1].

In Fig. 1 temperature dependencies of partial contributions to total pressure $\delta P/P_{tot}$ are shown. As can be seen, many corrections to the SVC are significant, taking into account high accuracy of the inversion in helioseismology problems.

In Fig. 2 deviation of the adiabatic exponent from the ideal-gas value, see (70), $(\Gamma_1 - 5/3) \cdot 10^3$ along the Sun trajectory is shown. Comparison of data calculated in complete physical model and chemical ones based on formulas Σ_{P-L}^{BS} and Σ_{SRM}^{BS} (with and without degeneracy account). In the same figure the S-model simulation results of "real" Sun with all present elements taken into account [1]. Even for evident inferiority of the pure hydrogen model it can be seen that the difference from the S-model value is not rather large, especially near the center of the Sun, where the physical model and the chemical model variant with Σ_{SRM}^{BS} are closer to experiment than the Σ_{P-L}^{BS}-chemical one.

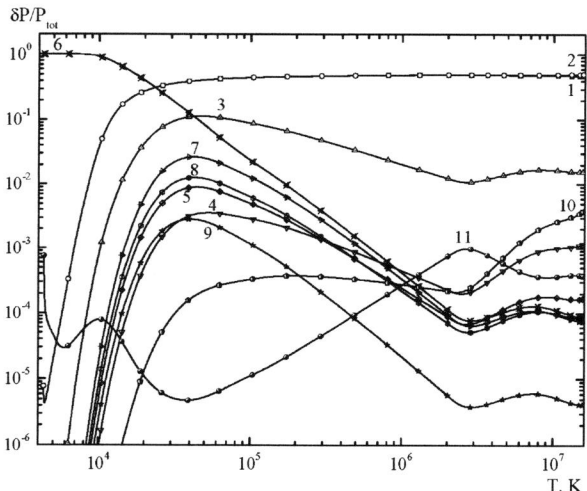

FIGURE 1. Specific partial contributions ($\delta P/P_{tot}$) to total pressure in dependence on temperature 1 — $P_{0,i}$; 2 — $P_{0,er}$; 3 — $\delta P_{\text{D-H}}$; 4 — $\Delta\Omega_{\text{diff}}/V$; 5 — $\delta\Omega^{\text{cl}}/V$; 6 — $-\delta\Omega^{\text{BS}}_{\text{SRM}}/V$; 7 — $\delta\Omega^{q}_{ep}/V$; 8 — $-\delta\Omega^{q}_{pp}/V$; 9 — $-\delta\Omega^{q}_{ee}/V$; 10 — $-\delta\Omega^{\text{exch}}_{ee}/V$; 11 — P_{R}.

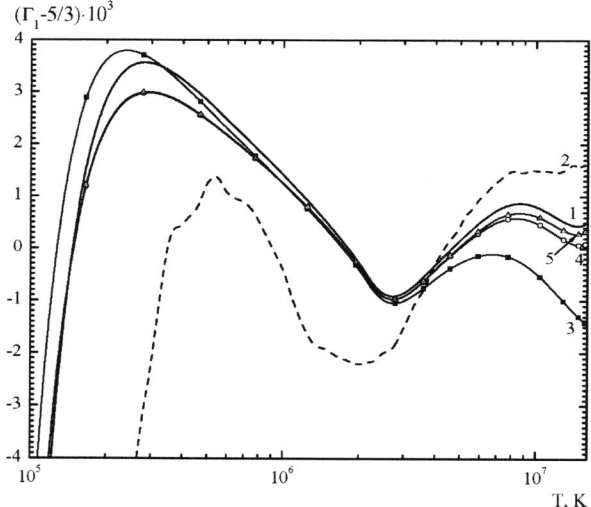

FIGURE 2. Relative deviations of the adiabatic exponent of the ideal-gas value $((\Gamma_1 - 5/3) \cdot 10^3)$ in dependence on temperature along the Sun trajectory: 1 — hydrogen physical model, 2 — S-model [1]; (chemical models with account of radiation) 3 — $\Sigma^{\text{BS}}_{\text{P-L}}$, 4 — $\Sigma^{\text{BS}}_{\text{SRM}}$, 5 — $\Sigma^{\text{BS}}_{\text{SRM}}$ without degeneracy accounting

8. CONCLUSION

The Sun is a unique scientific laboratory, since for weakly-nonideality of plasma it is possible to simulate a lot of processes starting from basic principles. Thus at the Sun center from condition

$$T \gg \max(e^2 \varkappa_\mathrm{D}, \hbar\omega_p, \mathrm{Ry}, E_\mathrm{st} d) \tag{71}$$

(where $E_\mathrm{st} = e n_e^{2/3}$ — Stark microfield in plasma, $d \sim e a_0$) it follows that the main limiting factor for atomic partition function is the temperature, since the bound states contribution is $\sim (\mathrm{Ry}/T)^2$ (see (35), (36)). In (71) the value $E_\mathrm{s} d$ — is the interaction energy of the Stark microfield and atomic dipole $d \sim e a_0$. The presence of large parameter (71) allows to start the EOS calculation from the ideal-gas approximation and to take into account all other effects subsequently. These effects include the influence of discrete spectrum on plasma pressure value, evaluated starting from undisturbed atomic states. Qualitative consideration composed of "atoms" destruction by screening, broadening, and microfield ionization (Unsöld mechanism [4]) etc., differs from the perturbation theory in activity series and even does not confirmed by experiment. (Compare [1], where [6] is closer to the inversion data than [4].) Similarly strict accounting of the Debye screening while calculating the atomic states (see, e.g. [13, 14]) leads to large shift of atomic levels energy as well as centers of spectral lines. This shift is not observed in weakly-nonideal plasma of arc discharge [38, 39].

In our opinion the account of broadening effects (see (42) and [18]) is more promising for spectral lines description as well as for weakly-nonideal dense plasma thermodynamics. In principle this approach joins problems of radiational gasdynamics and collisional-radiative kinetics, where "atoms" are represented in different way for calculation of pressure and radiation [18].

ACKNOWLEDGMENTS

Authors are thankful to S. V. Ayukov, V. A. Baturin, V. E. Fortov, V. K. Gryaznov, I. L. Iosilevskii, T. Kato, and R. More for many fruitful discussions. This work was supported by RFBR Grant #04-02-16775-a and the President of the Russian Federation Grant NSh-1257.2003.2.

REFERENCES

1. Christensen-Dalsgaard, J., Däppen, W., Dziembowski, W. A., and Guzik, J. A., "An introduction to helioseismology," in *Variable Stars as Important Astrophysical Tools, Proc. NATO Advanced Study Institute, Cesme, Turkey, Aug. 30 – Sept. 11 1998*, edited by C. Ibanoğlu, Kluwer Academic, Dordrecht, 2000, pp. 59–167, URL http://bigcat.ifa.au.dk/~jcd/solar_models/cptrho.15bi.d.15c.
2. Baturin, V. A., Däppen, W., Gough, D. O., and Vorontsov, S. V., *Mon. Not. R. Astr. Soc.*, **316**, 71–83 (2000).
3. Basu, S., Däppen, W., and Nayfonov, A., *Astrophys. J.*, **518**, 985–993 (1999).
4. Hummer, D. G., and Mihalas, D., *Astrophys. J.*, **331**, 794–814 (1988).
5. Mihalas, D., Däppen, W., and Hummer, D. G., *Astrophys. J.*, **331**, 815–825 (1988).
6. Rogers, F. J., Swenson, F. J., and Iglesias, C. A., *Astrophys. J.*, **456**, 902–908 (1996).
7. Ebeling, W., Hoffmann, H. J., and Kelbg, G., *Beitr. Plasmaphys.*, **7**, 233–248 (1967).
8. Ebeling, W., Kelbg, G., and Rohde, K., *Ann. der Phys.*, **21**, 235–243 (1968).
9. Kopyshev, V. P., *Zh. Éksp. Teor. Fiz.*, **55**, 1304–1310 (1968).

10. Krasnikov, Y. G., *Zh. Éksp. Teor. Fiz.*, **53**, 2223–2232 (1967).
11. Larkin, A. I., *Zh. Éksp. Teor. Fiz.*, **38**, 1896–1898 (1960), [Sov. Phys.-JETP **11**, 1363 (1960)].
12. Vedenov, A. A., and Larkin, A. I., *Zh. Éksp. Teor. Fiz.*, **36**, 1133–1142 (1959), [Sov. Phys.-JETP **9**, 806 (1959)].
13. Ebeling, W., Kraeft, W.-D., and Kremp, D., *Theory of Bound States and Ionisation Equilibrium in Plasmas and Solids*, Akademie-Verlag, Berlin, 1976.
14. Kraeft, W.-D., Kremp, D., Ebeling, W., and Röpke, G., *Quantum Statistics of Charged Particle Systems*, Plenum Press, New York, 1986.
15. Rouse, C. A., *Astrophys. J.*, **272**, 377–379 (1983).
16. Moore, C. E., Minnaert, M. G. J., and Houtgast, J., *The solar spectrum 2935 Å to 8770 Å*, vol. 61 of *NBS Monograph*, US Government Printing Office, Washington, 1966.
17. Norman, G. E., and Starostin, A. N., *Teplofizika vysokih temperatur*, **8**, 413–438 (1970).
18. Starostin, A. N., Roerich, V. C., and More, R. M., *Contrib. Plasma Phys.*, **43**, 369–372 (2003).
19. Vorobiev, V. S., and Khomkin, A. L., *Teor. Mat. Fiz.*, **8**, 109–118 (1971).
20. Beth, E., and Uhlenbeck, G. E., *Physica*, **4**, 915 (1937).
21. Uhlenbeck, G. E., and Beth, E., *Physica*, **3**, 729 (1936).
22. Levinson, N., *Kgl. Danske Videnskab. Selskab, Mat.-Fys. Medd.*, **25**, No.9 (1949).
23. Abrikosov, A. A., Gorkov, L. P., and Dzyaloshinskii, I. E., *Methods of Quantum Field Theory in Statistical Physics*, Dover Publications, New York, 1977, (transl. and ed. by R. A. Silverman).
24. Matsubara, T., *Progr. Theor. Phys.*, **14**, 351–378 (1955).
25. Kadanoff, L., and Baym, G., *Quantum Statistical Mechanics*, W. A. Benjamin, Inc., New York, 1962.
26. Keldysh, L. V., *Zh. Éksp. Teor. Fiz.*, **47**, 1515–1520 (1964), [Sov. Phys.-JETP, **20**, 1018–1022 (1965)].
27. Lifshitz, E. M., and Pitaevskii, L. P., *Physical Kinetics: (Course of Theoretical Physics, v. 10)*, Butterworth-Heinemann, Oxford, 1981, (translated by J. B. Sykes and R. N. Franklin).
28. Landau, L. D., and Lifshitz, E. M., *Statistical Physics: Part 1 (Course of Theoretical Physics, v. 5)*, Butterworth-Heinemann, Oxford, 1980, (translated by J. B. Sykes and M. J. Kearsley).
29. Galitskii, V. M., *Zh. Éksp. Teor. Fiz.*, **34**, 151–162 (1958), [Sov. Phys.-JETP, **7**, 104 (1958)].
30. DeWitt, H. E., *J. Math. Phys.*, **3**, 1216 (1962).
31. Montroll, E. W., and Ward, J. C., *Phys. Fluids*, **1**, 55–72 (1958).
32. Fradkin, E. S., "Metod funktsii Grina v teorii kvantovannyh polei i v kvantovoi statistike (diss. d.f.-m.n.)," in *Kvantovaya teoriya polya i gidrodinamika*, Nauka, Moscow, 1965, vol. 29 of *Trudy FIAN*, pp. 5–138.
33. Fock, V. A., *Z. Physik*, **98**, 145–154 (1935).
34. Landau, L. D., and Lifshitz, E. M., *Quantum Mechanics: Non-Relativistic Theory (Course of Theoretical Physics, v. 3)*, Pergamon, Butterworth-Heinemann, Oxford, 1981.
35. Akhiezer, A. I., and Berestetskii, V. B., *Quantum Electrodynamics*, Interscience Publishers, New York, 1965.
36. Landau, L. D., Lifshitz, E. M., and Pitaevskii, L. P., *Electrodynamics of Continuous Media (Course of Theoretical Physics, v. 8)*, Butterworth-Heinemann, Oxford, 1984, (translated by J. B. Sykes, J. S. Bell, and M. J. Kearsley).
37. Landau, L. D., and Lifshitz, E. M., *Fluid Mechanics (Course of Theoretical Physics, v. 6)*, Butterworth-Heinemann, Oxford, 1987.
38. Griem, H. R., *Phys. Rev.*, **131**, 1170–1176 (1963).
39. Wiese, W. L., Kelleher, D. E., and Paquette, D. R., *Phys. Rev. A*, **6**, 1132–1153 (1972).

Improved phenomenological equation of state in the chemical picture

Regner Trampedach

Mt. Stromlo Observatory, Australian National University

Abstract. I present an overview of an equation of state, being developed in the chemical picture, and based on the very successful MHD equation of state. The flexibility of the chemical picture combined with the free-energy minimization procedure, makes it rather straight-forward, albeit laborious, to include new effects in the model free-energy, simply by adding new terms.

The most notable additions to the original MHD equation of state, are relativistic effects, quantum effects, improved higher order Coulomb terms and a long list of molecules other than the H_2 and H_2^+ treated so far.

INTRODUCTION

Our understanding of stellar structure and evolution is the foundation for most astrophysical endeavors, whether related to clusters of stars, galaxies, the early universe or the distance scale. The quality of stellar modeling therefore has an impact beyond its own field. Our stellar models, in turn, depend crucially on the physics they are based on, of which there are two main categories; dynamical processes, *e.g.*, convection and rotation, and micro-physics, *e.g.*, the equation of state, opacities and nuclear reaction rates.

The late 80'es and early 90'es saw great advances in equation of state and opacity calculations for astrophysical plasmas and the published tables [1, 2, 3, 4] were immediately put to use. The improved atomic physics resolved some longstanding problems, *e.g.*, the disagreement between the theoretical evolutionary and observed dynamical masses of Cepheids and RR Lyraes [5, 6, 7].

At the same time, the increase in computer-power also enabled the emergence of realistic, 3-dimensional simulations of convection [8], convection being the most important of the dynamical processes. Due to the time-scales involved these simulations are constrained to the outer layers of stars. In the deeper layers, however, the stratification of the convection zone is adequately described by the mixing-length formulation [9]. The main parameter, α, determining the asymptotic adiabat of the convection zone, is not prescribed by this formulation. The 3D simulations can be used to calibrate α, and to describe the stratification in the upper layers where convective fluctuations are in the non-linear regime, and where radiation escapes the stellar surface.

The transition from diffusive to free-streaming radiative transfer (*i.e.*, the photosphere) is dealt with in great detail in conventional 1D stellar atmosphere calculations, where nowadays 10^5–10^6 wavelengths are used and non-LTE effects are included when necessary. Convection, however, is described in the mixing-length formulation which fails in the surface layers. The only alternative that can result in realistic and self-consistent stratifications, spectral line-shapes[10], abundances [11], *etc.*, are the convection simulations. It has recently been shown that a simplified radiative transfer scheme for the simulations, can reproduce the monochromatic solution, but for a small fraction of the computational cost [12]. Apart from non-LTE effects, the radiative transfer in the simulations can rival that of modern 1D atmosphere calculations, but with a realistic and consistent treatment of convection.

The convection simulations are deeper than normal atmosphere models, since, in general, convection turns adiabatic in deeper layers than where radiative transfer turns diffusive. As a consequence, the convection simulations straddle the regimes of a simple Saha equation of state that includes molecules, and a more sophisticated one that, however, can neglect molecules (other than H_2). The desire to accurately match the convection simulations with interior models, further promotes the concept of a stellar equation of state that includes everything from molecules at low temperature, interaction effects at high densities and relativistic effects at high temperatures. The present paper is an outline of such work in progress.

THE MHD EQUATION OF STATE

The MHD equation of state [13, 14] was conceived as part of the International Opacity Project (OP) [15, 16], of the late 1980's, to calculate precise and accurate opacities for stellar envelopes. The emphasis was on detailed quantum-mechanical calculations of the electron structure of all ions of the astrophysically most important elements. The resulting energy-levels and effective radii were used in the MHD equation of state and the transition probabilities were used for calculation of absorption coefficients for each state of each atom/ion. Combining the population of these states, as found from the equation of state, with the absorption coefficients, then results in the opacity; a self-consistent set of calculations.

The OPAL project [3, 4] is a parallel and similar equation of state and opacity effort, but independent of MHD and OP, and pursued in the physical picture.

The key concept introduced in the MHD equation of state, was that of occupation probabilities, w_i, of a particular state, i [17]. The Boltzmann-factors in conventional calculations of the population of states, are based on the assumption of an isolated particle and therefore ignore the presence of neighboring particles. This results in the well-known 24% un-ionized hydrogen at the center of the Sun, in this simplistic picture. It is obvious that the density at the Solar center leaves no room for hydrogen atoms. Two methods were introduced to cut-off the summation of the otherwise divergent partition function; one considered the available volume and the size of bound states and assumed dense packing, the other used a lowering of the ionization potential to spill bound states into the continuum. Both methods introduce discontinuities into the equation of state and most implementations have lacked internal consistency.

The MHD equation of state, on the other hand, includes a realistic model of a physical process that directly leads to ionization; ionization by repeated crossings of Stark-manifolds by fluctuating electric fields, as measured by [18]. This leads to ionization when the amplitude of the fluctuating field exceeds a threshold, F_{cr}, for given state. The probability of state i being present in a plasma is therefore

$$w_i = Q(F_{cr,i}) = \int_0^{F_{cr,i}} P(F) dF , \qquad (1)$$

where $P(F)$ is the micro-field distribution. This provides us with a smooth, self-consistent and physically plausible equation of state.

A further advantage over previous work is the analytical derivatives, which ensures smooth and consistent thermodynamic derivatives.

BEYOND THE ORIGINAL MHD

The improvements detailed below, are compared to the original MHD and to the OPAL equation of state[3] in Figs. 1–3. These plots show pressure differences in the left-hand panel and γ_1 differences in the right-hand panel. These quantities have been selected for their importance for the hydrostatic structure and the sound-speed profile, respectively. In these plots the solid line shows the difference (new MHD−original MHD) as the solid lines, (OPAL−original MHD) as the gray lines. The dotted lines show the original MHD (as published) with the τ-function that was introduced to curb the otherwise diverging (with density) Debye-Hückel term. This τ-factor was discussed by [19], disputing its physical foundation and showing that better agreement with OPAL is obtained using $\tau = 1$, as shown with the gray lines in Figs. 1–3. The solid, the dotted and the gray lines are the same in all three plots. The dashed lines show the effect of turning off some of the improvements, one at a time.

The OPAL equation of state is an important comparison case, since Solar models based on it show overall better agreement with helioseismic inversions [20], compared to MHD. Looking more carefully at the outer layers of the Sun, using high degree p-modes, models using the MHD equation of state seems to be closer to the Sun in the outer 7%, than ones using OPAL [21]. This region includes the H-, He- and some of the He^+-ionization zone, and covers the left half of each of the plots in Figs. 1–3. Deeper in the Sun we therefore want better agreement with OPAL (*i.e.*, the solid curve close to the gray curve in Figs. 1–3), whereas the situation is less clear further out.

Micro-field Distributions

The original MHD formulation used linear fits to the Holtzmark distribution, which assumes non-interacting particles. [22] introduced much improved fits to a micro-field distribution model that accounts for particle interactions through the Debye-Hückel theory. I have included this so-called Q-MHD formulation and the effect under Solar circumstances can be seen in Fig. 1. The effect is largest in the derivatives, where a lot of new structure is introduced. This change will most likely have a larger effect on the opacity since it changes the population of highly excited states that are often strong absorbers.

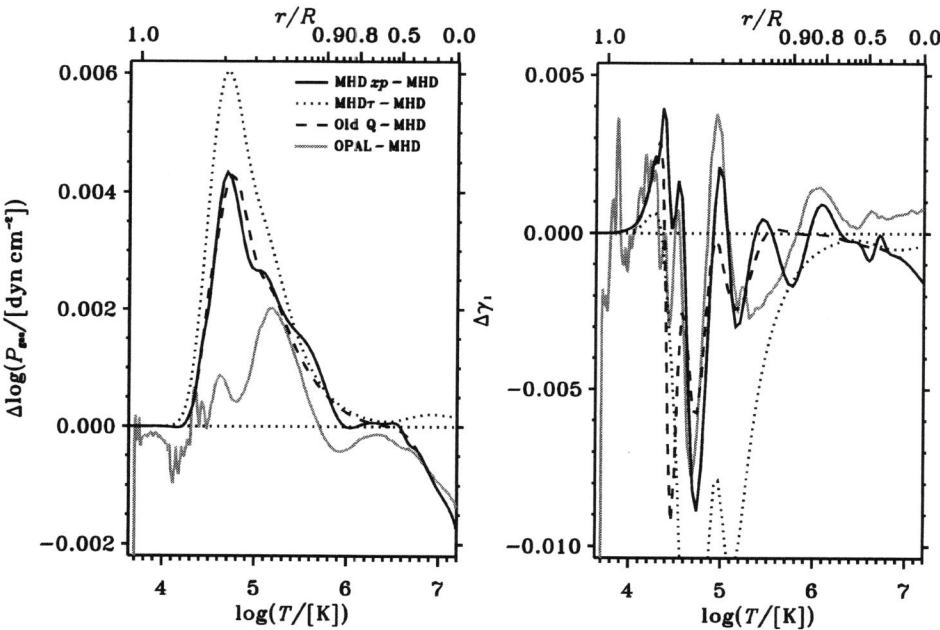

FIGURE 1. Differences between original MHD and: OPAL (gray), all changes (solid), all changes except old micro-field distribution (dashed), along a Solar ρ/T-stratification.

Quantum effects

Apart from the effect on particle statistics, in the form of degeneracy, there are at least two other effects of quantum mechanics that needs to be included; quantum diffraction and exchange interactions. The former is a consequence of Heisenberg's uncertainty relation and the realization that particles are wave-packets of finite size. This results in finite charge-densities and therefore avoids the short-range divergence of the Coulomb potential. This effect replaces the τ-factor mentioned above.

The exchange term arise from Pauli's exclusion principle between identical particles; the wave-functions of two identical particles will either overlap or repel each other, depending on their spin, thereby changing their interaction energy compared to that of differing particles. The effect of the first-order exchange-term [23] is shown in Fig. 2.

Interactions with Neutral Particles

In the original MHD equation of state neutral particles were treated as hard spheres, but only to first order in particle density, N_i. The second order term was included

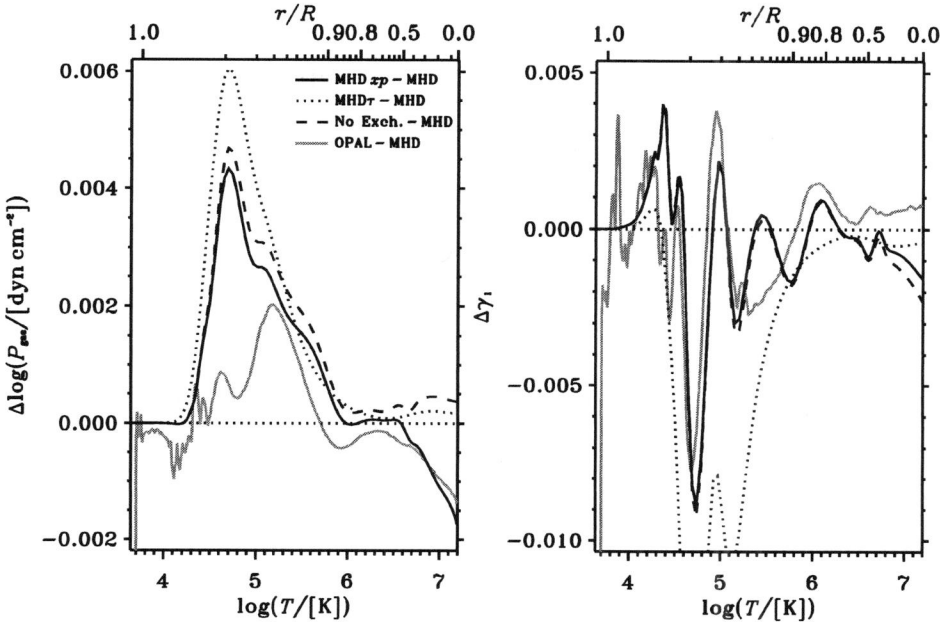

FIGURE 2. Differences between original MHD and: OPAL (gray), all changes (solid), all changes except for exchange effects (dashed), along a Solar ρ/T-stratification.

in an approximate way through the Ψ-function discussed by [24]. Apart from these approximations, the hard-sphere model has its problems. First of all, it is undefined for high densities and can disrupt the convergence of the free-energy minimization. Second, it is un-physical and ignores the underlying forces. A neutral atom has an extended electron distribution, and in close encounters, another particle will dip into this electron wave-function and begin to feel the charge of the nucleus which is no longer completely screened. At high densities net-neutral particles will therefore have a small effective charge. The nice thing about this model, is that it also applies to partially stripped ions, and all particles are treated on an equal footing. Under Solar circumstances this change has no discernible effect, but it most likely will for cooler stars.

Coulomb Interactions

By far the largest change—at least in the Sun—arise from including higher-order terms in the Coulomb interactions, beyond the Debye-Hückel term, F_{DH}. The original MHD had a τ-correction to F_{DH} that turned out to be un-physical [19], and is now replaced by a correction-factor based on Monte-Carlo simulations of the one-component-

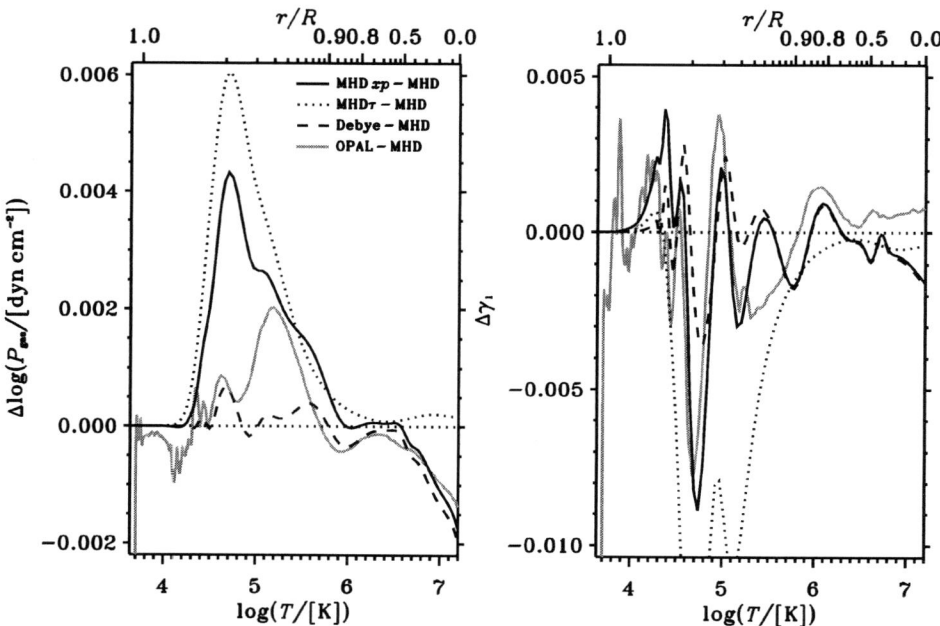

FIGURE 3. Differences between original MHD and: OPAL (gray), all changes (solid), all changes except pure Debye-Hückel (dashed), along a Solar ρ/T-stratification.

plasma [25], implicitly including many-body interactions. The effect of this change can be seen in Fig. 3; it is constrained to the convection zone and peaks at a depth of about 7 Mm, which is surprisingly close to the surface.

Additional Changes

Relativistically degenerate electrons are included as detailed by [26]. The effect is very small except in the adiabatic exponent, γ_1, which is lowered appreciably in the inner half of the Sun. Relativistic effects are not included in the version of OPAL we have used for comparison, but omitting it in the new MHD results in very good agreement with OPAL for $\log T > 6.7$. The new version of OPAL includes relativistic electrons, as well as a few other improvements [27].

Molecules are included by means of the fits to partition functions, compiled by [28], augmented by a parametrized pressure dissociation, based on the detailed treatment by MHD of the H_2- and H_2^+-molecules. Molecules are selected from a list of 315 di-atomic and 99 poly-atomic molecules, depending on the elements included in the equation of state calculation. This change will mostly affect the atmospheric opacities.

SUMMARY

A new equation of state is being developed, based on the MHD equation of state, but with a number of improvements that will hopefully bring it on par with the OPAL equation of state. Tables have not been computed yet, so calculations of Solar models with the new equation of state are not yet possible. Comparisons on a fixed ρ/T-track of a Solar model, as presented here, are encouraging, however.

ACKNOWLEDGMENTS

I am grateful to the organizers of the workshop, for the invitation to present and for a fruitful two weeks. I would also like to thank Werner Däppen for full access to the original MHD-code, including rights to and help with modifying it.

REFERENCES

1. Mihalas, D., Hummer, D. G., Mihalas, B. W., and Däppen, W., *ApJ*, **350**, 300–308 (1990).
2. Seaton, M. J., Yan, Y., Mihalas, D., and Pradhan, A. K., *M.N.R.A.S.*, **266**, 805–828 (1994).
3. Rogers, F. J., Swenson, F. J., and Iglesias, C. A., *ApJ*, **456**, 902–908 (1996).
4. Rogers, F. J., and Iglesias, C. A., *ApJS*, **401**, 361–356 (1992).
5. Andreasen, G. K., *A&A*, **201**, 72–79 (1988).
6. Andreasen, G. K., and Petersen, J. O., *A&A*, **192**, L4–L6 (1988.).
7. Yi, S., Lee, Y.-W., and Demarque, P., *ApJ*, **411**, L25–L28 (1993).
8. Nordlund, Å., and Stein, R. F., *Comput. Phys. Commun.*, **59**, 119 (1990).
9. Böhm-Vitense, E., *Zs. f. Astroph.*, **46**, 108–143 (1958).
10. Asplund, M., Nordlund, Å., Trampedach, R., Allende Prieto, C., and Stein, R. F., *A&A*, **359**, 729–742 (2000).
11. Asplund, M., Nordlund, Å., Trampedach, R., and Stein, R. F., *A&A*, **359**, 743–754 (2000).
12. Trampedach, R., and Asplund, M., "Radiative Transfer with Very Few Wavelengths," in *ASP Conf. Ser. 293: 3D Stellar Evolution*, 2003, pp. 209–+.
13. Mihalas, D., Däppen, W., and Hummer, D. G., *ApJ*, **331**, 815–825 (1988).
14. Däppen, W., Mihalas, D., Hummer, D. G., and Mihalas, B. W., *ApJ*, **332**, 261–270 (1988).
15. Seaton, M. J., editor, *The Opacity Project*, vol. 1, Institute of Physics Publishing, 1995.
16. Berrington, K. A., editor, *The Opacity Project*, vol. 2, Institute of Physics Publishing, 1997.
17. Hummer, D. G., and Mihalas, D., *ApJ*, **331**, 794–814 (1988).
18. Pillet, P., van Linden van den Heuvell, H. B., Smith, W. W., Kachru, R., Tran, N. H., and Gallagher, T. F., *Phys. Rev. A*, **30**, 280–294 (1984).
19. Trampedach, R., Däppen, W., and Baturin, V. A., *ApJ* (2004), (submitted).
20. Christensen-Dalsgaard, J., and Däppen, W., *A&AR*, **4**, 267–361 (1992).
21. Di Mauro, M. P., Christensen-Dalsgaard, J., Rabello-Soares, M. C., and Basu, S., *A&A*, **384**, 666–677 (2002).
22. Nayfonov, A., Däppen, W., Hummer, D. G., and Mihalas, D., *ApJ*, **526**, 451–464 (1999).
23. Kovetz, A., Lamb, D. Q., and Horn, H. M. V., *ApJ*, **174**, 109–120 (1972).
24. Trampedach, R., and Däppen, W., *ApJ* (2004), (in preparation).
25. Slattery, W. L., Doolen, G. D., DeWitt, H. E., and Slattery, W. L., *Phys. Rev. A*, **26**, 2255+ (1982).
26. Gong, Z., Däppen, W., and Zejda, L., *ApJ*, **546**, 1178–1182 (2001).
27. Rogers, F. J., and Nayfonov, A., *ApJ*, **576**, 1064–1074 (2002).
28. Irwin, A. W., *ApJS*, **45**, 621+ (1981).

Emulating the OPAL Equation of State in the Chemical-Picture Formalism

Aihua Liang

University of Southern California, Los Angeles, California, 90089-1342, U.S.A.

Abstract.
Helioseismic determination of solar interior is subject to the uncertainties in equation of states. Along with the technical improvements is the ever-increasing precision of helioseismic data. A more accurate equation of state that we can be more confident of is absoluately necessary. Among them, Mihalas-Hummer-Däppen (MHD) equation of state and a equation of state that was developed along with Opacity Project At Livermore (OPAL) are the two most popular ones. And fortunately MHD's lack of rigorous theoretical foundation is compensated by OPAL's conceptual conciseness, while OPAL's computational unwieldyness is compensated by MHD's flexibility and practicability. Since OPAL yields better match with solar data than MHD except at the top 2% of solar region, our research goal had been trying to find a modified MHD equation of state that emulates computational results of thermodynamic quantities from OPAL. So far this goal has been attained.

INTRODUCTION

In the field of stellar modeling, helioseismology in particular, the equation of state (EOS) is one of the three key material property ingredients that have to be hypothesized at the beginning based on our understanding of the steller interior. The other two ingredients are opacity and nuclear reaction rate at the solar center. And it is the most primary mission of helioseismology to probe inside the solar interior and see how well the model agrees with the real sun.

However omprehensive steller modeling is inherently a highly entangled problem that involves hydrodynamics, fluid dynamics, nuclear energy reaction, various energy transport scheme, electrodynamics, plus the three material properties mentioned above. Yet the situation is not entirely hopeless, because of one particularly convenient fact which states that as far as the deeper part of the convection zone is concerned, the equation of state can be isolated from the coupled solar model. The reason is that in the convection zone, energy is transported outwards mainly by convective motion, which is a highly efficient mechanism; as a result, the temperature gradient only needs to be slightly non-adiabatic to meet the observed solar luminosity requirement; the adiabatic condition, the equation of state, plus the hydrodynamic equilibrium and composition together can pretty much determined the seismic structure of the convection zone without invoking opacity. Therefore the deeper part of convection zone provides us with an ideal environment to study the equation of state, and in turn, with some elements fixed, the improvement of equation of state can be used to fine tune other elements that we are still uncertain of.

The technique that is used to extract solar interior properties from helioseimic data is called "differential inversion", where "differential" comes from the fact that, in practice, the real solar observational data is subtracted from thost computed from a good solar model, and the residue is believed to be small enough so that the linearization approximation, which greatly simplifies the inversion process, can be justified. Therefore apparently the determination of solar interior is subject to the uncertainties in solar models.

A typical illustration of the above statement is the thermodynamical determination of helium abundance in the 2nd helium ionization zone. As is well known, helium abundance can not be determined from spectroscopic observations as in the case of other elements; also another well known fact is that the ionization process lowers the adiabatic exponent $\gamma_1 \equiv (\partial \ln P/\partial \ln \rho)_s$, where s being the specific entropy at adiabatic conditions of the convection zone, from its ideal gas value $\frac{5}{3}$ to some lower value. And the abundance of helium abundance is obviously associated with how pronouced the "dip" is. The idea of determining helium abundance from the depth of the "dip" was proposed by [Gough(1984) 1].

MOTIVATION & GOAL

As is well known, the interior of the sun is a hot, dense, and yet weakly coupled plasma, the physical condition that is still attainable on earth. Thanks to the technical advances in observational field, the helioseismic data are now being obtained with ever-increasing precision, which in turn imposes tougher and tougher constrains on the accuracy of equation of state.

Among many prevailing equations of state in the past 20 years, the two most popular and sophisticated are MHD and OPAL equation of state. They are both at the base of two respective opacity computation efforts, the international Opacity Project (OP) [see, *e.g.*, 2], and the OPAL project pursued at Livermore [see, *e.g.*, 3, and references therein]. MHD equation of state [4, 5, 6, 7] is worked out in the framework of so-called "chemical picture" [8], where the notion of composite particles such as atoms and ions are assumed from the outset, and the ionization process is treated like a "chemical reaction". To assure thermodynamic consistency, all the thermodynamic properties need to be worked out from one particular thermodynamic potential. Since temperature T, density ρ, and volume V are the most natural control variables in such canonical systems, the free energy is consistently chosen as the thermodynamic potential. And so-called "free energy minimization" method is used to achieve thermodynamic equilibrium. In laying out an expression for free energy, another important assumption made by chemical picture formalism is the partition function's factorizability, which is essentially formulated as,

$$Z = Z_{trnsl} Z_{cnfg} Z_{int}, \qquad (1)$$

where, Z_{trnsl} stands for translational partition function, Z_{cnfg} for configurational partition function, and Z_{int} for internal partition function. This obviously leads to the modularity of free energy, which turns out to be the greatest strength of the whole chemical picture approach. To be more specific, taking advantage of the asymptotic nature of the free energy expression, intuitions can be used to bring in high order terms that correspond to finer physical effects. Apparently the thermodynamic consistency always remains intact. Although one should note that the introduction of new physical term is not all arbitrary due to the constrains of statistical consistency.

On the other hand, OPAL equation of state [9, 12, 11, 10] was developed within the framework of so-called "physical picture" formalism, where only fundemental particles such as electrons and nuclei are incorporated at the beginning of the formalism, and an activity (not density) expansion named ACTEX is carried out using grand canonical ensemble (GCE) instead of canonical ensemble, with the grand potential as the choice of thermodynamic potential. Composite particles enter the formalism as the effective couterparts of bound states, whereas the scattering states in the continuum are also kept. For partially ionized plasmas, an expansion with better convergence is achieved if an augmented set of activities is used that incorporates both fundamental particles and composite particles. The strength of OPAL lies in the fact that it is a systematic and rigorous approach so that it does not need to make as many artificial assumptions as MHD, and also it accounts for the many-body interaction effects from the plasma environment automatically while at the same time preserves the statistical consistency among all the elements in the plasma.

However when it comes down to numerical computations, MHD equation of state turns out to be much more convenient than OPAL. So far, the OPAL equation of state only exists in tabular form and the code is proprietary. The MHD equation of state program, on the other hand, is open-source, easy to operate and highly adaptable to a detailed chemical composition with many elements. These are desirable properties, especially in the light of new evidence that the heavy-element abundance of the Sun might have been over-estimated, and that new solar models will have to be re-computed soon.

As far as the convection zone excluding the top 2% is concerned, OPAL yields a closer match to the real sun than MHD, the discrepancy is well within the testing power of helioseismic data. Therefore our first step of research is to take advantage of chemical picture's modularity and physical picture's systematicness, identify those higher order physical terms that are present in physical picture but missing in chemical picture, and patch them to MHD so that the modified equation of state yields a closer match to OPAL. So far this step of our research has been attained. However the ultimate goal of our research would be to find the equation of state that matches the real solar observations. We note that there has been an earlier attempt the simulate aspects of the OPAL equation of state; it is the so-called SIREFF equation of state by Rogers, Swenson and Irwin. It is based on the Eggleton, Faulkner & Flannery [14] (EFF) equation of state [for details see 15].

PHYSICAL CONDTIONS OF HYDROGEN IONIZATION ZONE

So far we have studied the hydrogen ionization zone with the temperature and density profile matching the corresponding solar region. Listed in **Table 1** is a few dimensionless parameters that reveal the physical conditions of the corresponding region, which includes

- Plasma non-ideality parameter, $\Lambda_{ep} \equiv \frac{e^2}{kT\lambda_D}$;
- Coupling parameter, $\Gamma \equiv \frac{e^2}{kTr_{pp}}$;
- Qantum diffraction parameter, $\gamma_{ep} \equiv \frac{\lambda_{ep}}{\lambda_D}$;
- Electron degeneracy parameter, $\Theta_e \equiv \frac{\lambda_{ee}}{r_{ee}}$;

where, λ_D is the usual Debye-Hückel screening length defined as $\sqrt{\frac{kTV}{4\pi e^2 \sum_i Z_i^2 N_i}}$; r_{pp} is the mean distance between two neighbouring protons in the hydrogen plasma, so is r_{ee} defined; λ_j is the thermal de Broglie wavelength defined as $\frac{h}{\sqrt{2\pi m_{ij} kT}}$.

TABLE 1. The table is computed from our modified chemical picture equation of state that simulates the OPAL equation of state.

T in K	ρ in g/cm^3	Λ_{ep}	Γ	γ_{ep}	Θ_e
.812E+04	.220E-06	.142E-01	.323E-01	.572E-02	.129E-01
.115E+05	.496E-06	.490E-01	.737E-01	.235E-01	.354E-01
.145E+05	.184E-05	.102E+00	.120E+00	.553E-01	.650E-01
.181E+05	.721E-05	.185E+00	.178E+00	.111E+00	.107E+00
.231E+05	.264E-04	.293E+00	.243E+00	.199E+00	.165E+00
.309E+05	.843E-04	.388E+00	.292E+00	.304E+00	.229E+00
.433E+05	.226E-03	.413E+00	.305E+00	.384E+00	.283E+00
.636E+05	.498E-03	.358E+00	.277E+00	.403E+00	.312E+00
.948E+05	.103E-02	.288E+00	.240E+00	.396E+00	.329E+00
.141E+06	.214E-02	.229E+00	.206E+00	.384E+00	.345E+00
.212E+06	.418E-02	.174E+00	.171E+00	.358E+00	.353E+00
.320E+06	.793E-02	.129E+00	.141E+00	.327E+00	.356E+00

OUR RESEARCH APPROACH

As has been mentioned in the 2nd section, to accomplish our goal of research, we need to take advantage of the modularity and flexibility of chemical picture to model complicated mixture, and the systematicness and consistency of physical picture. So far even in the state-of-art MHD equation of state, the electrons, protons and the combined ions in the continuum are still modeled as free components, which is a gross simplification. More specifically, the interaction between electrons and protons beyond its bound states, which is supposed to be formulated in the 2nd virial coefficient, are not taken into consideration. With the heuristic nature, it is very difficult for the chemical picture itself to come up with a consistent scheme to account for such effects, while at the same time still preserve statistical consistency. On the other hand, such a scheme has readily existed in physical picture's activity expansion formalism ACTEX. Therefore the natural approach would be that we identify those terms that are present in OPAL and correspond to the scattering part of 2nd virial coefficient, and patch them up to chemical picture. What we have done in our research is to:

- transform the OPAL activity expansion to a density expansion, and compare it with the analogous density expansion as obtained from MHD;
- find the higher-order corrections that can be patched to the MHD's free energy formula to achieve a thermodynamically consistent theory, which will emulate the OPAL equation of state.

FORMULA FOR MODIFIED EQUATION OF STATE IN CHEMICAL PICTURE

Internal Partition Function

There exists a large body of literatures on chemical picture's free energy minization method [4, 5, 6, 7, 8]. Due to the assumed modularity, the free energy is usually expressed in a series expansion, each of which corresponds to a physical effect, *e.g.* translational free energy which is essentially the ideal gas term, configurational free energy which effectively modulates the influence cast on by the charged plasma environment, the internal partition function which describes a structured atom's internal degree of freedom, and also another term that is dedicated to the degeneracy effect of electrons. While MHD partition function is often regarded as the state-of-art partition function, here in our modified chemical picture formalism, Plank-Larkin partition function, which has been under intense debate for long time, is adopted as our choice of internal partition function,

$$\text{PLPF} = \sum_{nl}(2l+1)\left[\exp(-\frac{E_{nl}}{kT}) - 1 + \frac{E_{nl}}{k_BT}\right]. \quad (2)$$

Plank-Larkin partition function is a result of compensation of the bound state term by the scattering state term in two-body cluster coefficient [16]. Had the calculation on the the infi nite terms of full trace of two-body Hamiltonian been carried out effectively, such a compensation would not be necessary. But in reality the activity series has to be truncated at a certain point. Therefore it is accepted because of the faster convergence it brings to the formalism [12]. In the light of introducing the scattering state term into chemical picture, PLPF is both a necessary and natural choice for internal partition function, because they are statistically consistent. What is also worth pointing out here is that the internal partition function is conventionally regarded to relate to the occupation number of bound states directly. However since it has been shown very clearly that the Plank-Larkin partition function does not lead to the "real" occupations, but rather "effective" ones. Such a view must be transfered to chemical picture also. Therefore to obtain the "real" occupations, additional procedures have to be carried out for opacity computations. This has been prooved by people working on Opacity Project in Livermore.

Regarding the energy levels used in the formalism, since the plasma parameter Λ_{ep} is still much less than 1, it is reasonable to assume the deep bound states are not much affected by the screening effect, thus isolated energy levels are still kept intact [17, 19].

Electron-Proton Scattering Term

Under the condition that the quantum diffraction is not so large, $\gamma_{ep} < 0.4$, the perturbation expansion in weakly coupling plasma is used to calculate the scattering part of the electron-proton cluster coefficient [18]. In addition we can argue that since we are only interested in the scattering part and the bound state part is accounted for in Plank-Larkin partition function, there is no need to employ complicated pseudopotentials to deal with the region where $r < \lambda_{ep}$, which effectively reduces the quantum diffraction. The following formulae are all in natural units.

$$F_{ep} = 2N_e N_p \frac{T}{V} S^p_{ep}(T, V, N_e, N_p) \tag{3}$$

$$S^p_{ep} = -\frac{\pi}{3}\left(\frac{2}{T}\right)^3 \left(\ln\frac{\lambda_{ep}}{\lambda_D} + 0.86\right) \tag{4}$$

$$\lambda_{ep} = \frac{1}{\sqrt{T}} \tag{5}$$

$$\lambda_D = \sqrt{\frac{TV}{8\pi(N_e + Z_p^2 N_p)}} \tag{6}$$

where, λ_{ep} is the thermal de Broglie wavelength divided by $\sqrt{4\pi}$. λ_D is the Debye-Hückel screening length.

Proton-Proton & Electron-Electron Scattering Term

As we can see from **Table 1**, the degeneracy parameter for electrons is no greater than 0.356. Since the mass of a proton is about 1800 times that of an electron, the protons can be well regarded as classical Boltzmann particles, whereas a little marginal for electrons. Thus WKB approximation is used for electron-electron interaction, whereas simple classical definition of 2nd virial coefficient of proton-proton interaction will do [18, 20]

The following formulae are expressed in natural units,

$$F_{pp} = -N_p^2 \frac{T}{V} S^p_{pp}(T, V, N_e, N_p) \tag{7}$$

$$S^p_{pp} = 2\pi \int_0^\infty \left[\exp\left(-\frac{2\exp(-r/\lambda_D)}{rT}\right) - 1\right] r^2 \, dr \tag{8}$$

$$+2\pi\lambda_D^2\left(\frac{2}{T}\right) - \frac{\pi}{2}\left(\frac{2}{T}\right)^2\lambda_D$$

$$F_{ee} = -N_e^2\frac{T}{V}S_{ee}^p(T,V,N_e,N_p) \qquad (9)$$

$$S_{ee}^p = 2\pi\int_0^\infty\left[\exp\left(-\frac{2\exp(-r/\lambda_D)}{rT}\right) - 1\right]r^2\,dr \qquad (10)$$

$$-\frac{\pi}{6T^3}\int_0^\infty\exp\left(-\frac{2\exp(-r/\lambda_D)}{rT}\right)\left[-\frac{2}{r}\exp(-r/\lambda_D)\left(\frac{1}{r}+\frac{1}{\lambda_D}\right)\right]^2 r^2\,dr$$

$$+2\pi\lambda_D^2\left(\frac{2}{T}\right) - \frac{\pi}{2}\left(\frac{2}{T}\right)^2\lambda_D$$

a_0 is Bohr radius. The last two terms in the expressions of S_{pp} and S_{ee} are subtracted because they participate in the re-summation procedure that produces many-body screening effect, which has already been taken into account of.

NUMERICAL RESULTS & CONCLUSION

The following figures show the relative differences between, on the one hand, 4 different models,

- Simplified OPAL equation of state, +;
- MHD + (MHD PF), MHD equation of state with the usual MHD partition function, ×;
- MHD + PLPF, MHD equation of state with PLPF, *;
- Modified MHD, our latest MHD equation of state, intended to emulate OPAL, □;

and on the other hand, the OPAL results to be matched (horizontal zero line). The physical conditions of the comparison is given in **Table 1**, which roughly correspond to those found along a track through a part of the solar convection zone. As we see from the figures, our modified MHD model (□) works remarkably well not only for adiabatic exponent (γ_1), but also for pressure (p), specific heat at constant volume (C_v), isothermal compression coefficient (χ_ϱ), and the strain coefficient (χ_T). They are defined, respectively, as the following:

- $\gamma_1 = (\partial\ln p/\partial\ln\varrho)_s$;
- $C_v = T(\partial S/\partial T)_v$;
- $\chi_\varrho = (\partial\ln p/\partial\ln\varrho)_T$;
- $\chi_T = (\partial\ln p/\partial\ln T)_\varrho$;

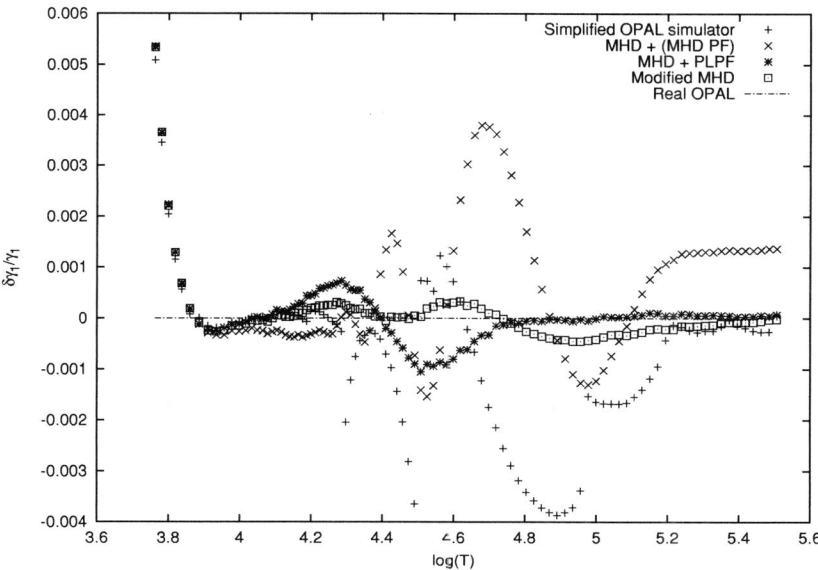

FIGURE 1. $\delta\gamma_1/\gamma_1$, where $\gamma_1 = (\partial \ln p / \partial \ln \varrho)_s$ (ϱ, s being density and specific entropy, respectively). See text for line styles and further details.

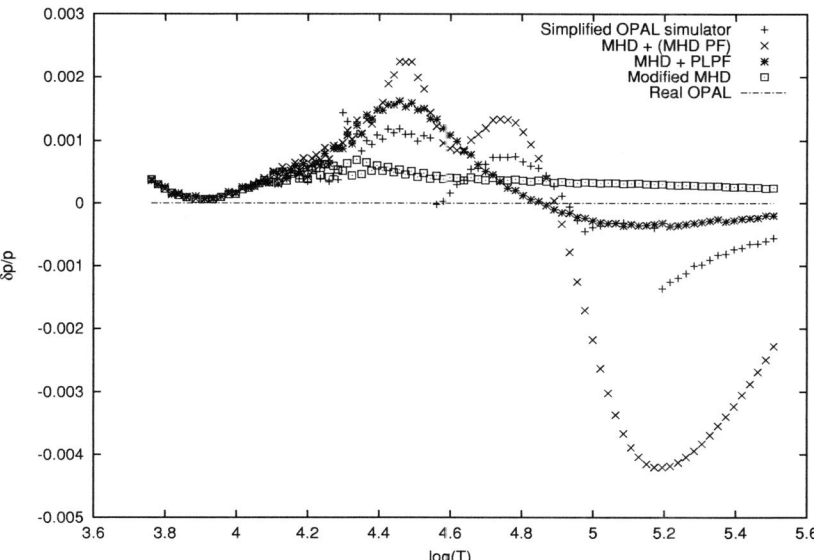

FIGURE 2. $\delta p/p$, where p refers to the pressure of 1 mole of pure hydrogen gas. See text for line styles and further details.

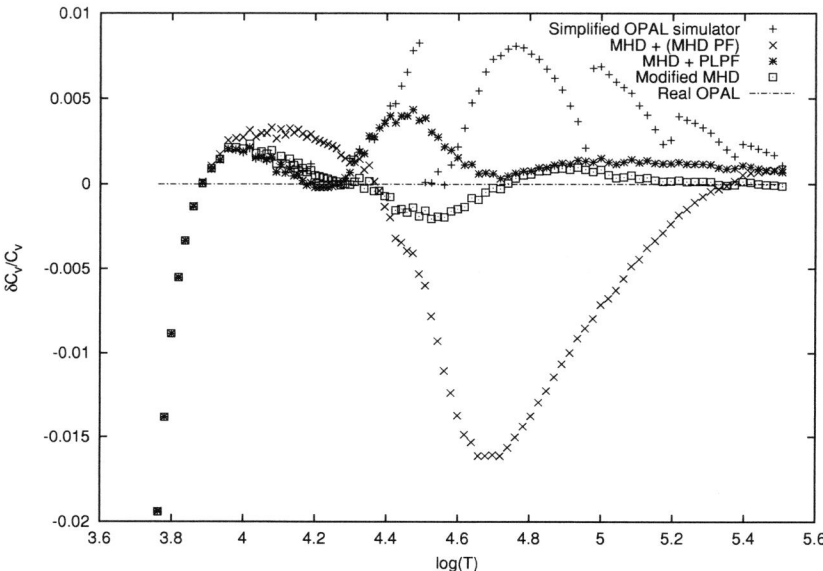

FIGURE 3. $\delta C_v/C_v$, where C_v refers to the specific heat at constant volume. See text for line styles and further details.

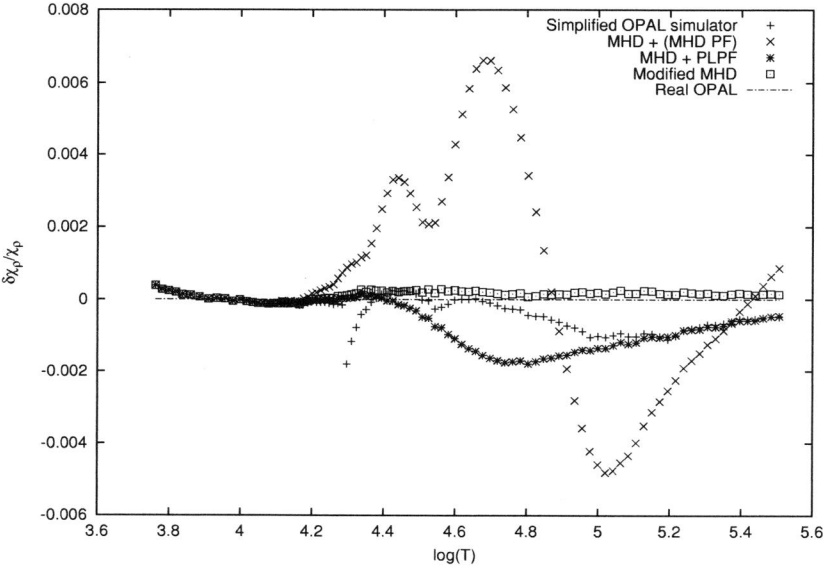

FIGURE 4. $\delta\chi_\varrho/\chi_\varrho$, where $\chi_\varrho = (\partial\ln p/\partial\ln\varrho)_T$. See text for line styles and further details.

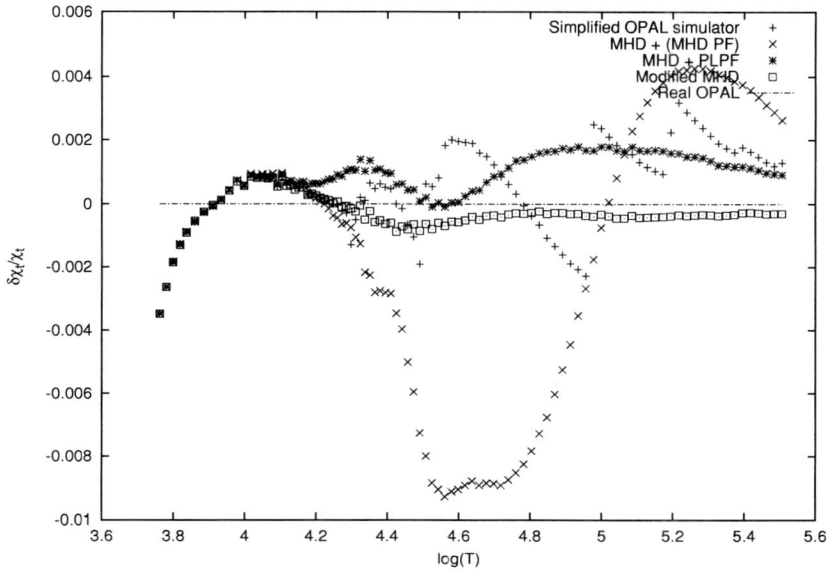

FIGURE 5. $\delta\chi_T/\chi_T$, where $\chi_T = (\partial \ln p/\partial \ln T)_\varrho$. See text for line styles and further details.

CONCLUSIONS

As we can see from the figures, in all cases, the remaining differences between the modified MHD equation of state have become much smaller than those between the original MHD equation of state and OPAL. Emulating OPAL has therefore been achieved.

ACKNOWLEDGMENTS

We thank Forrest Rogers for stimulating discussions. This work was supported by the grant AST-0307578 of the National Science Foundation.

REFERENCES

1. Gough, D. O. 1984, Mem. Soc. Astron. Ital., 55, 13
2. Seaton, M.J. 1995, *The Opacity Project* Vol. I, (Bristol: Institute of Physics Publishing)
3. Iglesias, C.A. & Rogers, F.J. 1996, ApJ, 464, 943
4. Däppen, W., Mihalas, D., Hummer, D. G. & Mihalas, B. W., 1988, ApJ, 332, 261
5. Hummer, D.G., Mihalas, D., 1988, ApJ, 331, 794-814
6. Mihalas, D., Däppen, W. & Hummer, D.G., 1988, ApJ, 332, 815
7. Nayfonov, A., Däppen, W., Hummer, D.G. & Mihalas, D.M. 1999,
8. Däppen, W., 1980, Astron. Astrophys. 91, 212-220
9. Rogers, F.J. & Iglesias, C.A., 1992, ApJS, 401, 361
10. Rogers, F.J. & Nayfonov, A. 2002, ApJ, 576, 1064
11. Rogers, F.J., Swenson, F.J., Iglesias, C.A., 1996, ApJ, 456, 902
12. Rogers, F.J., 1986, ApJ, 310, 723-728
13. Däppen W. 2004, *these proceedings*
14. Eggleton, P. P., Faulkner, J. & Flannery, B. P. 1973, A&A, 23, 325
15. Däppen, W. & Guzik, J. A. 2000, in Ibanoglu C., ed.,
16. Rogers, F. J., 1977, Phys. Letts. 61A, 6, 358
17. Rogers, F. J., 1981, Phys. Rev. A, 24, 3, 1531
18. Rogers, F. J., 1979, Phys. Rev. A, 19, 1, 375
19. Ebeling, W., Kraeft, W. D., Kremp, D. & Röpke, G., ApJ, 290, 24
20. Rogers, F. J., 1971, Phys. Rev. A, 4, 3, 1145

PART III

THE SUN AND THE STARS:
Detailed Techniques

The power of helioseismology to address issues of fundamental physics

Douglas Gough

Institute of Astronomy, Madingley Road, Cambridge, CB3 0HA, UK,
Department of Applied Mathematics and Theoretical Physics,
Wilberforce Road, Cambridge, CB3 0WA, UK
and Department of Physics, Stanford University, California 94305-4085, USA

Abstract. I first argue that, so far as macroscopic dynamical processes are concerned, to a high degree of accuracy the solar plasma, at least beneath the photosphere, may be regarded as being electrically neutral. I then turn to discuss, elaborating on Christensen-Dalsgaard's excellent but brief introduction in these proceedings, some of those aspects of helioseismology that are pertinent to determining those properties of the Sun that are pertinent to studies of the equation of state. I discuss one of the ways in which one can establish a representation of the structure of the Sun that is consistent with the helioseismic constraints, and I present some properties of the outcome. Finally, I describe some minor conclusions concerning opacity and the equation of state that can be drawn, mindful that in the not-too-distant future more penetrating insight will be gained.

INTRODUCTION

Jørgen Christensen-Dalsgaard (2004) has already provided us with an excellent basic summary of the principles of solar modelling and helioseismology. He has demonstrated how close the theoretical models now are to the helioseismological inferences of the actual structure of the Sun. For example, the sound speed differs from that of his Model S (Christensen-Dalsgaard *et al.*, 1996) by typically no more than about 0.1 per cent, the density, which is less well determined, by some ten times more. For many (e.g. Bahcall, 2001) this is an amazing success for the theory of stellar structure and evolution; it is also an amazing success for helioseismologists to have measured the inside of a star so precisely. But that is not to lead us into a state of complacency. There remain very significant differences between the theoretical models and the Sun. In some places the sound-speed discrepancy is more than ten formal standard errors in the helioseismic determinations. Moreover, there is evidence that Model S deviates from the current best theoretical opinion by even more. The reason is that a recent spectroscopic reassessment by Asplund *et al.* (2004) of chemical abundances in the solar atmosphere suggests that the abundances of the relatively common opacity-producing heavy elements C, N and O had previously been overestimated. As Christensen-Dalsgaard (2004) discusses in these proceedings, and as I discuss from a different viewpoint later in this report, the implied reduction in opacity below that which was used in the construction of Model S exacerbates considerably the discrepancy between theory and observation. We need to understand what factors might have caused the discrepancy, and, more importantly, what factors actually did cause it.

I hope it is hardly necessary for me to point out why it is that we wish to know the structure of the Sun so accurately. Aside from the intellectual challenge first of measuring the Sun's structure and then of providing a theoretical model that reproduces it to within the observational errors, we need to be able to check subtle aspects of the physics that is used in the construction of the theoretical models of the Sun. For some, that is important because the physics is needed for constructing models of other stars; for others, like those at this meeting, there is a deep interest also in the physics itself. In this presentation, therefore, my elaboration on what Christensen-Dalsgaard has provided will be confined to issues that relate closely to the physics of the dense plasma in the solar interior.

SOME PRELIMINARY REMARKS ON THE PHYSICS OF DENSE PLASMAS STRATIFIED IN A GRAVITATIONAL FIELD

As Christensen-Dalsgaard (2004) has already pointed out, different particle species in a plasma have a tendency to segregate when placed in an external force field. In making this statement I am viewing the plasma locally, on a scale of a few interparticle distances, so that in a star the so-called external forces are principally gravity and radiation pressure, even though those forces are produced (elsewhere) by the plasma itself. In the case of the Sun, gravity, on the whole, dominates radiation pressure, and heavy particles tend to settle through the lighter particles. That settling is opposed by diffusion, a process resulting from the random thermal motion. The diffusion coefficient increases with depth, and in practice it is only in the outer layers of the star that it is small enough for chemical segregation to be significant. But even there the process is slow: just beneath the base of the convection zone of the Sun, for example, the usual treatments of diffusion yield composition changes of only a few per cent over the lifespan of the Sun to date; in the core the change has been quite negligible.

Although electrostatic – and under some circumstances electromagnetic – forces are important, often dominant agents in producing the particle-particle interactions, on macroscopic (fluid) scales the plasma is normally regarded as being electrically neutral. But at this meeting Boris Vasiliev has challenged that tenet; he has argued that the electrical polarization produced by a separation between heavy ions (mainly nuclei) and electrons results in a structure of the Sun that is very different from current belief, producing, in particular, a much greater concentration of matter in a relatively small inner core. If correct, this argument must alter our view of the equation of state dramatically. It is therefore imperative that the arguments upon which our thinking is based be rehearsed, so that we are more conscious of the assumptions on which our solar models depend.

Because the modification that has been proposed is so striking, it is adequate here to present an approximate treatment of the equation of hydrostatic support, using, in particular, an equation of state based on nonrelativistic weakly interacting classical (i.e. not quantum-mechanical) point particles with no internal degrees of freedom, and, for added simplicity, neglecting the mass m_e of the electron compared with the masses of the heavy ions – indeed, I shall even consider all the ions to have the same mass, m_i.

This approximation is just what would straightforwardly yield the perfect gas law were the particles to be electrically neutral.

I presume, with others, that the Sun was born in a well mixed state. Any subsequent segregation of the particle species, based on the usual diffusive estimates, is likely to have been quite small. But I shall ignore that, in order to adopt, temporarily, an extreme view, and assume that the sole interaction between the electrons and the ions is to maintain locally a common temperature. If, in addition, electrostatic interactions were ignored too, then each species would come into its own hydrostatic equilibrium, being supported against gravity by its own partial pressure. Evidently the ions would be more concentrated towards the centre of the Sun than the electrons, the ratio of the pressure scale heights of the stratification of the two species being roughly $\varepsilon = m_e/m_i \lesssim 5 \times 10^{-4}$; the ratio of the density scale heights is similar, so the star would indeed be much more centrally condensed than are standard solar models. But this view is much too extreme: the electrostatic force resulting from the charge separation produced by the tendency to adopt such a configuration would act in opposition. What is the equilibrium state?

It is a straightforward matter to establish the equations of hydrostatic support for a spherical star. They are

$$k\frac{dnT}{dr} = -\frac{d\Phi}{dr}m_i n + \frac{d\phi}{dr}Zen \tag{1}$$

for the ions, and

$$k\frac{dnT}{dr} = \frac{nT}{1-v}\frac{dv}{dr} - \varepsilon\frac{d\Phi}{dr}m_i n + \frac{d\phi}{dr}en \tag{2}$$

for the electrons. Here r is the radius variable, $n = n_i$ is the number density of ions, and I have written the number density of electrons as $n_e = (1-v)Zn$, where Z is atomic number; T is the common temperature, e is the magnitude of the electronic charge and k is Boltzmann's constant; Φ and ϕ are respectively the gravitational and the electrostatic potentials, which satisfy the Poisson equations

$$\frac{1}{r^2}\frac{d}{dr}\left(r^2\frac{d\Phi}{dr}\right) = 4\pi G m_i[1 + \varepsilon Z(1-v)]n \simeq 4\pi G m_i n, \tag{3}$$

$$\frac{1}{r^2}\frac{d}{dr}\left(r^2\frac{d\phi}{dr}\right) = -\varepsilon_0^{-1}Zevn, \tag{4}$$

in which G is the gravitational constant and ε_0 is the dielectric constant of the vacuum.

I am merely going to discuss informally some properties of the solutions of these equations, rather than provide a detailed analysis. First, one should note that for there to be a balance between, on the one hand the mismatch between gravitational attraction and the gradient of partial pressure producing a tendency for ions and electrons to separate, and on the other hand the electrostatic force induced by the charge separation, the magnitudes of the gravitational and electrostatic forces acting on the (heavy) ions must be of the same order:

$$\left|\frac{d\Phi}{dr}\right|m_i n \sim \left|\frac{d\phi}{dr}\right|Zen; \tag{5}$$

this balance is expected to hold within a factor 2 or so. It follows that

$$\left|\frac{d\phi}{dr}\right| \bigg/ \left|\frac{d\Phi}{dr}\right| \sim \frac{m_i}{Ze}, \tag{6}$$

which can come about only if the respective sources of the gravitational acceleration and the electric field are also in that ratio:

$$\frac{Ze\overline{vn}}{4\pi\varepsilon_0 Gm_i\overline{n}} \sim \frac{m_i}{Ze}, \tag{7}$$

where the overbar denotes average over the spherical volume interior to r. Approximating $\overline{vn}$ by $\overline{v}\,\overline{n}$ yields

$$\overline{v} \sim \frac{4\pi\varepsilon_0 Gm_i^2}{Z^2 e^2} \simeq \frac{4\pi\varepsilon_0 Gm_p^2 A^2}{Z^2 e^2} \simeq 10^{-36}\left(\frac{A}{Z}\right)^2, \tag{8}$$

where A is the (mean) atomic number of the stellar material and m_p is the proton mass. It follows that even allowing for the possibility of a boundary layer in the net charge via a boundary layer in $v(r)$, the first term on the right-hand side of equation (2) is negligible. Consequently, subtracting equation (1) from equation (2) yields

$$m_i\frac{d\Phi}{dr} + (Z+1)e\frac{d\phi}{dr} \simeq 0, \tag{9}$$

because $\varepsilon \ll 1$, confirming the balance (5). Finally, by adding equations (1) and (2), and using equation (9) to eliminate $d\phi/dr$, one obtains

$$\frac{dp}{dr} = -\frac{d\Phi}{dr}\rho \equiv g\rho, \tag{10}$$

where p is the total gas pressure — the sum of the partial electron and ion pressures $p_e \simeq ZnkT$ and $p_i = nkT$ — and ρ is the mass density: $\rho \simeq m_i n$. This is the usual equation of hydrostatic support. Although there is electric polarization of the stellar material, the star being (slightly) positively charged throughout most of its interior and having a compensating shell of negatively charged material in an outer shell, the distribution of ions throughout the star is as though the material were considered to be electrically neutral, even though the ions are supported against gravity somewhat more by the electrostatic attraction to the electrons than by the gradient of p_i. In fact, because, in normal parlance, the ratio of gravitational to electrostatic forces is so small, the charge separation factor v is similarly small, and the distribution of the electrons is essentially the same as that of the ions: the fluid, treated macroscopically, is indeed essentially electrically neutral, at least when in hydrostatic balance. Furthermore, since the plasma frequency $\omega_p = (Zne^2/\varepsilon_0 m_e)^{1/2}$ is everywhere much higher than the characteristic frequency of any macroscopic fluid motion I shall consider here, equilibration between electrons and ions occurs very quickly in comparison with macroscopic fluid timescales, and therefore it is safe always to regard the fluid as being electrically neutral. The solution of equation (10) depends, of course, on the temperature distribution, which itself is determined by many additional physical considerations concerning mainly the generation and transport of heat; these have been summarized by Christensen-Dalsgaard (2004), and I shall not repeat that summary here. Suffice it to say that, whatever one considers the reliability of the physics of heat generation and transport to be, one can instead adopt the adiabatic sound-speed $c(r)$ as a quantity to close equation (10). In the Sun, c has been determined seismologically. It is therefore possible to infer the mass distribution throughout the Sun from the sound speed, provided, of course, that the equation of state relating c to p and ρ is known. Perhaps more reliably (although maybe less accurately) one can determine the density distribution, and thence the distribution of mass, directly by seismology, because the dynamics of acoustic modes is influenced by the density stratification, through buoyancy, through a reduction by density gradients in the propensity of sound waves to propagate, and through the wave perturbation to the gravitational field. How this is accomplished is briefly described by Christensen-Dalsgaard in these proceedings. I elaborate on that a little in the next section, in order to emphasize what the results mean, for it is extremely important to understand precisely the implications of the helioseismic inferences, and the assumptions on which they depend, before using them to make deductions about physics.

TOWARDS A SEISMIC MODEL OF THE SUN

Before proceeding, let us first establish what it is that one can learn from helioseismology. Except in the very outer layers of the star, seismic oscillations are adiabatic. Therefore wave propagation depends only on pressure p and density ρ, together with an equation of state relating the two under adiabatic change. The waves are also of very low amplitude, and can therefore be treated by linear theory. Consequently the equation of state enters only via the thermodynamic coefficient $\gamma_1 \equiv (\partial \ln p/\partial \ln \rho)_s$, the derivative being taken at constant specific entropy s. Only these three quantities can be determined by basic seismology, together, of course, with any function of them alone. (I speak here of basic helioseismology to be the inference of the spherically symmetrical component of the solar interior from the frequencies of seismic oscillation. These frequencies are influenced slightly in a recognizable way by interior fluid motion, mainly rotation, and by magnetic fields, which enables one to draw certain conclusions concerning the motion and the fields from measurements of the frequencies and the horizontal structure of the oscillations, but I refrain from discussing that here because it is not obviously immediately pertinent to the study of the equation of state.) Since the only seismic waves useful for diagnosing the solar interior that can be seen are acoustic, the most readily determined quantity is the sound speed c, given by $c^2 = \gamma_1 p/\rho$.

The most physically direct way of seeing how the sound speed is related to seismic frequencies is via the resonance conditions based on the simplified acoustic dispersion relation $\omega = kc$, valid for high k (and hence high ω), where ω is the oscillation frequency and k is the magnitude of the local wavenumber whose horizontal and vertical components are respectively $\mathbf{k}_h$ and k_v (e.g. Gough, 1993, 2003). As Christensen-Dalsgaard (2004) has pointed out, the resonance conditions may be written $|\mathbf{k}_h| = L/r$, where $L = l + \frac{1}{2}$, l being an integer called the degree of the mode, and

$$F(w) = (n + \alpha)\pi/\omega, \quad (11)$$

where $w = \omega/L$, the integer n being the order of the mode, and

$$F(w) = \int_{r_t}^{R} \left(1 - \frac{w^2}{a^2}\right)^{\frac{1}{2}} \frac{dr}{c} \equiv \int_{\tau_t}^{T} (1 - a^{-2}w^{-2})^{1/2} d\tau = \omega^{-1} \int_{r_t}^{R} k_v dr, \quad (12)$$

in which $a = c/r$ and r_t is the (lower) turning point, at which $a = w$, beneath which the waves cannot propagate. The independent variable τ in the second representation of the integral is acoustic radius, the natural variable for acoustic waves; $\tau_t = \tau(r_t)$ and $T = \tau(R)$, which I trust will not be confused with the same symbol when it stands for temperature. The upper limit R of the first and third integrals in equation (12) is the seismic radius of the star: it is the radius at which $c^2(r)$ — which itself decreases with r almost linearly in the outer adiabatically stratified regions of the convection zone beneath the upper superadiabatic boundary layer — would vanish if extrapolated linearly through the boundary layer and the lower atmosphere. Its value is quite well defined (e.g. Balmforth and Gough, 1990; Lopez and Gough, 2001), a property which is not universally appreciated (e.g. Bahcall and Ulrich, 1988). Finally, α, which can be regarded as a function of ω alone (that is to say, it does not depend separately on l, although it is certainly dependent on the stratification of the outer layers of the star where the scale heights of the background state are comparable with the wavelength of the acoustic modes). It takes into account principally the density variation of the Sun that is not recognized in the simple acoustic dispersion relation $\omega = kc$; the reason it is essentially independent of l (at least provided $l \lesssim 10^3$) is that in the outer layers the vertical scale of variation is very much shorter than the horizontal scale $L^{-1}r$.

The crucial property of equation (11) is that the left-hand side is a function of w alone (and the structure of the Sun, through $c(r)$, of course) and the right-hand side is a function of ω alone. In principle, this enables one to organize the helioseismic frequency data in such a way as to determine the function $F(w)$ from observation. One can then invert equation (12) to obtain $a(r)$, and consequently $c(r)$, implicitly in terms of F through the following relation:

$$\frac{r}{r_0} = \exp\left[-\frac{2}{\pi} \int_{w_0}^{a} (w^{-2} - a^{-2})^{-1/2} \frac{dF}{dw} dw\right], \quad (13)$$

where w_0 is the lowest measured value of w available (corresponding to the mode with $n = n_0$, $L = L_0$, say) and $r_0 = (1 - \Lambda)R$, Λ being a constant, depending on w_0 and the sound speed above the radius r_0 at which $a(r_0) = w_0$, whose value for the Sun is approximately $0.14(2n_0 + 3)/L_0$. With modern data sets, $\Lambda \ll 1$, and $r_0 \simeq R$. The major contributions to $\alpha(\omega)$ arise from the phase change $(\pi/4)$ at each caustic sphere between which the wave can propagate (the inner sphere is at $r = r_t$, the outer at approximately the level at which $\omega_c(r) = \omega$ — see immediately below) which is independent of ω, and from the density stratification of the outer layers of the star, which is represented in the acoustic wave equation by a displacement of ω^2 by an amount ω_c^2,

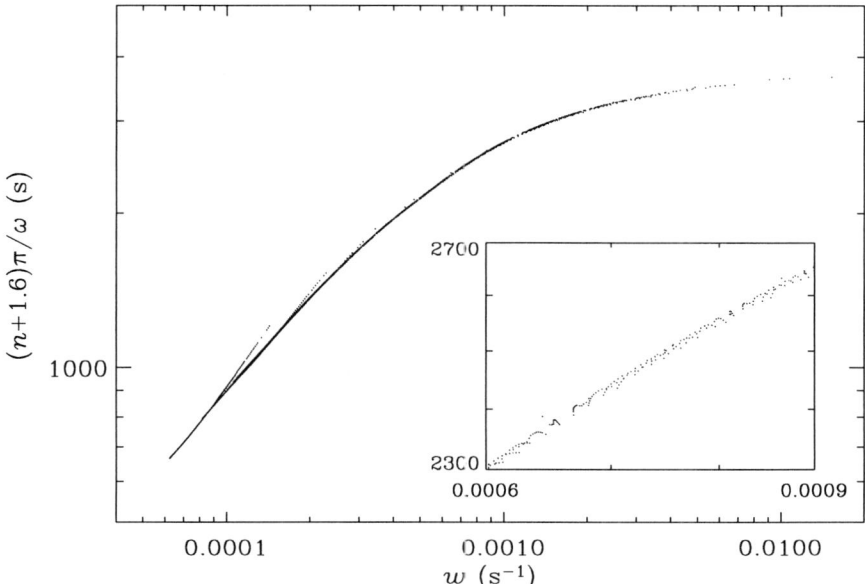

FIGURE 1. Solar acoustic frequencies measured by MDI (Scherrer *et al.*, 1995) plotted in the manner of equation (11). The branches near the lower end breaking away from the main band are respectively modes with $n = 1, 2, 3, ...$; for low w (high l) these low-order modes do not accord well with the asymptotic relation, which is formally valid only for large n. The structure of the band is more easily visible in the inset, which is a magnification of a portion of the main figure, and is largely accounted for by the expansion (14).

yielding $\omega^2 - \omega_c^2 = k^2 c^2$, where $\omega_c(r)$ is called the acoustic cutoff frequency (Lamb, 1908; Gough, 1993). In practice, the appropriate formula for $\omega_c(r)$ depends on how one chooses to cast the exact (linearized) wave equation for the seismic oscillations. But for a polytrope the outcome is merely to add a constant (whose value depends on the polytropic index) to α. (I must point out that below the superadiabatic boundary layer of the convection zone the stratification is approximately polytropic, because the first adiabatic exponent γ_1 does not deviate far from a constant, namely 5/3). Therefore the relative deviation of α from a constant is also small.

I have been through this argument in some detail in order to be able to illustrate how well the asymptotic formula represents the true state of affairs. In Figure 1, I plot $(n + \alpha)/\omega$ as a function of w for real solar data, having chosen a constant α whose value is such as to confine the data as well as is possible to a one-dimensional curve, rather than to lie on many distinct branches, as is the case, illustrated by Christensen-Dalsgaard (2004), when ω is plotted against l. In reality the curve has finite width, but it is quite clear from the figure that one can draw a curve through the middle of the distribution and so obtain an estimate of the function $F(w)$ on the right-hand side of equation (13).

The dispersion of the points in Figure 1 about a single curve gives (to the experienced[1]) some inkling of the accuracy with which one might determine $c(r)$.

Actually one can do much better than this very simple analysis suggests. One can start from the correct wave equation, which takes account of both the density stratification of the star and the perturbation to the gravitational potential Φ by the oscillations, and then carry out an asymptotic expansion. Including explicitly just the dominant corrections yields

$$F(w) = \pi \frac{n + \alpha_0(\omega) + w^{-2}\alpha_2(\omega) + \cdots}{\omega} + \omega^{-2} G(w) + \cdots, \quad (14)$$

which accounts for most of the structure evident in the inset to Figure 1. The expansion of $\alpha(\omega)$ recognizes that the thin subphotospheric region of the star in which the simple asymptotic wave equation is not as good as one would like is actually deep enough to sense the value of l through the horizontal variation of the mode; the function G takes account of buoyancy and the perturbation to the gravitational potential. Because sound propagation depends principally on c, the term $\omega^{-2} G(w)$ in equation (14) is small compared with $(n + \alpha)\pi/\omega$. One can take account of the explicit dependence of the terms in equation (14) on ω and w, and again extract $F(w)$ from the data. The rest is as before: one simply substitutes the resulting function F into equation (13) to determine $c(r)$. The procedure was first carried out by Gough and Vorontsov (1995) (who discuss how to overcome the ambiguities in extracting $\alpha_i(\omega)$ and $G(w)$), and has been pursued subsequently by Roxburgh and Vorontsov (2000a,b, 2001). In this report I shall discuss only how the method has been used to measure an aspect of the equation of state, namely γ_1, but I leave that until after I have finished discussing the construction of a seismic solar model.

Having determined $c(r)$, one can compute the variation of pressure and density by solving the hydrostatic equation in the form

$$\frac{d \ln p}{dr} = \frac{\gamma_1 g}{c^2}, \quad (15)$$

where $g = Gm/r^2$ is the magnitude of the acceleration due to gravity, $m(r)$ being the mass enclosed in a sphere of radius r:

$$m(r) = 4\pi \int_0^r r^2 \rho \, dr. \quad (16)$$

Equation (15) has the formal solution

$$p = p_0 \exp\left(-\int_{r_0}^r \frac{\gamma_1 g}{c^2} dr\right), \quad (17)$$

---

[1] Note that $F(w)$ must be differentiated, which magnifies the errors in the data. That is mitigated by the subsequent integration, but in view of the singular weight function $(1 - a^2/w^2)^{-\frac{1}{2}}$ this is effectively only the square root of integration, leaving one effectively to take only the square root of a derivative (Gough, 1990).

$$\rho = \frac{\gamma_1 p_0}{c^2} \exp\left(-\int_{r_0}^{r} \frac{\gamma_1 g}{c^2} dr\right), \qquad (18)$$

in which $p_0 = p(r_0)$ is determined by the mass constraint:

$$4\pi p_0 \int_0^R \frac{\gamma_1 r^2}{c^2} \exp\left(-\int_{r_0}^{r} \frac{\gamma_1 g}{c^2} dr\right) dr = M, \qquad (19)$$

where $M = m(R) = M_\odot$. Of course these equations are closed only if one knows the equation of state in the form

$$\gamma_1 = \gamma_1(p, \rho; \mathbf{X}), \qquad (20)$$

where $\mathbf{X}(r)$ represents the abundances of the chemical species comprising the solar plasma. Throughout all but the ionization zones of the abundant elements H and ^{4}He, $\gamma_1 \simeq 5/3$, and one can use that, together with an estimate of γ_1 in the ionization zones, as a basis from which to solve equations (16)–(20) by iteration. Alternatively, one can solve the equations in differential form directly.

An alternative procedure is to iterate from a reference solar model (although rarely is an iteration actually performed). One starts by assuming that the difference between the model and the Sun is everywhere small, so that $\delta\rho(x)/\rho(x) \ll 1$ etc, where $x = r/R$ and $\delta\rho = \rho_{\text{Sun}} - \rho_{\text{model}}$, and one linearizes those differences to obtain, as Christensen-Dalsgaard (2004) has explained, an integral formula for the differences $\delta\omega_{nl}$ between the observed and the theoretical frequencies of a mode of order n and degree l in terms of differences of any two thermodynamically independent seismic variables. For example, one can obtain

$$\frac{\delta\omega_{nl}}{\omega_{nl}} \simeq \int_0^{\tilde{X}} \mathscr{K}^{nl}_{c^2,\rho}(x) \frac{\delta c^2}{c^2} dx + \int_0^{\tilde{X}} \mathscr{K}^{nl}_{\rho,c^2}(x) \frac{\delta\rho}{\rho} dx + I_{nl}^{-1} \mathscr{G}(\omega_{nl}), \qquad (21)$$

where the kernels $\mathscr{K}$ depend on the structure and the unperturbed oscillation eigenfunctions of the model, I_{nl} is the inertia of the mode in question and $\mathscr{G}$ is an arbitrary function of ω which is introduced to account for the inevitable unknown errors in the kernels $\mathscr{K}$ in the uncertain outer layers of the star where the simple adiabatic eigenfunctions do not adequately represent the oscillations. The upper limit of integration is usually taken to be high in the atmosphere, typically near the temperature minimum, which renders the eigenfunctions beneath the photosphere insensitive to the mechanical boundary condition that is imposed there. Thus, $\tilde{X} \gtrsim 1$. These constraints are supplemented with the linearized mass-conservation constraint, which may be written in the form (21) with $\mathscr{K}^{nl}_{c^2,\rho} = 0$, $\mathscr{K}^{nl}_{\rho,c^2} = 4\pi R^3 x^2 \rho/M$ and $\mathscr{G} = 0$. The constraints (21) can usefully be generalized to take into account differences between the masses and the seismic radii of the Sun and the reference solar model, if that is necessary, but I shall not describe those elaborations here.

One now adopts some procedure for combining these data in such a way as to make the relations (21) more easily digestible. One way of achieving this is to represent the unknown functions $\delta c^2/c^2$ and $\delta\rho/\rho$ as expansions in some set of basis functions of x, expand $\mathscr{G}$ in a set of basis functions of ω, and determine the expansion coefficients in such a way that the constraints (21) applied to all the modes are satisfied as well

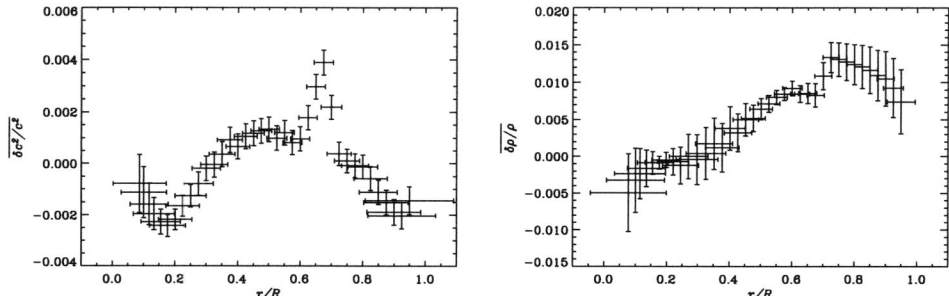

FIGURE 2. Optimally localized averages of the relative differences $\delta c^2/c^2$ and $\delta\rho/\rho$ between the Sun and Model S of Christensen-Dalsgaard *et al.* (1996), inferred by M. Takata from the frequencies plotted in Figure 1.

as possible. This, and any other way of expressing a representation of functions such as $c^{-2}\delta c^2$ and $\rho^{-1}\delta\rho$ in terms of the data $\omega_{nl}^{-1}\delta\omega_{nl}$, is called inversion. If the errors in the data were independent, one would minimize the sum χ^2 of the squares of the differences between the left and right sides of the constraints (21), weighted by the reciprocals of the variances of the observational errors. This leads to a set of linear equations relating the expansion coefficients to the data. It is necessary to 'regularize' the procedure by adding to χ^2 a penalty function $\mathscr{P}$, which penalizes undesirable features of the solution that are introduced by the errors. Typically one penalizes excessively-small-scale erratic behaviour, because it is known that such behaviour is characteristic of the effect of random errors. This is often accomplished by setting $\mathscr{P}$ to be λ times the sum of the integrals of the squares of the derivatives of $c^{-2}\delta c^2$ and $\rho^{-1}\delta\rho$. The parameter λ is called a regularization parameter, and choosing it well is part of the art of the subject. Having determined δc^2 and $\delta\rho$, one can obtain δp from the linearized difference between the hydrostatic equation (10), together with $g = Gm\rho/r^2$, applied to the Sun and to the model. The quantities δc^2, $\delta\rho$ and δp can then be added to the values of c^2, ρ and p in the original reference model to obtain a better estimate of the seismically accessible component of the structure of the Sun. In principle the whole process could be repeated, but it rarely is because, as is evident in Figure 2 (which was obtained by the OLA method described below, rather than by the method I have just described), theoretical models are so good that the adjustments are very small.

It should be realized that the outcome of the inversion is dependent not only on the data errors but also on the form of the penalty function $\mathscr{P}$ and on the regularization parameter λ; it is one of the infinite set of functions that formally satisfy the data (one hopes) to within the likely errors. This procedure for carrying out the inversion is called regularized least-squares (data fitting), or RLS(F) (the parts in parentheses usually being omitted).

Perhaps a more useful procedure is to consider a linear combination of the relations (21):

$$\sum_{n,l} a_{nl}(\bar{x}) \frac{\delta\omega_{nl}}{\omega_{nl}} \simeq \int_0^{\tilde{X}} \left(\mathscr{A}_{c^2,\rho} \frac{\delta c^2}{c^2} + \mathscr{A}_{\rho,c^2} \frac{\delta\rho}{\rho} \right) dx + \mathscr{S}, \tag{22}$$

where the averaging kernels $\mathscr{A}_{c^2,\rho}(x;\bar{x})$ and $\mathscr{A}_{\rho,c^2}(x;\bar{x})$ and the surface term $\mathscr{S}(\bar{x})$ are linear combinations of $\mathscr{K}^{nl}_{c^2,\rho}$, $\mathscr{K}^{nl}_{\rho,c^2}$ and $I_{nl}^{-1}\mathscr{G}(\omega_{nl})$ with the same coefficients a_{nl} as those on the left-hand side of equation (22). One then tries to choose those coefficients in such a way as to make $\mathscr{S}$ small, and simultaneously $\mathscr{A}_{c^2,\rho}$, say, unimodular and concentrated and positive in the vicinity of $x=\bar{x}$ for some $\bar{x}$, and $\mathscr{A}_{\rho,c^2}$ everywhere small.

The data combination is then approximately a localized average $\overline{c^{-2}\delta c^2}$ of $c^{-2}\delta c^2$ about $x=\bar{x}$. Procedures for accomplishing that are summarized, for example, by Gough and Thompson (1991); as with the data-fitting method described above, some regularization is required by trading localization of the averaging kernels against magnification of the data errors. One then repeats the procedure for different $\bar{x}$. In addition one can find another set coefficients to localize $\mathscr{A}_{\rho,c^2}$ with $\mathscr{A}_{c^2,\rho}$ everywhere small to obtain averages $\overline{\rho^{-1}\delta\rho}$.

This procedure is called optimally localized averaging (OLA). The averages obtained are correct values (to within the data errors) of the averages of $c^{-2}\delta c^2$ and $\rho^{-1}\delta\rho$ weighted by the averaging kernels $\mathscr{A}$, and can therefore be more reliably interpreted than the raw results of RLSF. Unlike RLSF in its usual form (i.e. as I have described it above), averages obtained using different forms of regularization are not contradictory; they are merely different because they have different averaging kernels and possibly different error estimates. Of course, because RLSF constructs functional representations that are linear combinations of the data, one can construct corresponding averaging kernels as similar combinations of the data kernels $\mathscr{K}$ and interpret the results correctly as weighted averages. Then the inferences drawn are comparable. Not surprisingly, RLSF averaging kernels are not as localized as those that have been expressly tailored to be so, and they tend to have negative sidelobes; therefore the averages are more difficult to interpret.

One can make further headway with OLA if one makes an assumption. If one assumes that there is no rapid variation in $c^{-2}\delta c^2$ and $\rho^{-1}\delta\rho$ (if there were, and its form were presumed known, one could incorporate it into the procedure, as did Elliott and Gough, 1998), one can regard the averages as estimates of the corresponding point values of the functions at $x=\bar{x}$. Moreover, provided that $\mathscr{S}$ is very small in all cases, one can then correct the estimate of $c^{-2}\delta c^2$ (and $\rho^{-1}\delta\rho$) by subtracting from the data combination on the left-hand side of equation (22) the estimate of the second (first) integral obtained by substituting $\overline{\rho^{-1}\delta\rho}$ ($\overline{c^{-2}\delta c^2}$) for $\rho^{-1}\delta\rho$ ($c^{-2}\delta c^2$), and iterating. This procedure has been adopted by Takata and Gough (2005). The averaged differences $\overline{c^{-2}\delta c^2}$ and $\overline{\rho^{-1}\delta\rho}$ from model S of Christensen-Dalsgaard *et al.* (1996) so obtained are illustrated in Figure 2. The horizontal bars represent the standard widths (approximately 0.7 of the full-width at half maximum) of the (Gaussian-like) averaging kernels, the vertical bars the standard errors (resulting from the errors in the data). Not surprisingly, because the dynamics of acoustic waves depends more strongly on c than on ρ, $c^{-2}\delta c^2$ is determined more accurately than $\rho^{-1}\delta\rho$. It is important to realize that even though the original data errors may be independent, the errors in neighbouring averages in Figure 2 are correlated with each other, and with averages elsewhere in the star, particularly near the surface (cf. Howe and Thompson, 1996).

There are various options for proceeding from here. For example, one could reformulate equation (22) in terms of, say, $p^{-1}\delta p$ and $\gamma_1^{-1}\delta\gamma_1$, to complete the determination of the localized averages of the seismic variables. Alternatively, one could accept the averages of $c^{-2}\delta c^2$ and $\rho^{-1}\delta\rho$ as estimates of the point values, and then determine δp by integrating the perturbed hydrostatic equation (10). That is the procedure I have adopted here for determining the seismic representation of the Sun listed in the first five and the ninth columns of Table I. Then γ_1 is determined by the relation $\gamma_1 = \rho c^2/p$.

A PHYSICALLY MOTIVATED AND SEISMICALLY CALIBRATED REPRESENTATION OF THE SUN

There are several seismic representations of the Sun in the literature (e.g. Gough, 1990; Shibahashi and Takata, 1996; Turck-Chièze et al., 2001; Couvidat, Turck-Chièze and Kosovichev, 2003). Most have, in addition to the seismically accessible variables, quantities such as temperature that require non-seismic information for determining them. Different authors have used different information, and choosing between them is a matter of judgement. Here I describe the construction of the representation presented by Gough and Scherrer (2001). It is based on a determination of the seismic variables c^2 and ρ along the lines described by Takata and Gough (2001), using solar model S of Christensen-Dalsgaard et al. (1996) as the reference, and then obtaining p and γ_1 by integrating the hydrostatic equation as I described in the previous section. The remaining variables were obtained by accepting that the Sun is in global thermal balance — that is to say, that the luminosity $\tilde{L}(R(t_\odot), t_\odot)$ at the surface, whose value is observed today (at $t = t_\odot$), is equal to the rate of generation of energy by nuclear reactions in the core, and assuming that the formulae for the nuclear reaction rates and the equation of state are correct. Note that local thermal balance, which requires that $\partial \tilde{L}/\partial m = \tilde{\varepsilon}$ throughout the star, where $\tilde{\varepsilon}$ is the rate of generation of energy per unit mass, and a thermal transport equation providing a relation between $\tilde{L}$ and the stratification of the star, has not (yet) been assumed.

The first task was to determine the chemical composition, which was represented by $\mathbf{X} = (X, Y, Z)$, where X, Y and Z are respectively the abundances of hydrogen, ^{4}He and the sum of all the remaining heavy elements. In constructing the reference model these had been assumed to have been uniform at the start of the main-sequence evolution, but had subsequently developed variations partly as a result of gravitational settling and partly by nuclear transmutation. Account of mixing in the tachocline, a thin well mixed rotational shear layer immediately beneath the convection zone, had not been taken, which is why there is a sharp hump in $\delta c^2/c^2$ between $x \simeq 0.6$ and $x \simeq 0.7$ evident in Figure 2. So first Y and Z (and consequently $X = 1 - Y - Z$) were homogenized with the convection zone through a layer of thickness $\Delta x = 0.02$ representing the tachocline, that value of Δx having been taken from the seismic calibration of Elliott, Gough and Sekii (1998). That is the only change that was made to Z. Then a constant, δY, was added to Y throughout, to be used later to ensure that the luminosity $\tilde{L}(r,t)$ is consistent with the rate of generation of energy by nuclear reactions. The justification for taking this route, rather than, say, scaling the helium abundance profile by a constant

TABLE 1. Seismic model of the Sun: the total acoustic radius T of the Sun is 3523 s. Density ρ is in g cm^{-3}, sound speed c in km s^{-1}, temperature T in K, pressure p in dyne cm^{-2} and opacity κ in cm^2 g^{-1}.

τ/T	r/R	m/M	c	ρ	X	$100Z$	T	p	κ
0.000	0.000	0.000	5.042E+02	1.530E+02	0.338	2.03	1.57E+07	2.33E+17	1.24
0.010	0.026	0.002	5.065E+02	1.451E+02	0.361	2.03	1.54E+07	2.23E+17	1.27
0.020	0.051	0.013	5.098E+02	1.266E+02	0.415	2.03	1.48E+07	1.97E+17	1.33
0.030	0.077	0.040	5.108E+02	1.049E+02	0.483	2.02	1.40E+07	1.64E+17	1.43
0.040	0.103	0.082	5.068E+02	8.488E+01	0.550	2.01	1.30E+07	1.31E+17	1.55
0.050	0.128	0.137	4.977E+02	6.807E+01	0.605	2.00	1.19E+07	1.01E+17	1.70
0.060	0.153	0.201	4.848E+02	5.443E+01	0.644	1.99	1.10E+07	7.67E+16	1.85
0.070	0.177	0.269	4.697E+02	4.344E+01	0.671	1.98	1.01E+07	5.75E+16	2.03
0.080	0.201	0.338	4.541E+02	3.462E+01	0.687	1.98	9.34E+06	4.28E+16	2.21
0.090	0.223	0.405	4.387E+02	2.755E+01	0.697	1.97	8.66E+06	3.18E+16	2.42
0.100	0.245	0.468	4.240E+02	2.192E+01	0.703	1.97	8.06E+06	2.36E+16	2.64
0.125	0.297	0.601	3.918E+02	1.249E+01	0.711	1.96	6.86E+06	1.15E+16	3.26
0.150	0.345	0.702	3.655E+02	7.302E+00	0.714	1.95	5.96E+06	5.85E+15	4.01
0.175	0.389	0.775	3.438E+02	4.427E+00	0.716	1.94	5.27E+06	3.14E+15	4.87
0.200	0.432	0.828	3.254E+02	2.789E+00	0.717	1.94	4.72E+06	1.77E+15	5.83
0.225	0.472	0.867	3.095E+02	1.820E+00	0.719	1.93	4.27E+06	1.05E+15	6.88
0.250	0.510	0.897	2.955E+02	1.228E+00	0.720	1.93	3.89E+06	6.43E+14	8.02
0.275	0.547	0.919	2.829E+02	8.522E-01	0.720	1.92	3.56E+06	4.09E+14	9.29
0.300	0.582	0.936	2.713E+02	6.072E-01	0.721	1.92	3.27E+06	2.68E+14	10.71
0.325	0.615	0.949	2.604E+02	4.428E-01	0.721	1.92	3.01E+06	1.80E+14	12.31
0.350	0.648	0.959	2.497E+02	3.302E-01	0.723	1.93	2.76E+06	1.24E+14	14.31
0.375	0.679	0.968	2.388E+02	2.518E-01	0.731	1.92	2.52E+06	8.61E+13	16.95
0.388	0.694	0.971	2.328E+02	2.222E-01	0.740	1.88	2.38E+06	7.23E+13	18.44
0.400	0.708	0.974	2.263E+02	1.977E-01	0.740	1.81	2.25E+06	6.08E+13	20.42
0.404	0.713	0.976	2.237E+02	1.903E-01	0.740	1.81	2.19E+06	5.71E+13	21.39
0.425	0.736	0.980	2.114E+02	1.607E-01	0.740	1.81	1.96E+06	4.30E+13	
0.450	0.762	0.984	1.973E+02	1.306E-01	0.740	1.81	1.71E+06	3.05E+13	
0.475	0.786	0.988	1.841E+02	1.060E-01	0.740	1.81	1.49E+06	2.15E+13	
0.500	0.808	0.991	1.716E+02	8.573E-02	0.740	1.81	1.30E+06	1.51E+13	
0.525	0.829	0.993	1.597E+02	6.907E-02	0.740	1.81	1.12E+06	1.06E+13	
0.550	0.849	0.995	1.484E+02	5.534E-02	0.740	1.81	9.71E+05	7.30E+12	
0.575	0.867	0.996	1.375E+02	4.401E-02	0.740	1.81	8.35E+05	4.99E+12	
0.600	0.883	0.997	1.270E+02	3.470E-02	0.740	1.81	7.14E+05	3.35E+12	
0.625	0.899	0.998	1.169E+02	2.705E-02	0.740	1.81	6.06E+05	2.21E+12	
0.650	0.913	0.999	1.071E+02	2.080E-02	0.740	1.81	5.10E+05	1.43E+12	
0.675	0.926	0.999	9.761E+01	1.573E-02	0.740	1.81	4.24E+05	8.97E+11	
0.700	0.938	0.999	8.833E+01	1.165E-02	0.740	1.81	3.49E+05	5.45E+11	
0.725	0.948	1.000	7.924E+01	8.417E-03	0.740	1.81	2.83E+05	3.17E+11	
0.750	0.958	1.000	7.033E+01	5.894E-03	0.740	1.81	2.26E+05	1.76E+11	
0.775	0.966	1.000	6.154E+01	3.965E-03	0.740	1.81	1.76E+05	9.14E+10	
0.800	0.973	1.000	5.291E+01	2.528E-03	0.740	1.81	1.35E+05	4.40E+10	
0.825	0.980	1.000	4.498E+01	1.504E-03	0.740	1.81	1.02E+05	1.93E+10	
0.850	0.985	1.000	3.793E+01	8.213E-04	0.740	1.81	7.32E+04	7.38E+09	
0.875	0.989	1.000	3.004E+01	3.956E-04	0.740	1.81	4.99E+04	2.33E+09	
0.900	0.992	1.000	2.309E+01	1.551E-04	0.740	1.81	3.47E+04	5.93E+08	
0.925	0.995	1.000	1.813E+01	4.692E-05	0.740	1.81	2.50E+04	1.19E+08	
0.950	0.997	1.000	1.444E+01	1.046E-05	0.740	1.81	1.86E+04	1.77E+07	
0.975	0.999	1.000	1.144E+01	1.647E-06	0.740	1.81	1.38E+04	1.80E+06	
1.000	1.000	1.000	7.893E+00	1.998E-07	0.740	1.81	5.78E+03	7.61E+04	

factor, as is sometimes done, is that the luminosity evolution $\tilde{L}(R,t)$ of solar models that are correctly calibrated to $R(t_\odot)$ and $\tilde{L}(R(t_\odot),t_\odot)$ is quite insensitive to any mild alterations to the physics adopted for computing those models and to the initial chemical composition. Therefore so is $\int \tilde{L}(R,t)\mathrm{d}t$, which determines the amount of ^{4}He produced. Therefore it is the value of $\int [Y(r,t) - Y(r,0)]\mathrm{d}m$ integrated over the energy-generating core (in which gravitational settling may be neglected), that is more-or-less constrained, and which remains unaltered if a constant δY is added to $Y(r,t)$. Note that adopting this procedure is to accept, as do standard solar model computations, that the heavy products of the nuclear reactions remain in situ, and are not transported by any putative macroscopic flow in the core. Having obtained (X,Y,Z), the temperature T can then be determined, in terms of the unknown constant δY, by the equation of state. This determination was carried out together with the application of the thermal-balance constraint $\int \tilde{\varepsilon}\mathrm{d}m = \tilde{L}(R,t_\odot)$, yielding both $T(r)$ and δY simultaneously. The outcome is presented in Table I. Note that the result depends on having accepted the heavy-element abundance $Z(r,t_\odot)$ from the reference model (aside from homogenizing the tachocline with the convection zone), which presumes that the gravitational settling is not altered significantly by the small corrections to the other variables. These are reasonably safe assumptions, because both the equation of state and the solar nuclear energy generation rate are much less sensitive to Z than they are to Y, but provided, of course, that the Sun supports no other mechanism of abundance change.

Having thus obtained a representation of the Sun, one can now address local thermal balance. If one assumes that beneath the convection zone heat is transported only by radiation, as is the case in standard solar models — once again I am assuming no transport by macroscopic material motion other than convection — then one can use the heat transport equation $\kappa = -(16\pi a \tilde{c} r^2 T^3/3\tilde{L}\rho)\mathrm{d}T/\mathrm{d}r$ to determine the opacity κ. (Here a is the radiation density constant and $\tilde{c}$ is the speed of light.) That too is included in Table I.

ADDRESSING PHYSICS

What might now come to mind in the light of the manner in which I have presented the seismic model is to ask how the inferred value of κ compares with theoretical expectation. It was through opacity that helioseismology first addressed microphysics, from an early determination of the sound speed (Christensen-Dalsgaard et al., 1985). It was found that the solar sound speed substantially exceeded that of the preferred theoretical model of the time (Christensen-Dalsgaard, 1982) between $x \simeq 0.25$ and $x \simeq 0.6$; this could be explained if the opacity, which had been provided to astrophysicists as tables by the Los Alamos National Laboratories, were too low by some 20 per cent or so at temperatures between about 10^6 (actually 2×10^6) and 4×10^6K, and densities in the vicinity of 1g cm^{-3}. After much discussion, spearheaded by Werner Däppen, Forrest Rogers and Carlos Iglesias at the Lawrence Livermore National Laboratory were persuaded to carry out a few independent spot calculations. Their results were in reasonable agreement with the prediction from seismology, and after subsequent careful comparison with the Los Alamos calculations it was revealed that, amongst other things, the Los Alamos calculations contained an error in the treatment of spin-orbit coupling in radiative transitions. In consequence, more extensive calculations were carried out by Carlos, Forrest and their colleague B.G. Wilson to provide new tables (Iglesias, Rogers and Wilson, 1987).

I have included the opacity in Table I. What I could try to do in addition is to present the intrinsic errors $\delta_{int}\kappa$ in those values. But unfortunately I cannot reliably do that. The reason is that there is considerable uncertainty in the abundances of the heavy elements, which are seismically inaccessible. As I mentioned earlier, and as Christensen-Dalsgaard (2004) and Trampedach (2004) discuss in more detail in these proceedings, the recent redetermination by Asplund et al. (2004) of the atmospheric abundances of C, N and O reduces substantially the opacity in the standard solar model, because, in the standard solar model, photospheric abundances are indicative of interior abundances. Nevertheless, I present in Figure 3 the apparent intrinsic opacity error computed from Model S, which is the difference between the opacity listed in Table I and the theoretical opacity used to construct the reference solar model evaluated at the temperature, density and chemical composition of the representation of the Sun listed in Table I. I emphasize that this result is not the true intrinsic error, but includes the effect of errors in the chemical composition that was adopted for the table (both through the direct dependence of κ on X, and indirectly through the errors that arise in the inferred temperature), and also any error that might have been introduced by the assumption that all the heat is transported outside the convection zone by radiation alone. The cumulative effect of these additional errors is likely to be substantially greater

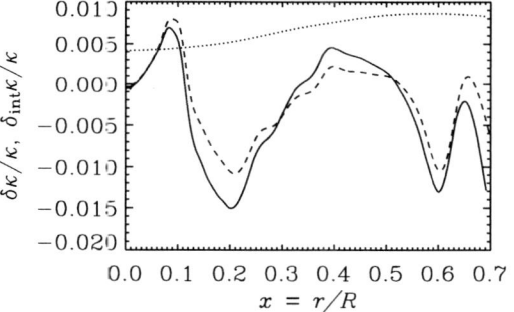

FIGURE 3. The dashed curve is the estimated relative actual difference $\delta\kappa/\kappa$ between the opacity in the radiative zone in the Sun and that in Model S of Christensen-Dalsgaard et al. (1996). The solid curve has been 'corrected' for the differences in temperature, density and chemical composition (represented by Y and Z), and is therefore an estimate of the intrinsic difference. However, it should not be taken literally, because there is evidence that some of the heavy-element abundances assumed for the Sun are incorrect. The dotted curve is $0.01\,(\partial\ln\kappa/\partial\ln Z)_{\rho,T,X}$.

than the apparent intrinsic error plotted in Figure 3. The abundance reassessment by Asplund et al. (2004), for example, implies, broadly speaking, a reduction in Z by about 30 per cent. Since $\partial\ln\kappa/\partial\ln Z$ is of order unity (it is depicted, as a dotted curve, in Figure 3 to help one assess the result of changing Z), this implies a contribution to the apparent value of $\delta\kappa_{int}/\kappa$ of typically about -20 per cent (assuming that one is correctly confining consideration to models that are in accord with the helioseismic data), which swamps the values plotted in the figure.

I should also point out that there are other routes by which one can obtain estimates of $\delta_{int}\kappa$. Elliott (1995), for example, has cast the frequency constraints (21) in terms of $\delta\ln\kappa$ and $\delta\ln\gamma_1$, supplementing them with the energy transport equation valid in the radiative zone and, of course, the equation of state to relate $\delta \ln \kappa$ and $\delta \ln T$ to the seismically accessible variables; he then inverted them directly for $\delta \ln \kappa$. He did not use the constraint that $\tilde{L}(r,t)$ is produced by the nuclear reactions, although in principle that would have been possible, but instead assumed that $\delta \tilde{L} = 0$. As with the method I have adopted here, any conclusion drawn from the results concerning $\delta_{\text{int}} \ln \kappa$ is contaminated by an unknown contribution from the errors in the chemical composition.

A quantity of greater pertinence to this workshop is the first adiabatic exponent, γ_1. It has been employed as an indicator of the partition function by Nayfonov and Däppen (1998), Basu, Däppen and Nayfonov (1999) and Däppen and Nayfonov (2000) and is discussed further by Däppen (2004) in these proceedings. In principle it too could be determined directly from the integral constraints used by Elliott to infer $\delta_{\text{int}} \ln \kappa$. But instead, Elliott (1996) used constraints on $\delta \ln \gamma_1$ and $\delta \ln A^*$, where $A^* = \gamma_1^{-1} d \ln p / d \ln r - d \ln \rho / d \ln r$ is the convective stability parameter. This is a more robust route because one need then confine consideration to only seismic variables, using the equation of hydrostatic support and the mass continuity equation but not the equation for energy transport. Once again, there is uncertainty arising from uncertainty in the chemical composition. Basu and Christensen-Dalsgaard (1997) subsequently tried to obviate that defect by seeking combinations of frequency constraints that are independent of δY (whilst ignoring the dependence of γ_1 on the abundances of the heavy elements). At first sight this might seem a remarkable thing to have tried to do, given that the relation between $\delta \ln \gamma_1$ and $\delta_{\text{int}} \ln \gamma_1$ depends on the seismically inaccessible chemical abundances yet no non-seismic constraint was adopted to determine δY; it appears to be formally possible because the dependence on chemical composition is introduced explicitly through the linearized equation of state for γ_1, and the resulting influence of its contribution to the frequency constraints is eliminated in a manner analogous to the elimination of the uncertain surface term $\mathscr{G}$ in equation (21). One might also wonder at using the theoretical equation of state to determine an error in itself. That is formally possible by virtue of the reliance of the procedure on the principle that the equation of state is used only through the partial derivatives of γ_1 with respect to p, ρ and the chemical abundances, which merely determine the function subspace in which abundance variation influences γ_1 and, implicitly, where in that subspace the Sun lies, and that any error in those derivatives influences the results of the procedure only to second order. Therefore, if the reference model does not deviate too far from the Sun, one can hope that those results approximate the intrinsic error being sought. But does it? It is unfortunate that the results presented appear not to be consistent with each other, so it is difficult to interpret them.

Some further progress can be made in the convection zone, by comparing γ_1 at different locations (i.e. by looking at the spatial derivative of γ_1), because there the chemical composition is uniform, and equal to that in the visible photosphere. Moreover, well beneath the upper boundary layer the stratification is isentropic and the Reynolds stresses of the turbulence are negligibly small compared with the gas pressure gradient. In principle, this additional condition (which, I emphasize, is not seismic) enables one to eliminate the explicit dependence of the γ_1 gradient on the density, a quantity which is seismically less well determined than the sound speed. One can show that

$$W(r) \equiv \frac{r^2}{Gm}\frac{dc^2}{dr} = 1 - \gamma_1\left[1 + \left(\frac{\partial \ln \gamma_1}{\partial \ln p}\right)_s\right] \equiv \Theta \tag{23}$$

(cf. Gough, 1984), where s is specific entropy and, as before, m is the mass enclosed in the sphere of constant r. Of course, this equation is not completely independent of ρ — any corresponding equation that is would be much more complicated — but it is nearly so: the convection zone contains only 2 per cent of the mass of the Sun, and so a useful simple diagnostic of the equation of state is obtained by replacing m by $M_\odot$. For a perfect monatomic gas, $\gamma_1 = 5/3$, and $\Theta = -2/3$ is constant. Deviations of Θ from $-2/3$ are measures of deviations from the perfect-gas law.

The original motivation for using the diagnostic $\tilde{W} = (r^2/GM_\odot)dc^2/dr$ was to obtain a measure of the helium abundance Y from the deviation of Θ from its perfect-gas value due to the second ionization. (The zone of first ionization overlaps with the much more potent zone of ionization of hydrogen; using these is actually more delicate, however, because the stratification in the hydrogen ionization zone is substantially different from being isentropic. It is worth noting that in the middle of the second helium ionization zone $M_\odot - m \simeq 4 \times 10^{-5}$, so $\tilde{W}$ is an excellent approximation to W, and hence of Θ.) But it was found subsequently that Y could not be determined as precisely as was hoped, because the equation of state that was used to convert the form of Θ to a helium abundance was not accurate enough for the purpose (Kosovichev et al., 1992). The studies set a new standard for equations of state, and opened up the possibility of using helioseismology to investigate microscopic physics. It motivated the refinements to the equation of state that are discussed at this workshop. I shall address here just one example.

In Figure 4a is plotted $\tilde{W}$ through the convection zones of models of the solar convection zone constructed by Baturin et al. (2000) with two versions of the MHD equation of state (Mihalas, Däppen and Hummer, 1988; Däppen et al., 1988). In one, labelled MHD-2, all the heavy elements are presented by oxygen only; in the other, MHD-5, C, N, O and Fe were explicitly considered, with the same total heavy-element abundance $Z = 0.02$. The dashed curves are the actual functions, the solid curves are corresponding seismological deductions of them, carried out asymptotically using equation (13) with $F(w)$ determined from theoretical eigenfrequencies represented by equation (14). The inversions are poor at the low and high ends of the radius range, beyond the limits indicated by the horizontal bar drawn near the bottom of the panel, because only modes in a limited range of w were employed. Included also is the seismological inference of $\tilde{W}$ in the Sun (solid curve with error bars), obtained in an identical manner from observed frequencies of the same set of modes. The differences between theory and observation are substantially greater than the formal errors in the latter (and are comparable with the changes produced by redistributing the abundances of the heavy elements), which indicates that there is still much room for improvement.

As an educative exercise, Baturin et al. invented a toy model of the equation of state derived from a factorized partition function $\mathscr{Z}$ incorporating electrons with the usual Fermi-Dirac statistics and heavy particles with the usual integrals over momentum and configuration space, multiplied by an internal component $\mathscr{Z}_{int}$ which is the Boltzmann-

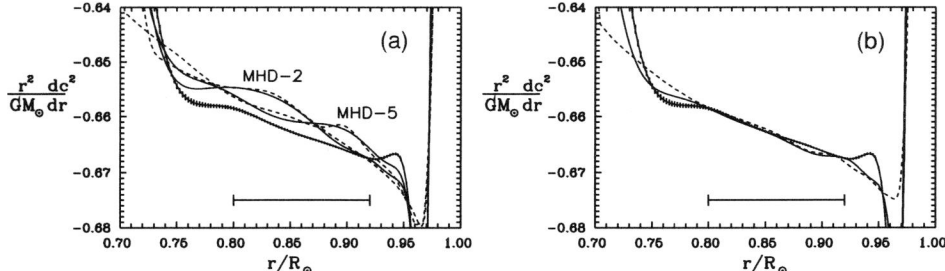

FIGURE 4. (a) The dashed curves are $\tilde{W}(r)$ plotted through the convection zones of two solar models with MHD equations of state with different abundances of some heavy elements, but with the same value of Z. The corresponding solid curves are the values inferred by inverting a limited set of oscillation eigenfrequencies. The inversions are tolerable where the dashed and the corresponding solid curves more-or-less agree, in the range indicated by the horizontal bar near the bottom of the figure. The solid curve with error bars estimates $\tilde{W}$ in the Sun; it was computed from the same mode set as were the theoretical curves. (b) The dashed and the corresponding solid curve are $\tilde{W}$ and its value inferred by frequency inversion (of the same mode set as before) for a model with an equation of state with a van der Waals exclusion term, as explained in the text. As before, the solid curve with error bars is the value inferred for the Sun (from Baturin et al., 2000).

weighted sum over bound states computed with Debye-Hückel screening, into which a van der Waals-like term was incorporated by multiplying each term in the internal partition function of species i by an exclusion term $\exp(-\zeta R_i/\langle r \rangle)^3$, where R_i is the isolated-atom rms radius, $\langle r \rangle$ is the mean distance between particles and ζ is a free parameter, common to all species in the model, and expected to be of order unity. The diagnostic $\tilde{W}$ was compared with the solar value for various values of ζ. Figure 4b shows the outcome for $\zeta = 1.67$, which provided the best fit; when $\zeta = 2.5$, for example, the fit is as poor as in Figure 4a. Therefore it is evident that helioseismology has the power to measure the effective radii of atoms and ions when they are immersed in a dense plasma. It is not claimed that the true radii have actually been measured, because there may be other factors in the partition function that have not been taken into account (and, indeed, the partition function is surely not actually factorizable). Nevertheless, it demonstrates that helioseismology has adequate diagnostic power to address the kinds of issue concerning the equation of state that are currently under study.

CONCLUDING REMARKS

There are other areas of physics where helioseismology has played an important role, and no doubt will continue to do so. But these do not concern the equation of state, which is the reason I have not yet mentioned them. Now I merely record some of them for completeness.

There are two issues which were at the forefront of attention from the very beginning: the solar neutrino problem and the oblateness of the external gravitational field. Early sound-speed inversions were inconsistent with solar models that generated neutrino fluxes low enough to be consistent with observation; this result forced the community to consider more seriously than it had previously the possibility of neutrino transitions, with the implication that neutrinos are not massless. Measurements from SuperKamiokande have confirmed that transitions do occur, and, coupled with measurements at the Sudbury Neutrino Observatory, the fluxes have been found to be consistent with the theory. That is not to presume that future detailed measurements of the energy spectrum will also be consistent; it would be dull if they were. A similar issue is the possibility that weakly interacting massive particles (wimps) may be captured by the Sun from a cosmic sea, and contribute to the transport of heat in the radiative zone. This would modify the sound speed and the density stratifications; therefore seismology can put a constraint on the properties of the wimp sea.

Seismological measurement of the internal angular velocity of the Sun from rotational splitting data provides the most accurate measure by which to determine the Sun's gravitational quadrupole moment; that result, coupled with the observed value of the precession of the perihelion of the orbit of the planet Mercury (and from measurements of the circumsolar orbits of other bodies) confirm General Relativity, which in the post-Newtonian approximation satisfies the data better than any other theory.

Helioseismology has also raised many issues concerning the internal dynamics of the fluid flow within the Sun. But now I risk digressing too far from fundamental physics, so I bring my discussion to a close.

ACKNOWLEDGMENTS

I am very grateful to Di Sword for preparing the LATEX file and to Günter Houdek for his help with the diagrams.

REFERENCES

1. Asplund, M. *et al.*, 2004, *Astron. Astrophys.*, **417**, 751-768
2. Bahcall, J.N., 2001, *Nature*, **412**, 29
3. Bahcall, J.N. and Ulrich, R.K., 1988, *Rev. Mod. Phys.*, **60**, 297-372
4. Balmforth, N.J. and Gough, D.O., 1990, *Astrophys. J.*, **362**, 256–266
5. Basu, S. and Christensen-Dalsgaard, J., 1997, *Astron. Astrophys.*, **322**, L5
6. Basu, S., Däppen, W. and Nayfonov, A., 1999, *Astrophys. J.*, **518**, 985
7. Baturin, V.A., Däppen, W., Gough, D.O. and Vorontsov, S.V., 2000, *Mon. Not. R. Astron. Soc.*, **316**, 71
8. Christensen-Dalsgaard, J., 1982, *Mon. Not. R. Astron. Soc.*, **199**, 735
9. Christensen-Dalsgaard, J., 2004, these proceedings
10. Christensen-Dalsgaard, J. *et al.*, 1985, *Nature*, **315**, 378
11. Christensen-Dalsgaard, J. *et al.*, 1996, *Science*, **272**, 1287–1292
12. Couvidat, S., Turck-Chièze, S. and Kosovichev, A.G., 2003, *Astrophys. J.*, **599**, 1434
13. Däppen, W., 2004, these proceedings
14. Däppen, W. *et al.*, 1988, *Astrophys. J.*, **332**, 261
15. Elliott, J.R., 1995, *Mon. Not. R. Astron. Soc.*, **277**, 1567
16. Elliott, J.R., 1996, *Mon. Not. R. Astron. Soc.*, **280**, 1244
17. Elliott, J.R. and Gough, D.O., 1999, *Astrophys. J.*, **516**, 475

18. Elliott, J.R., Gough, D.O. and Sekii, T., 1998, in *Structure and dynamics of the interior of the Sun and Sun-like stars* (ed. S.G. Korzennik & A. Wilson, European Space Agency, SP-418, Noordwijk), 763
19. Gough, D.O., 1984, *Mem. Soc. Astron. Italiana*, **55**, 13
20. Gough, D.O., 1990, in *Progress of seismology of the Sun and stars*, (ed. Y. Osaki & H. Shibahashi, Springer, Heidelberg), *Lecture Notes in Physics*, **267**, 283
21. Gough, D.O., 1993, in *Astrophysical fluid dynamics*, (ed. J-P. Zahn & J. Zinn-Justin, Elsevier, Amsterdam), *Les Houches Session XLVII*, 399–560
22. Gough, D.O., 2003, *Astrophys. Sp. Sci.*, **284**, 165–185
23. Gough, D.O. and Scherrer, P.H., 2001, in *The Century of Space Science*, (ed. J.A.M. Bleeker, J. Geiss & M.C.E. Huber, Kluwer, Dordrecht), 1035
24. Gough, D.O. and Thompson, M.J., 1991, in *Solar interior and atmosphere* (ed. A.N. Cox, W.C. Livingston & M.S. Matthews, Univ. Arizona Press, Tucson), 519
25. Gough, D.O. and Vorontsov, S.V., 1995, *Mon. Not. R. Astron. Soc.*, **273**, 573
26. Howe, R. and Thompson, M.J., 1996, *Mon. Not. R. Astron. Soc.*, **281**, 1385
27. Iglesias, C. A., Rogers, F. J. and Wilson, B. G., 1987, *Astrophys. J.*, **322**, L45
28. Kosovichev, A.G. *et al.*, 1992, *Mon. Not. R. Astron. Soc.*, **259**, 536
29. Lamb, H., 1908, *Proc. London Math. Soc.*, **7**, 122
30. Lopes, I. and Gough, D.O., 2001, *Mon. Not. R. Astron. Soc.*, **322**, 473
31. Mihalas, D., Däppen, W. and Hummer, D.G., 1988, *Astrophys. J.*, **331**, 815
32. Nayfonov, A. and Däppen, W., 1998, *Astrophys. J.*, **499**, 489
33. Roxburgh, I. W. and Vorontsov, S. V., 2000a, *Mon. Not. R. Astron. Soc.*, **317**, 141
34. Roxburgh, I. W. and Vorontsov, S. V., 2000b, *Mon. Not. R. Astron. Soc.*, **317**, 151
35. Roxburgh, I. W. and Vorontsov, S. V., 2001, *Mon. Not. R. Astron. Soc.*, **322**, 85
36. Scherrer, P.H. *et al.*, 1995, *Solar Phys.*, **162**, 129
37. Shibahashi, H. and Takata, M., 1996, *Publ. Astron. Soc. Japan*, **48**, 377
38. Takata, M. and Gough, D.O., 2001, in *Proc SOHO 10 / GONG 2000 Workshop: Helio- and astero-seismology at the dawn of the millennium*, (ed. A. Wilson, ESA SP-464, Noordwijk) 543
39. Takata, M. and Gough, D.O., 2005, in preparation
40. Trampedach, R., 2004, these proceedings
41. Turck-Chièze, S. *et al.*, 2001, *Astrophys. J.*, **555**, L69

The State of ^{7}Be in the Sun

Nir J. Shaviv * and Giora Shaviv[†]

*Racah Institute of Physics, Hebrew University of Jerusalem, Jerusalem 91904, Israel
[†]Department of Physics, Israel Institute of Technology, Haifa 32000, Israel

Abstract. The ionization state of ^{7}Be in the solar core affects the prediction of the solar ^{8}B neutrino flux. Recently, Shaviv & Shaviv (2003) examined the theory and the effect of pressure ionization on the ionization state of ^{7}Be and all elements with $12 \geq Z \geq 4$. It was shown that under the conditions prevailing in the solar core, the ionization state is determined mainly by the nearest neighbor and not by the Debye potential created by the entire plasma.

As pressure ionization is determined mainly by the nearest neighbor, we first redefine the ionsphere for a mixture and then solve the Schrödinger and later the Kohn-Sham equations for an ion immersed in a dense plasma under conditions for which the mean interparticle distance is smaller than the Debye radius.

Contrary to previous estimates showing that Beryllium is partially ionized, we find that it is fully ionized. As a consequence, the predicted rate of the ^{7}Be $+ e^-$ reaction is reduced by 20-30%, quite insensitive to the exact solar model. Since ^{7}Be is a trace element, its total production is controlled by the unchanged ^{4}He $+^3$He reaction rate, and its destruction is determined by the rate of electron capture. As the latter rate decreases when the Beryllium is fully ionized (relative to the case of partially ionized Be), the estimate for the abundance of ^{7}Be increases and with it the ^{8}B neutrino flux. The increase in $\phi_\nu(^8B)$ is by about $20-30\%$. The neutrino flux due to Be^7 electron capture remains effectively unchanged because the change in the rate is compensated for by a change in the abundance. The prediction for the ratio of $\phi_\nu(^8B)/\phi_\nu(^7Be)$ changes as well.

INTRODUCTION

Although ^{7}Be is a trace element in the Sun, it is very important for the Solar neutrino problem since it gives rise to the most energetic neutrinos. ^{7}Be is destroyed in the Sun via two competing processes, electron capture and proton capture. Here we discuss the first one.

Electron capture takes place predominantly from the continuum. As early as 1967 Iben, Kalata and Shwartz, using the Saha equation, discovered that ^{7}Be keeps its K-shell electron at least part of the time. If so, electron capture would also proceed via the s-wave of the bound electron and increase the rate of capture and affect significantly the predicted neutrino flux.

Hitherto, based on Iben et al (1967), the ^{7}Be was assumed to be partially ionized. However, if the Beryllium is indeed not fully ionized, then so are the CNO elements, implying that the nuclear screening in reactions like $^{12}C + p$ should be calculated assuming partial ionization and not with a charge $Z = 6$. The same is true with the ^{7}Be $+ p$ reaction. Similarly, the electron pressure and the speed of sound depend on the number of free electrons which in turn depends on the ionization. This is important when a very accurate speed of sound in the Sun is required, for example, for helioseismology.

The first correction for the plasma effect was carried out by Iben et al (1967) who assumed a Debye-Hückel potential, instead of the Coulomb potential. They calculated the new energy levels and then used the new energy levels in the Saha equation. The energy levels were calculated under the assumption of vanishing electron wave functions at infinity, as if the atom is isolated.

At the high densities found in the solar core, the ions are too close to each other for the assumption of an isolated atom to be correct. Hence, our main purpose here is to assess the role of the nearby atom on the bound states and ionization.

The paper is structured as follows: We examine the thermodynamic variables at the solar core and show that the mean inter particle distance is of the same order as the Debye radius. We then redefine the ion sphere for a mixture and calculate the energy levels for an ion inside its ion sphere. Namely, instead of demanding the wave function to vanish at infinity we demand the derivative to vanish at a finite range (the boundary of the ion sphere). We then solve for the energy levels assuming the Schrödinger and then the Kohn-Sham equations. The latter includes electron screening and exchange. Once the ground state is found we assess the degree of ionization in the Sun and the effect on the solar neutrino flux prediction.

THE CONDITIONS IN THE SUN

The classical approach of Iben et al (1967) to the problem is as follows: First, solve for the energy level at $T = 0$. Then apply the Saha equation with the energy levels to find the occupation numbers at finite T and ρ. Iben et al. assumed a Debye potential and the boundary condition on the electronic wave function: $\psi(\infty) = 0$, to find the energy levels when the atom is immersed in the plasma. However, the nearest ion is much closer than the Debye radius, which is the range of the potential. If the nearest ion is well inside the Debye radius, then its effect on the electrons dominates over that of the plasma.

For the Debye theory to be valid, the number of particles in a Debye sphere must be much larger then unity. If so, whenever the Debye theory is valid, the distance between particles is much smaller than the Debye radius and hence the Debye theory is inappropriate for the calculation of energy levels. The condition for the validity of the Debye theory is therefore:

$$R_D \gg \langle r \rangle \quad \rightarrow \quad \rho \ll \rho_{crit} = 1.57(1+3X)^2 \left(\frac{T_6}{3+X}\right)^3. \qquad (1)$$

For the conditions in the Solar core ($T_6 = 15.5, \rho = 155 g/cc$, $X = 0.35$ and $\rho_{crit} = 1451 g/cc$). In fig. (1) we plot ρ_{crit} as a function of temperature and composition. Also plotted is the structure line for the Sun. We see that the critical density is not sensitive to the composition. More important, the Sun is very close to the critical density and in the outer parts, its structure line even crosses the critical density. Hence, we have to treat the ionization problem under the assumption that $\langle r \rangle < R_D$, implying that we cannot assume the wave function to vanish at infinity.

The ion under consideration is a trace ion. Hence, we need a definition of the ion sphere for an inhomogenous composition. We define the radius of the unit atomic cell as

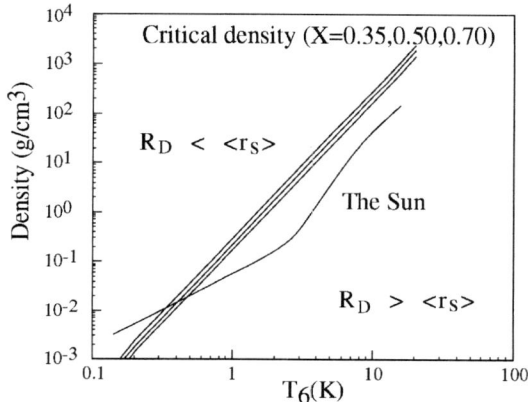

FIGURE 1. The density-temperature plane and the relation between the Debye radius and the mean inter particle distance. The three straight lines (in a log-log scaling) correspond to three different compositions and describe the location $R_D = \langle r \rangle$.

the location where the force between two neighbouring ions vanishes. Define:

$$\langle r_Z \rangle = \frac{\delta}{1+\delta} \langle r \rangle, \quad \delta = \sqrt{\frac{Z \sum X_i Z_i / A_i}{\sum X_i / A_i}}, \qquad (2)$$

where the sum is carried over all species in the mixture. The new definition reduces to the canonical one in the case of homogenous plasma, namely $\langle r_Z \rangle = 0.5 \langle r \rangle$.

We have now to replace the boundary condition $\psi(\infty) = 0$ with a boundary condition on $\langle r_Z \rangle$. The Bloch condition on the electronic wavefunction in the case of periodic boundary conditions is given by:

$$\begin{aligned} \psi_k(r) &= e^{2ik\cdot r} \psi_k(-r) \\ \frac{\partial}{\partial r} \psi_k(r) &= -e^{2ik\cdot r} \frac{\partial}{\partial r} \psi_k(-r). \end{aligned} \qquad (3)$$

The implementation of these boundary conditions leads to band structure. In spherical packing the conditions becomes (cf. Lai et al. 1991):

$$\frac{\partial}{\partial r} \psi_k(r = \langle r_Z \rangle) = 0. \qquad (4)$$

We solved therefore the Schrödinger equation in a unit cell $\langle r_Z \rangle$ for the trace elements Be, C, N, C and Mg assuming a Coulomb potential. We stress that the use of the Debye potential is not justified and if used will not have a significant effect when $\langle r_Z \rangle \ll R_D$.

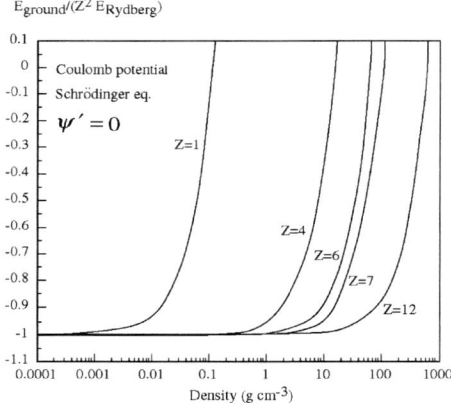

FIGURE 2. The solution of the Schrödinder equation for the ion sphere and several trace elements. (The solution for the Hydrogen did not assume this assumption). The boundary condition is the vanishing of the derivative on the boundary of the ion sphere.

ASSUMING THE SCHRÖDINGER EQUATION

We first assume the Schrödinger equation and solve it inside the ion sphere to find the ground state energy. We defined complete ionization when the ground state merges into the continuum.

The results for the ground state of the trace ions (and Hydrogen) are given in fig. (2). We find that Beryllium is fully ionized in the Sun long before the density and temperature reach those of the solar core. In fact, even Nitrogen is fully ionized under conditions found in the core of the Sun.

The pressure ionization due to the merging of the ground state into the continuum, is rather fast and takes place over a narrow range of densities. Usually the temperature rises inward and hence in reality the ionization takes place even faster. From Table 1 we see that the critical ion sphere at which the ground state merges into the continuum is, when measured in units of inter particle distance, practically independent of Z

ASSUMING THE KOHN-SHAM EQUATION

Two major deficiencies exist in the implementation of the Schrödinger equation and statistical theories. First, the potential is not self consistent. Second, the Thomas-Fermi like equations do not lead to high kinetic energy when the density is high. Third, the shape of the N-body function is not fully anti-symmetric. The solution comes in the form of the Kohn-Sham equation:

$$-\frac{\hbar^2}{2m}\nabla^2\psi + \left[-\frac{Ze}{r} + e^2\int\frac{n(r')dr'}{|r-r'|} - e^2\left(\frac{3}{\pi}n(r)\right)^{1/3}\right]\psi(r) = E\psi(r) \qquad (5)$$

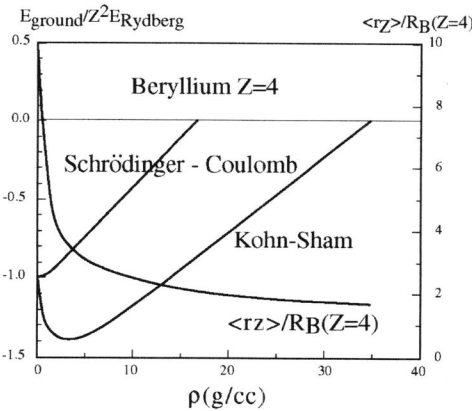

FIGURE 3. The comparison between the ground state calculated à la Schrödinder and à la Kohn-Sham. Due to electron pressure, ionization takes place according to the KS equation at a higher density.

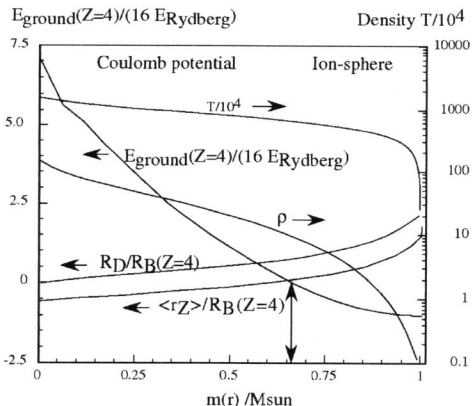

FIGURE 4. The variation of the ground state of Beryllium along the Sun. Also shown are the Debye radius and the ion-sphere as well as the density and temperature.

where ψ is the single electron wave function such that $n(r) = \sum_{i=1}^{N} |\psi_i|^2$. The first term under the integral sign is the Coulomb potential, the second is the screening by electrons and the third is the exchange term.

In our particular case, the elements of interest are trace elements. Hence the exchange term is dominated by the electrons contributed by the fully ionized Hydrogen and no complicated iteration is needed. The KS equation was solved for the same trace ions as before. In fig. (3) we see a typical behavior of the two approximations. While in the Schrödinger approximation the ground state rises monotonically, in the KS it first decreases and the ion becomes more bound. This is due to the exchange term. Eventually, as the density increases and the ion-sphere radius decreases, the energy of the ground

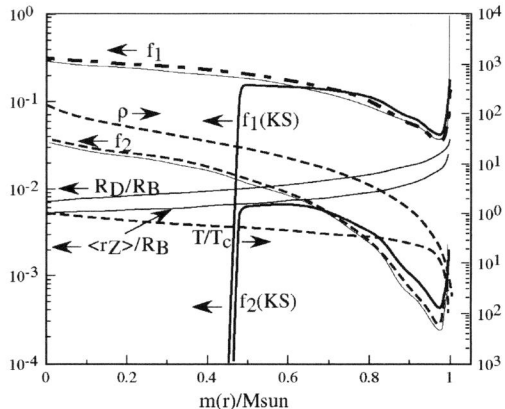

FIGURE 5. A comparison between the ionization level calculated à la Iben et al and the present calculation. f_1 and f_2 are the per cent occupancy of one and two electrons in the K-shell.

state rises and leads to complete pressure ionization at a higher density.

The results for the critical density at which the ground state merges into the continuum are given in table 1. The table contains the results in the two approximations used. A notable difference, which is a consequence of the higher densities required to ionize ions with higher Z is the change in $\langle r_Z \rangle / \langle r \rangle$ in the KS case.

COMPARISON WITH OTHER METHODS

Perrot (1990-1993) applied the Neutral-Pseudo-Atom method to calculate the complete ionization of pure metallic Beryllium. The results of Perrot is a critical ionization density of $\rho = 92.5 g/cc$. If we use simple spherical packing, as one expects in stellar plasma, the result for metallic Beryllium is $56 g/cc$. If however, we consider an fcc lattice (with a phase transition at a compression of 4) we find $89.5 g/cc$—in good agreement given all assumptions in the calculations. We just mention here that the Neutral-Pseudo-Atom method does not strictly apply above compression ratios of about 40 and the method Perrot used was based on band calculations. Our result should pertain to the bottom of the band.

THE EFFECT ON THE SOLAR NEUTRINO FLUX PREDICTION

The revised results for the solar neutrino flux are summarised in Table 2. We find that irrespective of the detailed solar model, the effect of full ^{7}Be ionization is an enhancement of the ^{8}B flux and an enhancement of the ^{8}B flux relative to the ^{7}Beneutrino flux. Detailed analysis can be found in Shaviv & Shaviv (2003).

TABLE 1. The critical densities and atomic cell radii in units of mean interparticle distance in the two approximations. The ion-spheres of complete ionization are given in units of $\langle r \rangle$.

Ion	$\rho_{crit}(Coul)$	$\langle r_Z \rangle$	$\rho_{crit}(KS)$	$\langle r_Z \rangle$
^{7}Be	16.33	1.921	36.18	1.474
^{12}C	65.25	1.923	86.25	1.752
^{14}N	110	1.927	132.5	1.806
^{16}O	171	1.927	196.5	1.840
^{24}Mg	660	1.925	698.8	1.890

TABLE 2. The effect of treating the ^{7}Be as fully ionized on the solar neutrino flux

S(0) in MeV barn	Cas97		DS96		RMP98		DS96S	
$S_{pp}(0)/10^{-25}$	3.89		4.01		4.00		4.01	
$S_{34}(0)/10^{-4}$	5.10		4.50		4.5		5.1	
$S_{33}(0)/10^{0}$	5.10		5.1		5.1		5.32	
$S_{17}(0)/10^{-6}$	22.4		17		22.4		22.4	
Ionization	partial	full	partial	full	partial	full	partial	full
$\phi(^8B)/10^6 cm^{-2}s^{-1}$	5.33	6.36	3.13	3.74	5.19	6.20	4.68	5.58
$\phi(^8B)10^3/\phi(^7Be)$	1.19	1.42	0.783	0.935	1.05	1.25	1.094	1.306
Ga(SNU)	133	133	121	121	132	132	129.5	130.0
Cl(SNU)	7.80	9.14	5.10	5.90	7.65	8.96	7.06	8.24

CONCLUSION - THE STATE OF BERYLLIUM IN THE SUN

In this work, we were interested in the internal structure of a trace ion in a high density plasma, and not in the general properties of the plasma. The relevant physical quantity for the latter is the Debye radius and potential. On the other hand, the electronic structure depends on the distance to the nearest neighbour if $\langle r \rangle < R_D$, and on the Debye potential in the opposite case. Hence, the method to calculate the energy levels and the boundary conditions should conform with the physical conditions dictated by the situation.

Under the conditions prevailing in the Sun (and Main Sequence stars), the ionization state of a given ion depends mostly on the distance to the nearest neighbor rather than the distance scale of the screened potential. The treatment of ionization should be based on the ion-sphere with the proper boundary condition on the wave function.

We have calculated the critical density of complete ionization assuming at first the Schrödinger equation and a Coulomb potential. Next, we implemented the Kohn-Sham equation which takes into account the exchange interaction due to the free electrons and the screening by them. Comparison of our results with the NPA method yields a very good agreement for the case of metallic Beryllium.

The improvements in the treatment of pressure ionization of trace elements show that all species with $Z \leq 8$ are fully ionized in the core of the Sun. The frequently applied correction to the ^{7}Be electron capture rate due to bound electrons, is not needed. The revised prediction for the $\phi_v(^8B)$ from the Sun is higher by about 20-30%. The agreement between the predictions of the standard solar model (assuming the standard set of nuclear cross sections) and the SNO result for the ^{8}B neutrino flux is improved.

The discussion here was centered around the high density effects. However, one should take into account collective effects as well. The energy of the electron plasma oscillations is

$$\hbar\omega_e = 3.7 \times 10^{-11} \sqrt{n_e} \, eV. \qquad (6)$$

At an electron number density of $10^{26}/cc$ the energy of the plasma oscillations amounts to 370 eV which is higher than the ionization potential of ^{7}Be. Thus the bound states of ^{7}Be have been broadened into continuum states long before.

ACKNOWLEDGMENTS

GS would like to thank the Asher Space Research Fund for partial support of this research.

REFERENCES

Adelberger, E. G., et al. 1998, Rev. Mos. Phys. 70,1265 (RMP)
Castellani, V., degl'Innocenti, S. Fiorentini G., Lissa, M., & Ricci, B. 1997, Phys. Rep. 281, 309 (Cas97)
Dar, A. & Shaviv, G. 1996, ApJ 468, 933 (DS96)
Iben, I. J., Kalata, K., Schwartz, J. 1967, ApJ 150, 1001
Lai, D., Abrahams, A. M., & Shapiro, S. L. 1991, ApJ 377, 612
Kohn, W., Sham, L. J. 1965, Phys. Rev. 140, 1133
Perrot, F. 1990, Phys. Rev. A 42, 4871
Rogers, F. J., Graboske, H. C., Harwood, D. J. 1970, Phys. Rev. A. 1,1577
Shaviv, N. J., Shaviv G. 2003, MNRAS 341, 119

SAHA-S model: Equation of State and Thermodynamic Functions of Solar Plasma

V.K.Gryaznov[1], S.V.Ayukov[2], V.A.Baturin[2], I.L.Iosilevskiy[3], A.N.Starostin[4], V.E.Fortov[1]

[1] *Institute of Problems of Chemical Physics RAS, Chernogolovka, Russia*
[2] *Sternberg Astronomical Institute, Moscow, Russia*
[3] *Moscow Institute of Physics and Technology, Dolgoprudnyi, Russia*
[4] *Troitsk Institute for Innovation and Fusion Research, Troitsk, Russia*

Abstract. A thermodynamic model SAHA-S for solar plasma is presented. Effects of Coulomb interaction, exchange and diffraction effects, free electron degeneracy, relativistic corrections, radiation pressure contributions are taken into account. Contribution of bound states of atoms and ions is corrected in comparison with Plank-Brillouin-Larkin formula. Calculations of equation of state of solar plasma with different element composition are carried out. Contribution of various plasma effects and chemical element abundance to thermodynamic functions and in particular Γ_1 is discussed.

INTRODUCTION

A great amount of high precision observation data on eigenfrequencies of solar oscillations obtained during the last decade [1, 2] provides detailed information on physical conditions in the Sun interior and permits to develop more precise models of its inner structure. To build an adequate model of the Sun one needs a very accurate equation of state (EOS hereafter) of solar plasma together with other microphysical parameters (opacity, nuclear reaction rate et al.). A comparison of computed model frequencies with observational data [1, 2] gives us an opportunity to improve and refine the EOS of weakly coupled plasmas on the high level of accuracy – better then 10^{-4} Modern EOS models of solar plasma now in use [3, 4] reproduce sound velocity with high accuracy over the whole temperature range, but there are some questions which have to be clarified. Among them is the level of accuracy of theoretical estimations for sound speed and adiabatic exponent Γ_1 over solar model profile, and especially in the lower part of the convection zone. Also, what is possible effect in EOS of nonideal plasma corrections on the profile of Γ_1 in the solar conditions? These questions are actual in relation with real discrepancy between observation and model frequencies caused by sound speed deviations in the lower part of the convection zone.

This paper presents a thermodynamic model of weakly coupled solar plasma named SAHA-S. The SAHA-S EOS inherits basic plasma physical effects, which have traditionally been concerned with solar EOS, but also proposes generalizations of some of them. We are intended to discuss the adiabatic exponent profile inside the Sun, using comparison of the values for different EOS-models, which successively include different plasma effects

THERMODYNAMIC MODEL

The SAHA-S EOS is based on so-called "chemical picture" of plasmas [5, 6, 7] which starts from the representation of the free energy as a sum of the zero approximation term $F^{(id)}$, corresponding to the "ideal–gaseous" mixture with varying composition of a wide spectrum of simple and complex particles – (electrons and ions, atoms and molecules and so on), and of further contributions stemming from interaction between these particles. Historically, the free energy formulation in astrophysical EOS procedure has been written in [8]. Generally complex particles have internal degrees of freedom - excited states, being in thermodynamic equilibrium with the system as a whole, i.e. with its translational degrees of freedom.

$$F(\{N_i\},V,T) = \sum_i F_i^{(id)} + F_e^{(id)} + F^{(rad)} + \Delta F_{ii,ee,ie}^{(int)}(\{N_i\},V,T) \tag{1}$$

Here the first term on the right-hand-part side is the contribution of the ideal-gas of "heavy" particles, atoms, ions and molecules. The second term is the ideal-gas contribution of electrons (which may be partially degenerate). The third term represents contribution of radiation and the forth term is responsible for the inter-particle interactions between all particles and includes contribution of Coulomb interactions between charged particles, neutral-neutral interactions, and also charge-neutral interactions.

We suppose that the electrically neutral system (multi-component plasma) consists of L components (atoms, molecules, atomic and molecular ions and electrons) within a fixed volume V and at a temperature T. All particles of the system are constructed of K sorts of nuclei and electrons. It is well known [9] that at fixed V, T, and number of nuclei of each species $\{N_k^0\}, k = 1,2,...,K$ the free Helmholtz energy is minimal in this volume. The minimum necessary condition gives the equations of chemical and ionization equilibrium:

$$\mu_j = \sum_{k=1}^{K+1} v_k^j \mu_k; \quad j = K+2,...,L; \quad \mu_j = \frac{\partial F(\{N_i\},V,T)}{\partial N_j} \tag{2}$$

where μ_j is the chemical potential of the j-th component, v_j^k is the number of nuclei of species k in a particle of species j.

To complete the equilibrium system of equations one has to add to the equations (2) the equations

$$\frac{\sum_{j=1}^{L} v_j^k n_j}{\sum_{l=1}^{K}\sum_{i=1}^{L} v_i^l n_i} = c_k^0, \quad k = 1,2,...,K \tag{3}$$

of chemical proportions, the equations

$$\sum_{p=1}^{L} n_p Z_p^{(+)} = \sum_{m=1}^{L} n_m Z_m^{(-)} \quad (4)$$

of electrical neutrality and the equation

$$\sum_{j=1}^{L} m_j n_j = \rho_0 \quad (5)$$

of mass conservation.

Here $n_j = N_j / V$ is the particle density of species j, $Z_i^{(+)}$ and $Z_i^{(-)}$ are the charges of particles with positive and negative charges respectively, m_j is the mass of a particle of species j. The abundance of the chemical element of species k, c_k^0 and mass density ρ_0 are given constants. So we have $L+1$ equations (5-8) with respect to L variables n_j. Notice that due to the condition

$$\sum_{k=1}^{K} c_k^0 = 1$$

only K-1 equations of the equations (3) are linearly independent. Solving the system of equations (2-5) we can calculate the component composition $\{n_j\}$.

The first term of the Helmholtz free energy (1) represents the Boltzmann ideal gas expression for a mixture of particles of various species

$$F_i^{(id)} = \sum_{j=1}^{L} N_j k_B T \left(\ln \frac{n_j \lambda_j^3}{Q_j} - 1 \right) \quad (6)$$

Here k_B is Boltzmann constant, $\lambda_j = \sqrt{\dfrac{2\pi\hbar^2}{m_j k_B T}}$ is the thermal De Broglie wavelength, Q_j is the partition function for a particle of species j.

$$Q(\{n_j\}, \tau, T) = \sum_i \omega_i(\{n_j\}, T) g_i e^{-\varepsilon_i / k_B T}. \quad (7)$$

Here ε_i, g_i are the excited energy levels and statistical weights for an isolated particle. For atoms and ions we use data [10, 11], for diatomic molecules the partition functions correspond to the approximation of the non-rigid rotator – the non-harmonic oscillator with the data from [12,13]. The factor ω in general is dependent on particle numbers and temperature, but in the present work only the dependence on temperature is taken into account. For molecules

$$\omega_i(T) = \begin{cases} 1, & \varepsilon_i \leq D - k_B T \\ 0, & \varepsilon_i > D - k_B T \end{cases} \quad (8)$$

where D is the dissociation energy of diatomic molecule.

For atoms and ions two versions are used. One of them is the Planck-Larkin partition function [14],

$$\omega_i^{PL}(T) = 1 - (1 + \tilde{E}_i)e^{-\tilde{E}_i}; \quad \tilde{E}_i = \frac{\varepsilon_i}{k_B T} \qquad (9)$$

another one uses a more rigorous [15] expression (for the contribution due to the bound states only) than that given by the Planck-Larkin formula [14] representing in fact some mixture of contributions of bound and scattering states. In this case, the factor ω is given by the expression [15]

$$\omega_i^{SR}(T) = 1 - e^{-\tilde{E}_i}\left[4 - \frac{6}{\sqrt{\pi}}(\tilde{E}_i)^{1/2} + \frac{4}{\sqrt{\pi}}(\tilde{E}_i)^{3/2}\right] + \frac{\Gamma(1/2, \tilde{E}_i)}{\sqrt{\pi}}\left[3 - 4\tilde{E}_i + 4\tilde{E}_i^2\right] \qquad (10)$$

Notice that from (9) and (10) and at $\varepsilon_i/k_B T \ll 1$, i.e in the very high temperature limit, we have

$$\omega_i^{SR}(T) \Rightarrow 4\omega_i^{PL}(T) \qquad (11)$$

In Fig.1 the dependence of the ratio $\omega_i^{SR}(T)/\omega_i^{PL}(T)$ on temperature is represented. It can be seen that for low temperature the difference between two expressions for the partition function is negligible, but at the maximum solar temperature $\approx 1000\ eV$ is close to the asymptotic limit (11).

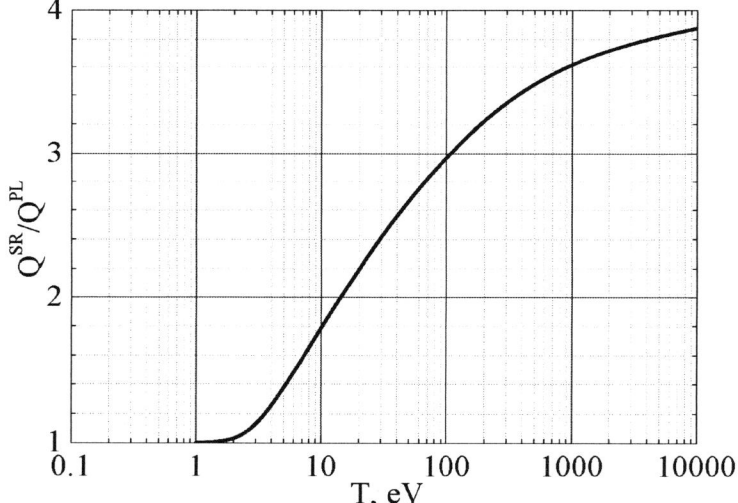

FIGURE 1. Ratio of Q^{SR} to Q^{PL} at various values of temperature for hydrogen atoms.

The contribution of the electronic ideal gas to the Helmholtz free energy expression (1) is represented by the second term $F_e^{(id)}$. The ideal gas of free electrons is considered as a partially degenerate Fermi gas. In accordance with [9]

$$F_e^{id} = N_e k_B T \left[\alpha - \frac{4}{3\sqrt{\pi}} \frac{2}{n_e \lambda_e^3} \int_0^\infty \frac{x^{3/2}}{1+\exp(x-\alpha)} dx \right]; \quad \alpha = \frac{\mu_e}{k_B T}, \quad (12)$$

where the electronic particle density n_e and electronic chemical potential are related by the expression

$$\frac{n_e \lambda_e^3}{2} = 2\pi^{-1/2} I_{1/2}(\alpha); \quad I_t(\alpha) = \int_0^\infty \frac{y^t dy}{1+\exp(y-\alpha)} \quad (13)$$

From (12) expressions for pressure, internal energy and chemical potential can be obtained by simple differentiation. The interpolation formulas for the Fermi-Dirac functions $I_t(\alpha)$ [16] are used, they have the relative accuracy not worse than 10^{-6} over the whole range of degeneracy parameter $n_e \lambda_e^3$.

The third term of the Helmholtz free energy (1) is responsible for inter-particle interaction. In this work the main inter-particle interaction for solar plasma is the Coulomb interaction between charged particles in the framework of the Debye approximation in the grand canonical ensemble, generalized for the case of multi-stage ionization (DGCE)[17].

$$-\frac{\Omega^{(Coul)}}{Vk_B T} \equiv -\frac{F^{(Coul)} - \sum_{i=1}^L N_i \mu_i^{(Coul)}}{Vk_B T} \equiv \frac{P^{(Coul)}}{k_B T} = \sum_{i=1}^L n_i - \frac{\Gamma_D^3}{24\pi f^3}. \quad (14)$$

Here $\Omega^{(Coul)}$, $F^{(Coul)}$, $\mu^{(Coul)}$, $P^{(Coul)}$ are the thermodynamic potential, the Helmholtz free energy, the chemical potential and the pressure of charged particles, $f = e^2/k_B T$ is the Coulomb scattering amplitude and Γ_D is a real root of the equation

$$\Gamma_D^2 = 4\pi f^3 \sum_{i=1}^L \frac{Z_i^2 n_i}{1 + Z_i^2 \frac{\Gamma_D}{2}}, \quad (15)$$

and Z_i is the particle charge number. In (14,15) the summation is extended over the charged particles only. Due to (14) the dimensionless correction to the ideal gas pressure and energy are

$$\frac{\Delta P^{(Coul)}}{n_e k_B T} = \frac{\Delta E^{(Coul)}}{3N_e k_B T} = -\frac{\Gamma^3}{24\pi n_e f^3}, \quad (16)$$

and for the chemical potential one has:

$$\frac{\Delta \mu_i^{diff}}{k_B T} = -\ln\left(1 + Z_i^2 \frac{\Gamma}{2}\right). \quad (17)$$

Certainly, the properties of the DGCE approximation differ from the well known Debye-Hückel (DH) one. In the DH approximation the correction to the ideal gas pressure has the form

$$\frac{\Delta P^{(DH)}}{\sum_i Z_i^2 n_i k_B T} = -\frac{\Gamma_D}{6}, \qquad (18)$$

where Γ_D is the Debye coupling parameter

$$\Gamma_D = f\sqrt{4\pi f \sum_i Z_i^2 n_i}. \qquad (19)$$

Notice that the DH pressure loses stability along the isotherm at $\Gamma_D=4$ and becomes negative at $\Gamma_D=6$ so that

$$\frac{\Delta P^{(DH)}}{\sum_i Z_i^2 n_i k_B T} \Rightarrow -\infty \quad \text{when } \Gamma_D \Rightarrow \infty,$$

while the DGCE is thermodynamically stable at any value of the coupling parameter and

$$\frac{\Delta P^{(DGCE)}}{\sum_i n_i k_B T} \Rightarrow -\frac{1}{3} \quad \text{when } \Gamma_D \Rightarrow \infty.$$

To make the EOS of the solar plasma more precise, the electron exchange and diffraction corrections to the Coulomb interaction are taken into account. In accordance with [18], the Coulomb correction to the ideal gas pressure (thermodynamic potential Ω) with the exchange and diffraction effects has the form

$$\frac{\Delta P^{(CoulExDif)}}{n_e k_B T} = -\frac{\tilde{\Gamma}^3}{24\pi n_e f^3} - \frac{1}{4}\zeta_e \lambda_e^2 f \frac{\zeta_e}{n_e} + \frac{\lambda_e \zeta_e f^2}{4} \frac{\tilde{\Gamma}^2}{4 n_e f^3} +$$
$$+ \frac{\pi \lambda_e f^2 \zeta_e}{4} \frac{\zeta_e}{n_e} \left\{ \frac{(1-\sqrt{2})}{\sqrt{2}} + \frac{1}{2}\sum_{i=1}^{L}\sum_{k=1}^{L} Z_i^2 Z_k^2 \frac{\zeta_i}{\zeta_e} \frac{\zeta_k}{\zeta_e} \frac{\lambda_{ik}}{\lambda_e} \right\}, \qquad (20)$$

where λ_e, ζ_e are the thermal De Broglie wavelength and activity ($\zeta_e = \exp(\mu_e/k_B T)/\lambda_e^3$) for the electrons and the same λ_i, ζ_i for the ions, λ_{ik} being the thermal De Broglie wavelength with the reduced mass $m_{ik}=m_i m_k/(m_i+m_k)$, the summation in (20) is over ions only. In (20) all activities have to be expressed in terms of the particle densities using equations [18]

$$\zeta_e = \frac{n_e}{1 + \frac{\tilde{\Gamma}}{2} + \frac{1}{4}\zeta_e \lambda_e^2 f - \frac{\lambda_e}{f}\frac{\tilde{\Gamma}^2}{16} - \frac{\pi \lambda_e f^2 \zeta_e}{4}[\sqrt{2}-1]}$$

$$\zeta_i = \frac{n_i}{1 + Z_i^2 \frac{\tilde{\Gamma}}{2} - \frac{\pi \lambda_e f^2 \zeta_e Z_i^2}{4}\left(1 + \sum_{k=2}^{I} Z_k^2 \frac{\zeta_k}{\zeta_e} \frac{\lambda_{ik}}{\lambda_e}\right)} \qquad (21)$$

By virtue of (21) the corrections to the ideal gas chemical potentials of electrons and ions have the form

$$\frac{\Delta\mu_e^{diff}}{k_B T} = -\ln\left(1 + \frac{\Gamma}{2} + \frac{1}{4}\zeta_e\lambda_e^2 f - \frac{\lambda_e}{f}\frac{\Gamma^2}{16} - \frac{\pi\lambda_e f^2 \zeta_e}{4}\left[\sqrt{2}-1\right]\right)$$

$$\frac{\Delta\mu_i^{diff}}{k_B T} = -\ln\left(1 + Z_i^2\frac{\Gamma}{2} - \frac{\pi\lambda_e f^2 \zeta_e Z_i^2}{4}\left(1 + \sum_{k=2}^{I} Z_k^2 \frac{\zeta_k}{\zeta_e}\frac{\lambda_{ik}}{\lambda_e}\right)\right)$$
(22)

and the same for the internal energy:

$$\frac{\Delta E^{(CoulExDif)}}{N_e k_B T} = -\frac{\Gamma^3}{8\pi n_e f^3} - \frac{1}{2}\zeta_e\lambda_e^2 f \frac{\zeta_e}{n_e} + \frac{5}{2}\frac{\lambda_e f^2 \zeta_e}{4}\frac{\Gamma^2}{4n_e f^3} +$$
$$+ \frac{5}{2}\frac{\pi\lambda_e f^2 \zeta_e}{4}\frac{\zeta_e}{n_e}\left[\left(\frac{1}{\sqrt{2}}-1\right) + \frac{1}{2}\sum_{i=2}^{I}\sum_{k=2}^{I} Z_i^2 Z_k^2 \frac{\zeta_i}{\zeta_e}\frac{\zeta_k}{\zeta_e}\frac{\lambda_{ik}}{\lambda_e}\right]$$
(23)

Due to the high temperature values of plasma in the inner part of the Sun, the relativistic effects in the electron motion and radiation pressure can affect thermodynamics of the solar plasma. To understand how important these effects are, the relativistic corrections [9] were accounted for. It has to be noticed that the first order correction to the pressure vanishes exactly,

$$\frac{\Delta P_e^{rel}}{k_B T} = 0,$$
(24)

but there are corrections to the internal energy of the same order:

$$\frac{\Delta E^{rel}}{k_B T} = \zeta_e \frac{15 k_B T}{8 m_e c^2},$$
(25)

and the electronic chemical potential

$$\frac{\Delta\mu_e^{rel}}{k_B T} = -\ln\left[1 + \frac{15 k_B T}{8 m_e c^2}\right],$$
(26)

where c is the light velocity.

The radiation pressure and the internal energy contributions are taken in the form [9]

$$\Delta P^{(Rad)} = \frac{\Delta E^{(Rad)}}{3V} = \frac{4\sigma_{SB}}{3c}T^4; \quad \sigma_{SB} = \frac{\pi^2 k_B^4}{60\hbar^3 c^2}.$$
(27)

In the context of the structure of the Helmholtz free energy expression described above, all corrections to the ideal gas thermodynamic values, the expression for the chemical potential of any particle can be split into two parts. They are Boltzmann ideal gas contribution and the correction to its value

$$\frac{\mu_j}{k_B T} = \ln\frac{n_j \lambda_j^3}{Q_j} + \Delta\mu_j^{(Corr)}(\{n_i\}, T).$$
(28)

Notice that if the second term in (28) is put equal to zero, the partition function is reduced to the ground state statistical weight, equations (2) are transformed to the Saha equations [19] for the of ionization equilibrium.

RESULTS AND DISCUSSION

On the basis of the chemical picture and using the model (1-28), massive computations of composition and thermodynamic properties of the solar plasma have been carried out. All the calculations were performed by specially designed computer code SAHA-S. This code continues the SAHA code line [6, 20, 21, 22, 13] used to compute the EOS and thermodynamic functions of multi-component plasmas with strong inter-particle interactions. Two types of computations were performed. The first one is the calculation of comprehensive tables for different helium abundance Y, fixed heavy elements abundance Z (2%), densities ranging from 10^{-8} g/cm^3 to 10^4 g/cm^3, and temperatures ranging from 10^3 to 10^8 K. On the basis of these tables a full model of the Sun was thus constructed and its results for the calculated values of Γ_1 were compared to the data obtained from other solar models.

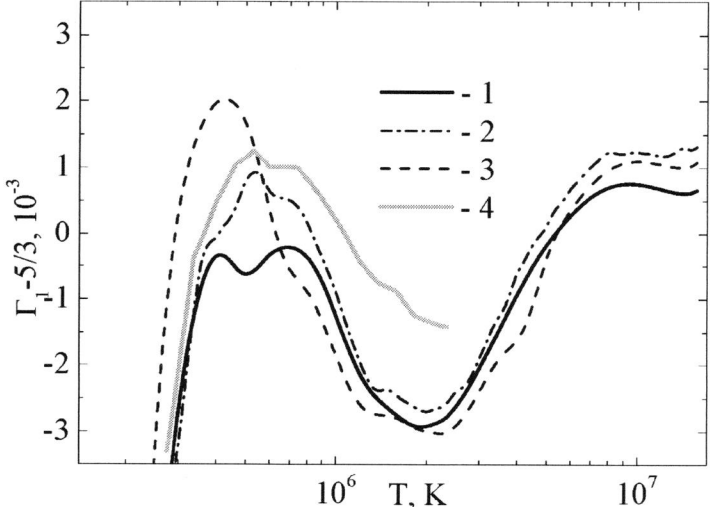

FIGURE 2. Adiabatic exponent along the solar trajectory, 1 - present work, 2 – OPAL[3], 3 – MHD [4], 4 – the result of helioseismic inversion [1].

The "adiabatic exponent" (AE) Γ_1

$$\Gamma_1 = \left(\frac{\partial \ln P}{\partial \ln \rho}\right)_S, \qquad (29)$$

which, strictly speaking, has to be called the isentropic exponent, one of the key thermodynamic value in helioseismology. This object will be paid the main attention.

In these calculations the heavy element composition (in mass) was the following: carbon fraction was 0.1906614, nitrogen - 0.0558489, oxygen - 0.5429784, neon - 0.2105114. In these calculations plasma consisted of 54 components, including electrons, atoms, all ions for each chemical element, and the diatomic molecules. From the model described above the following elements of the approach were taken into account. Corrections due to the Coulomb interactions of charged particles were included in the framework of the DGCE (14-17), the effects of degeneracy of free electrons were described according to (12,13), and the partition functions were calculated in the form (10-11), effects of the radiation pressure by (27). At the first step in the solution of equations (2-5), we obtained the component composition, then the thermodynamic functions, the dimensionless parameters, and then the differential characteristics such as heat capacities, sound velocity, adiabatic compressibility were calculated. One can see that the model SAHA-S demonstrates a good agreement with other [3, 4] solar models. Nevertheless it is seen that the SAHA-S and the other models demonstrate growing discrepancy with the results of helioseismic inversion [1] to temperatures, corresponding to the bottom of the convection zone. Understanding of possible reasons of this discrepancy is impossible without answering the question on the influence of various plasma effects on the behavior of thermodynamic values along the solar trajectory.

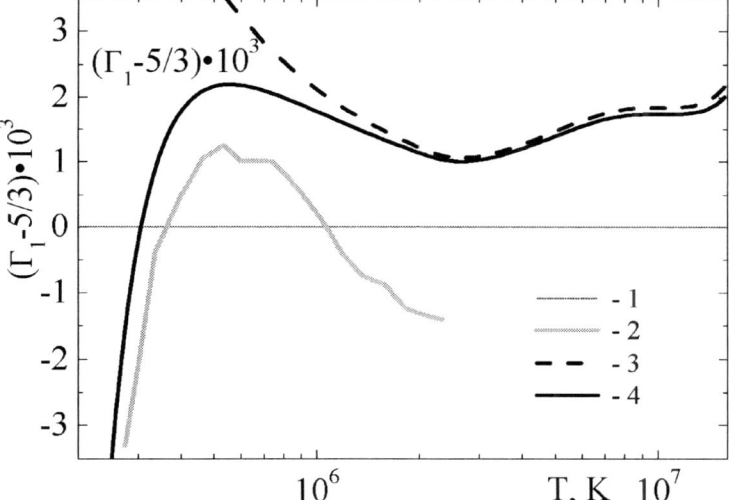

FIGURE 3. Adiabatic exponent of ideal plasma of protons and alpha-particles (curve 1) is the present calculation, the results of inversion [1] -2, 3 is the fully ionized plasma of hydrogen and helium with Coulomb interaction in the framework of the DGCE (14-18), 4 – the contribution of bound states of H and He to AE.

From this point of view of the second type of computations of thermodynamics of solar plasmas was carried out. The idea of these calculations consisted in successive switching of the plasma effects to the EOS on to analyze the dependence of the thermodynamic functions on these effects. In this case we used data of the model S for the solar trajectory, namely the dependence of the density on temperature, and the helium and heavy elements abundance. The first step was the simplest one.

The fully ionized plasma of hydrogen and helium was considered as the ideal Boltzmann gas (no corrections of (7-28) were taken into account), X corresponded to its value in the model-S. In this case Γ_1 has to be equal to 5/3 exactly, that is natural for the ideal Boltzmann gas of particles without any internal structure. In the next step the fully ionized plasma of hydrogen and helium was considered with the Coulomb interactions in accordance to the DGCE (14-17). Fig. 3 demonstrates the shift of Γ_I to the higher values due to the contribution of the charged particle interactions.

Larger shift at lower temperatures corresponds to higher values of the Coulomb coupling parameter (19). It is seen in Fig.4, where the Coulomb coupling parameter and degeneracy parameter vs. the solar radius are represented.

In the next step the calculations of plasma properties were carried out in which together with the Coulomb interactions the recombination of free charges was allowed. The contribution of bound states using the Planck-Larkin [14] formula to calculate the partition functions was considered. The results of such calculations are demonstrated also in Fig.3. It is seen that at region of lower temperature where the contribution of bound states is noticeable, the recombination effects lead ΔE to a steep falling down. This effect is important up to the temperature of the convection zone bottom, where the helium ionization goes to finish. Addition of degeneracy effects of free electrons does not change this picture, since the adiabatic exponent of the ideal electron gas is exactly equal to five thirds, which is the strict result for the ideal Fermi gas at any value of the degeneracy parameter [9].

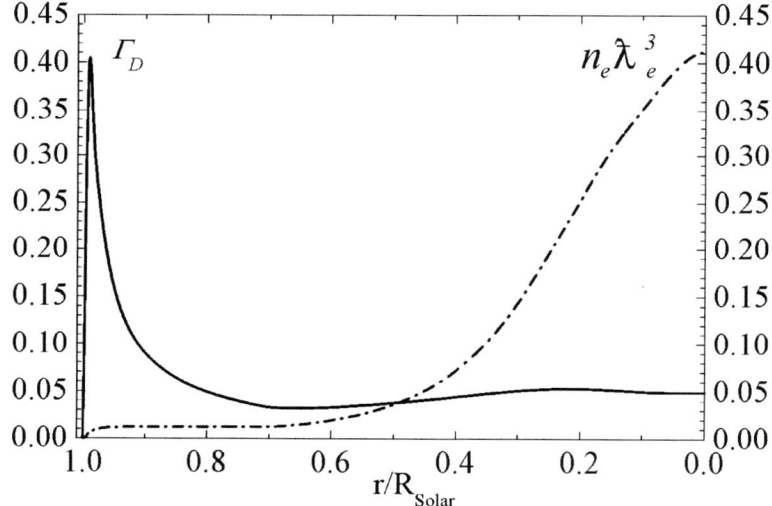

FIGURE 4. Coulomb coupling parameter and degeneracy parameter along the solar radius.

The next step in our calculations consisted in the inclusion of heavy elements. The abundance of heavy elements corresponded to the model S and the distribution of mass fractions was the same, as in the calculations represented in Fig.2 (C - 0.1906614, N - 0.0558489, O - 0.5429784, Ne - 0.2105114). All degrees of ionization for each chemical element were taken into account so as the diatomic molecules H_2, H_2^+, C_2, N_2, O_2, CH, CN, CO, NH, NO. The partition functions of excited states of atoms were calculated using the Planck-Larkin formula; the partition functions of molecules were calculated in the approximation of the non-rigid rotator - non-harmonic oscillator with the cutoff procedure (8). Results of these calculations are presented in Fig.5. The maxima of AE in the temperature range 400-700 kK are due to the ionization of inner electronic shells of heavy elements. The difference between the values of AE calculated with the Planck-Larkin (9) and Starostin - More - Rörich (10) formulae for the partition functions of atoms and ions can be seen in the same Figure.

It is clear from (11) and Fig.1 that the maximal difference between the calculation results with two partition functions is of factor 4 at very high temperature, but for the solar plasma the maximal effect occurs at the temperatures of about 500kK, which is due to the contribution of the abundance of bound states in this temperature region.

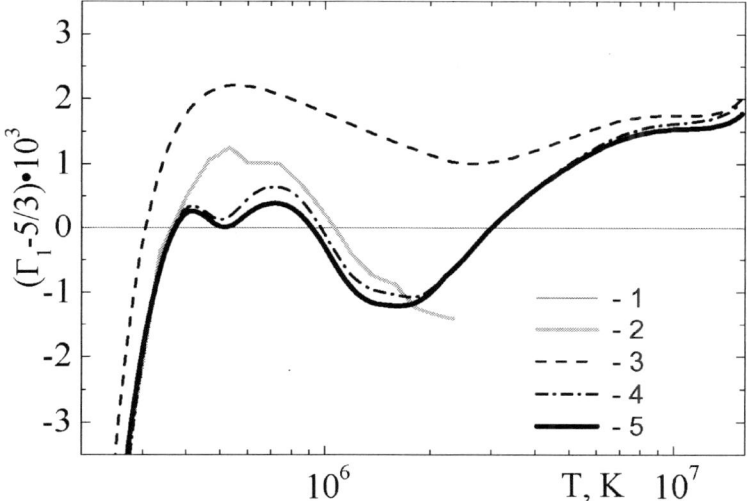

FIGURE 5. Contribution of heavy elements and bound states with the Plank-Larkin (4) and Starostin-More-Rörich (5) partition functions; 1, 2, 3 as 1,2,4 in Fig.4.

Whereas in the center of the Sun where temperature reaches maximum, this difference is not so high due to the full ionization of the main chemical elements of the solar plasma, H and He. The contribution of the radiation pressure and energy to AE can be seen in Fig.6.The difference between AE with and without the radiation effects is practically constant beginning from the temperature of about 1000kK, which can be explained by the increase of the solar plasma density in the direction to the center of the Sun and, respectively, the growth of the kinetic pressure and energy. A smaller effect is connected with additional quantum corrections (exchange and diffraction), those are given by (24-27), where the Coulomb corrections in the framework of the DGCE combine with the exchange and diffraction corrections. The contribution of these corrections is represented by the curve 5 in Fig.6. The last effect under consideration was the relativistic one. As it is seen from (28-30), the first order pressure relativistic correction equals zero exactly, but nevertheless and due to a very high temperature of the solar plasma in the zone deeper than the bottom of the convection zone, these effects are quite noticeable. The curve 6 in Fig.6 for AE represents the calculation with all plasma effects considered above including the Coulomb interaction effects with quantum corrections, degeneracy of free electrons, radiation effects, relativistic effects and the partition functions of atoms and ions according to the Starostin-More-Rörich taken into account.

The last numerical experiment in the present work consisted of the variation of the composition heavy elements, namely it was enriched with iron and silicon. The variation of fractions consisted mainly in the addition of small fractions of Fe and Si and corresponding decreasing of the Ne fraction.

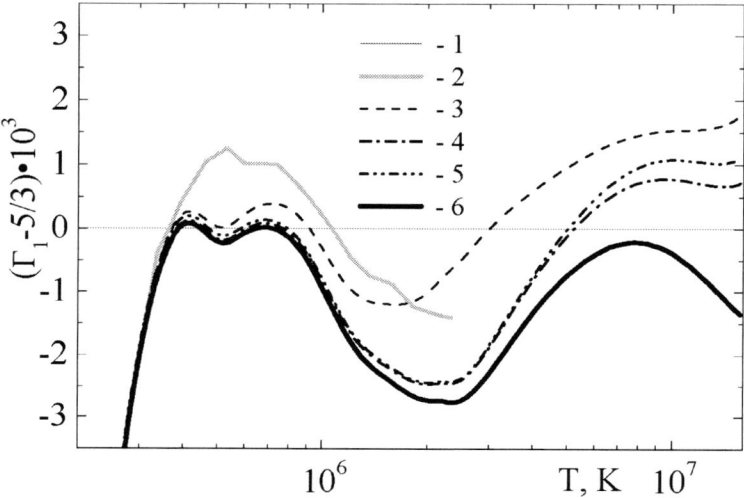

FIGURE 6. Contribution of radiation effects – curve (4), exchange and diffraction effects (5) and relativistic effects (6) to AE. 1, 2, 3 are the same as 1, 2, 5 in Fig.5.

The element composition of heavy elements (in the limits of the abundance of heavy elements within the model S) was the following: C – 0.1762, N – 0.0516, O – 0.5018, Ne – 0.0967, Fe – 0.0982, Si –0.0755. Other parameters of calculation coincided with those demonstrated in Fig.6, that is all plasma effects were taken into account.

The results of such calculations are represented in Fig.7, where the comparison with the same calculations but with usual (more poor) element composition is carried out.It is seen that the enriched composition of heavy elements does not lead to the radical change in the AE behavior and becomes apparent by removing the two peaks structure in the 400-700kK temperature range.

The last step in this type of calculations was the variation of the heavy element abundance. In this case the heavy element abundance was equal to Z/2, where Z was that corresponding to the model S value. The results of these calculations are shown also in Fig.7. One can see this more than satisfactory accordance of these results with the data obtained from the procedure of heliosesmic inversion [1], but to our opinion, it has to be appreciated as occasional at the moment. In Fig 8 the distribution of mean ion charges of each chemical elements versus temperature is represented. We must note that ionization degree for H and He reaches its maximum in the convection zone, for C it occurs in the core only, whereas for other chemical elements it does not reach the maxima even in the center of the Sun.

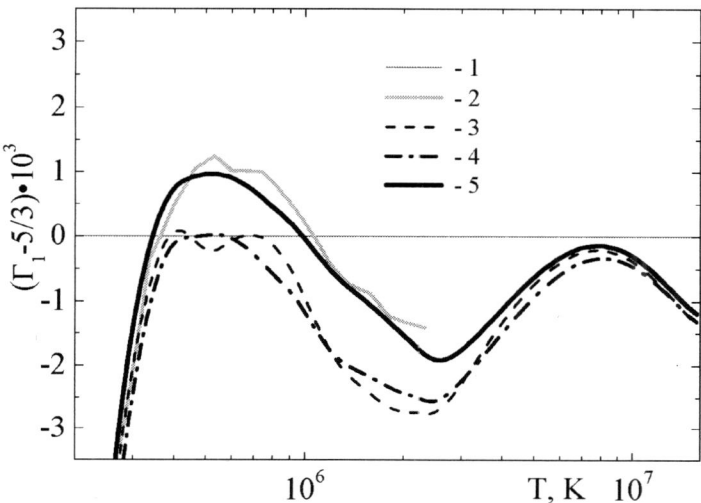

FIGURE 7. Influence of extension (4) of heavy element composition on the behavior of ΔE; 1, 2, 3 are the same as 1, 2, 6 in Fig.6, 5 – calculation with extended composition and double diminished Z.

It points on a significant role of bound states of atoms and ions in the solar plasma and the knowledge of their energy level structure as an adequate description of contribution to thermodynamic functions and equation of state.

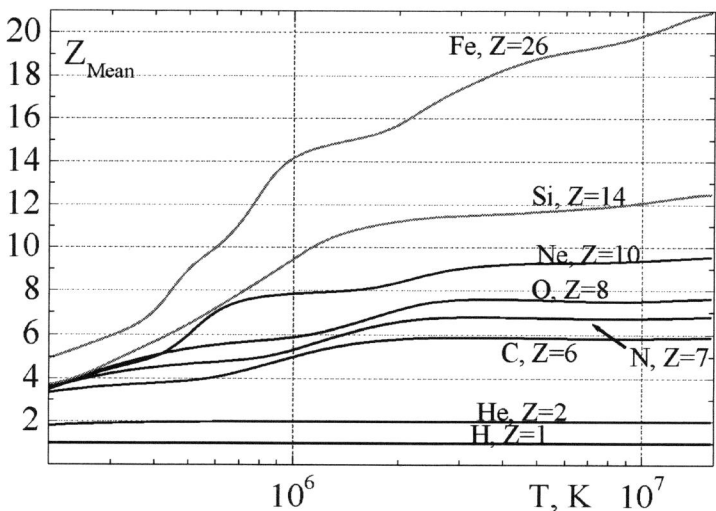

FIGURE 8. Mean charge of ions of chemical elements in the solar plasma. Names of chemical elements and charges of their nuclei are pointed out.

CONCLUSIONS

In this work we have proposed the thermodynamic model SAHA-S of the solar plasma. The model is based on the chemical picture of plasma and includes the Coulomb correlations in the frame of the Debye approximation in the grand canonical ensemble with the exchange and diffraction corrections, free electron degeneracy effects, relativistic effects, new approach for the calculation of atomic and ionic partition functions, the radiation contribution to the plasma pressure and energy taken into account. The results of this model are in good agreement with the solar thermodynamic models now in use. To isolate discrepancies in Γ_1 between the theoretical data and the data of the helioseismic inversion contribution of various plasma effects has been analyzed. The most important factors which affect the behavior of the adiabatic exponent in the convection zone are the Coulomb correlations and the bound states of atoms and ions. In the region of the solar radiation zone and core the relativistic and radiation effects are of great importance. Extending the set of heavy elements with Fe and Si leads to the removal of additional peaks of the adiabatic exponent because of the ionization of Ne inner shells. Double decrease of extended heavy element abundance makes it possible to reach improved agreement between the helioseismic and calculated data.

ACKNOWLEDGMENTS

We are grateful to Sergey Vorontsov for very interesting and helpful discussions and new data on helioseismic inversion and to Werner Däppen for helpful discussion and permanent attention to this work.

This work was supported by the grants of the President of Russian Federation № Ssc-1257.2003.2. and Ssc-1938.2003.2, contract no. 40.009.1.1.1192 of the Ministry of Education and Science of the Russian Federation, the Program for Scientific Research in Thermal Physics and Mechanics of Extreme States of Matter of the Presidium of the Russian Academy of Sciences, and grant № 04-02-16775-a of the RFBR.

REFERENCES

1. Vorontsov S.V., In Proc. SOHO 11 Symposium, 'From Solar Min to Max: Half a Solar Cycle with SOHO', Davos, Switzerland 11-15 March 2002, ed. A.Wilson, ESA SP-508, Noordwijk, The Netherlands, p.107-110, (2002); in this Proc. Workshop "Equation-of-State and Phase Transition Issues in Ordinary Astrophysical Matter", Leiden, The Netherlands (2004).
2. J. Christensen-Dalsgaard, W.Däppen, S.V.Ayukov et al., *272*, *Science*, **272**, 1286-1292 (1996) ; J. Christensen-Dalsgaard, *Rev.Mod.Phys* **74**, 1073-1129 (2002)
3. F.Rogers, F.Swenson, G.Iglesias, *Astrophysical J.* **456**, 902-908 (1996); F.J.Rogers, A.Nayfonov, *Astrophysical J.* **576**, 1064-1074 (2002)
4. D.Mihalas, D.Hummer, W.Däppen: in *Astrophysical J.*, **331**, 794; ibid, **331**, 815 (1988); ibid **332**, 261 (1988)
5. W.Ebeling, *Physica* **43**, 293-303 (1969)
6. V.K.Gryaznov, I.L.Iosilevskiy, et al, *Thermal physics of gas-core nuclear reactor*, ed. by V.M.Ievlev, Atomizdat, Moscow, 1980, 304 p.
7. V.K.Gryaznov, I.L.Iosilevskiy, V.E.Fortov, *Thermodynamic properties of shock compressed plasma based on chemical picture* In: High-Pressure Shock Compression of Solids VII, Shock Waves and Extreme States of Matter, Springer, Hannover , 2004, 531p.
8. G.M.Harris, J.E.Roberts, and J.G.Trulio, *Phys. Rev.*, **119**, 1832-1841 (1960)
9. Landau L.D., Lifshits E.M., Statistical physics, Moscow, Nauka, 1976, 584p.
10. Moore C. E., Atomic Energy Levels, Nat. Bur. Stand. (U.S.), Circ. 467, 1949, 309 p.
11. S.Bashkin; J. O Stoner Atomic energy levels and Grotrian diagrams, Amsterdam : New York : North-Holland Pub. Co. ; American Elsevier Pub. Co., 1975-1981
12. Huber K. P. and Herzberg G. *Constants of Diatomic Molecules*. New York: Van Nostrand Reinhold, 1978.
13. *Thermodynamic properties of individual substances.* Handbook 4vol. Ed by L.V.Gurvich, G.B.Khachkuruzov, V.A.Medvedev, v.1, Moscow: 1978.
14. Planck M., *Ann. der Phys.* **75**, 673 (1924); Larkin A.N., *Zh. Eksp. Teor. Fiz.* **38**, 1896 (1960)
15. A.N. Starostin, V.C. Rörich, R.N. More, Contrib. Plasma Phys., **43**, 369 (2003)
16. Kalitkin N.N., Kuzmina L.V *Interpolation formulas for Fermi-Dirac functions*, Preprint no.62 of Inst Appl Mat, 1972, 22p.
17. A.A.Likalter, *Zh. Eksp. Teor. Fiz.* **56**, 240 (1969); V.K. Gryaznov, I.L.Iosilecskiy,V.E.Fortov., *Zh.Prikl.Meh.Teh.Fiz*, **3**, 70-76 (1973)
18. Ebeling W., Kraeft W., Kremp D., *Theory of Bound States and Ionization Equilibrium in Plasmas and Solids*, Moscow, Mir, 1979,
19. Saha M.M., Phil. Mag., **40**, 472 (1920)
20. W. Ebeling, A.Förster, V. Fortov, V.Gryaznov, A. Polishchuk, *Thermophysical Properties of Hot Dense Plasmas*, Stuttgart-Leipzig , Teubner , 1991, 315p.
21. V. K. Gryaznov, V.E. Fortov, M.V.Zhernokletcv, G.V.Simakov, R.F.Trunin, L.I.Trusov, I.L Iosilevskii, *JETP* **87**, 678-690 (1998)
22. V.E.Fortov, V,Ya.Ternovoi., M,V,Zhernokletov, M.A.Mochalov,A.L.Mikhailov,A.S.Filimonov, A.A.Pyalling, V.B.Mintsev,V.K.Gryaznov, I.L.Iosilevskii, *JETP* **97**, 259-278 (2003)

Thermodynamics of the Convection Zone Through Adiabatic Exponent

V. A. Baturin

Sternberg Astronomical Institute, Moscow, Russia

Abstract. Methods of comparison of the plasma equations of state (EOS) in solar conditions are discussed. The most useful quantity in helioseismic applications is the adiabatic exponent, and its profile is investigated, involving geometric illustration in natural variables – pressure and density. The amount of information which can be obtained about EOS from a profile of adiabatic exponent on an adiabatic curve is studied. The comparison of modern versions of EOS is presented from the point of view of adiabatic profiles.

DEFINITION OF THE EQUATION OF STATE

Let us start from a definition of the equation of state (EOS hereafter). Consider five primary thermodynamic variables – pressure P, volume V, temperature T, entropy S, internal energy U. Between every three of them exist a unique relation, which describes possible equilibrium states of system and can be called an equation of state. The number of possible combinations of 3 things out of 5 is 10, but three appropriate relations are already enough to get all others. It means, that there are no states outside of this surface, defined by three relations. The EOS surface in any coordinate space of the primary variables is two dimensional.

Generally speaking, one can call EOS a complete set of the relations between any thermodynamic variables, described all equilibrium states of the system. Naturally, we are interested in most compact description. Taking into account the thermodynamics laws, we can expect that two relations will be enough. In fact, even two relations may contain excessive information. It means, that they are not fully independent, for example it is not possible "suggest" arbitrary EOS for internal energy $U(P,V)$ and copulate it with perfect gas relation $P(T,V)$.

It is well known, that exist functions of two thermodynamics parameters, which is enough to describe all the EOS set. One example of such function is a free energy, expressed as function of temperature and volume $F(T,V)$ (equivalently, function of density and temperature). All other values can be obtained as corresponding partial derivatives. Such functions used to be called a thermodynamic potential, and its main characteristics is a pair of independent variables. Among the primaries, the only thermodynamic potential is internal energy as function of volume and entropy $U(S,V)$.

Let us make a remark on a number of necessary functions. In view of equilibrium thermodynamics, internal energy and entropy defined with arbitrary constant. More strictly, one needs to know differentials dU and dS of these functions. But differential is 1-form and defined (in specific coordinates) by two functions instead of one. So formally, a number of functions increase, but again between these new functions some relations exist. Usually we believe that if a function is known, then its differential is known too. In practice it may be not so easily, because expression for function can be complicate and implicit. Inversely, a function can be integrated from differentials only with arbitrary constant, but in case of thermodynamic equilibrium it is enough.

Thermodynamic laws for the equilibrium EOS can be written as relations between differential 1-forms (or differentials, see for example [1])

$$T\tilde{d}S = \tilde{d}U + P\tilde{d}V \qquad (1)$$

Obviously, internal energy is a thermodynamic potential as function of two extensive variables, which appeared in (1) as differential forms. To get a thermodynamic for other pair, like temperature and volume for example, one need to involve other function. For example, using a substitution $T\tilde{d}S = \tilde{d}(TS) - S\tilde{d}T$, one can get $-S\tilde{d}T = \tilde{d}F + P\tilde{d}V$ where $F = U - TS$ is the already mentioned free energy, which is a thermodynamic potential in desirable variables. Not every pair of primary values can be arguments for some thermodynamic potential. These exceptions are appeared to be conjugated in terms of (1), namely (P,V) and (T,S).

So, besides of the thermodynamic potential $U(V,S)$, all others combinations of any three primary parameters do not produce a complete EOS. Thus a thermal EOS in form $P(T,V)$ even in simple case of perfect gas, is not enough to calculate all other values, like specific capacity $c_V = \left.\dfrac{\partial U}{\partial T}\right|_V = T\left.\dfrac{\partial S}{\partial T}\right|_V$. Another important example is the internal energy as a function of pressure and volume $U(P,V)$.

Examples of incomplete EOS provide us with restricted information on physical properties of the system. But generally available data is also restricted. In this case, study of incomplete EOS give a measure of information principally available from the data. If we can restore only incomplete relation, there are a lot of physical system may satisfy to the data.

GEOMETRY OF EOS

As we have seen from previous section, one surface (or manifold) together with its differentials in corresponding coordinates provides all necessary thermodynamic information. However, essential part of information presented by differentials, so strictly speaking, a set of functions to be used. We are most interested in values, which are important in helioseismic analysis, sound speed and adiabatic exponent. Starting from internal energy $U(S,V)$ we can calculate adiabatic sound speed as second derivative

$$c^2 \equiv -V\left.\frac{\partial P}{\partial V}\right|_S = V\left.\frac{\partial^2 U}{\partial V^2}\right|_{SS} \tag{2}$$

It is clear, that adiabatic exponent $\gamma_1 = \left.\frac{\partial \lg P}{\partial \lg \rho}\right|_S$ related with sound speed $\gamma_1 = c^2/(PV)$. To calculate γ_1 from a free energy $F(T,V)$, one needs two derivatives of pressure $\chi_T = \left.\frac{\partial \lg P}{\partial \lg T}\right|_\rho$ and $\chi_\rho = \left.\frac{\partial \lg P}{\partial \lg \rho}\right|_T$ (represent of a differential of pressure dP) together with specific capacity c_V (all these values are second derivatives of free energy). Then one can write

$$\gamma_1 = \chi_\rho + \frac{P}{\rho T}\frac{\chi_T^2}{c_V} \tag{3}$$

It is not a best way for calculation and analysis, because expression includes three derivatives together pressure $\pi(T,\rho) \equiv P/(\rho T)$. Due to partial ionization of element, for example, both χ_T and c_V increase roughly in similar manner, so their ratio is compensated by decreased χ_ρ. If we try to estimate effect of some disturbances of EOS, we need to calculate disturbances in four derivatives. Even in case of relatively simple disturbances, like a Debye correction to a perfect gas EOS, it is difficult to predict at sight even a sign of γ_1 variation. (In fact, in this specific case, adiabatic exponent increases).

It maybe instructive to express a adiabatic exponent γ_1 in more simple way, from incomplete EOS in form (here denote $U_A \equiv (\partial U/\partial A)_B$ if $U(A,B)$ is a function of A and B)

$$\gamma_1 = \frac{1 + P^{-1}U_V}{V^{-1}U_P} = \frac{PV + U_{\ln V}}{U_{\ln P}} \tag{4}$$

It may look strange, why it is possible to get adiabatic derivative from expression $U(P,V)$, which does not contain entropy and temperature. That is because even on an incomplete thermodynamic manifold we able to plot levels of constant entropy (adiabatic reversible processes) based only the geometry of the manifold – utilize thermodynamic laws.

In formal way we can explain this fact as follow procedure. Let i_s is a vector field which is tangent to adiabatic curve. Then an equation for this field can be written after interior production by 2-form $i_s(dT \wedge dS) = i_s(dT) \wedge dS - dT \wedge i_s(dS) = dS$ (we save formal exterior production even when production of i_s and 1-form is just a number. See [2] for details). But after taking exterior differential from (1) we obtain $dT \wedge dS = dP \wedge dV$ and it becomes obvious, that equation for adiabatic field can be written in (P,V)-plane as well.

From view of geometry expression (1) maybe interpreted in follow way. For fixed values P and V there is exist a tangent plane to the manifold $U(P,V)$. This tangent plane defined by two crossed lines, according to derivatives U_V and U_P. Line of the adiabatic curve can be deduced from the manifold, because they determined by zero left part of (1). The adiabatic exponent is a tangent to internal energy surface in direction of the line of adiabatic curve. These system of adiabatic curves and its tangents can be projected on (P,V)-plane, where they equivalent to one function $\gamma_1(P,V)$.

Using of the incomplete form of EOS $U(V,P)$ reduce a number of parameters directly affect the adiabatic exponent. Moreover, internal energy itself contains all information to calculate derivatives and adiabatic levels. So it is reasonable to compare of different version of EOS as two manifold of internal energy, if we interested only in adiabatic exponent. Of course, mapping of $U(P,V) \to \gamma_1(P,V)$ is irreversible, so it is not possible to restore an internal energy on base of a field of adiabatic exponent. To get this we will need absolute values of internal energy on some line do not belong of an adiabatic curve and cover desirable rage of pressure and volume. In practice, however, we have only values of exponent on some unknown adiabatic curve.

Thus geometry of 2-dimential surface provides compact and convenient description of hydrodynamic properties of the EOS. It is important for theoretical analysis, but not to easy to plot. Moreover, it includes the absolute value of internal energy that makes it less physically valid. Of course, arbitrary constant does not affect of internal geometry of surface. But if we try to exclude this constant, we will need to analyze already two functions, which determine differential of internal energy. As we mostly interested in adiabatic exponent, it is clear, that a map $dU(P,V) \to \gamma_1(P,V)$ exists and it does not depend on a constant in internal energy itself. So, it is quite reasonable to use a projection all adiabatic structure on the plane of pressure and density.

A function $\gamma_1(P,V)$ is equivalent to a vector field in the (P,V)-plane, which is tangent to levels of constant entropy (that is projection of adiabatic curves on the plan). But two points should be mentioned in connect with this picture. First, we have geometrical representation for adiabatic field, but not a coordinate representation of 1-form dS. To find a distance between adiabatic levels, we should induce a temperature scale, what we try to avoid in this approach. As we will see below, this incompleteness is principal, and but can be avoided only partially in complete EOS. For practical reason we may use other value, instead of temperature – isothermal sound speed, $u = P/\rho$. For example for EOS of perfect gas in solar conditions, these values T and u differ by a mean molecular weight.

Second point connected with a lack of natural parameterization along an adiabatic curve. This fact looks rather technical, but it reveals a serious problem when we try to compare two different EOS, based on adiabatic exponents.

This problem is a central point of this article. From the discussion above, the comparison of two EOS means comparison of two systems of adiabatic levels, covered the region.

Main point of the paper is a comparison of different EOS based on profiles of adiabatic exponent. All previous discussion provides evidence that natural variables for such comparison are pressure and density.

To explain, why P and ρ is more convenient variables, give only one example. Using traditional variables - temperature and density, the function $\gamma_1(T,\rho)$ does not allow plotting a system of adiabatic levels. To do this one needs additionally a differential of pressure in form of χ_T and χ_ρ, to get $\left.\dfrac{\partial \ln T}{\partial \ln \rho}\right|_S = \dfrac{\gamma_1 - \chi_\rho}{\chi_T}$.

THERMODYNAMIC COMPARISON OF EOS AT FIXED POINTS

The comparison of adiabatic exponent of different EOS even in (P,ρ)-plane is still nontrivial task. Generally, we have to compare two close but not identical surfaces in some region of arguments. Such problem cannot be resolved without additional restrictions. Specific method of providing for such restrictions depends on aim of the comparison. Simplest approach is thermodynamic comparison, when we look for difference in given arguments, like given pressure and density. In this case we can consider whole surface $\delta\gamma_1(P,\rho)$. In every fixed point differences in adiabatic exponent means differences in inclination of tangents to adiabatic curve.

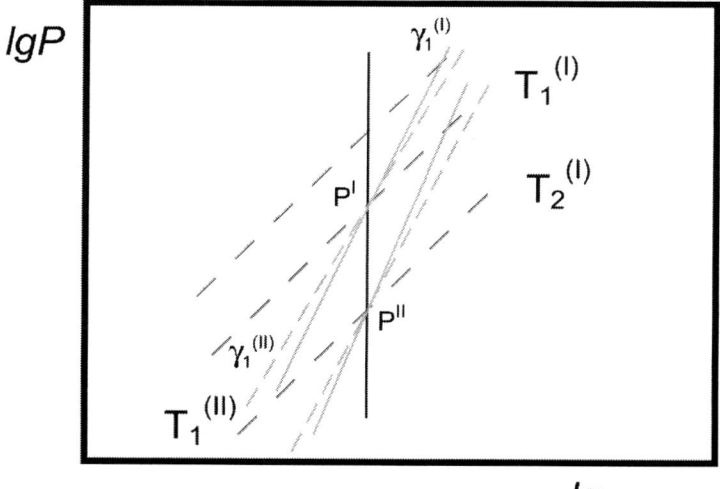

FIGURE 1. Adiabatic and isothermal curves in pressure-density plane for two EOS.

It is worth to note, that the differential surfaces will differ, if comparison performed with given T and ρ instead. The source of this divergence may look not very obvious, but it revealing is important, because (T,ρ)-arguments used much more often.

Let us consider illustrative Figure 1. Steep lines are parts of the adiabatic curves, described by functions $\gamma_1^I(P,\rho)$ in EOS type I and by $\gamma_1^{II}(P,\rho)$ in EOS type II. For both EOS we also have own levels temperature, $T^I(P,\rho)$ and $T^{II}(P,\rho)$, and isothermal lines plotted as well. These temperature levels are not matched generally, in forms and values. When we fix temperature and density in comparison procedure, the pressure $P^{II} = P^I + \delta_T P$ in EOS II is changed to make temperature equal $T^I(P,\rho) = T^{II}(P^{II},\rho)$. As result, a cumulative difference consists of two terms — one is difference for fixed pressure and density, and other is shift of comparison point. In linear approximation the difference of adiabatic exponent will be written as a sum

$$\delta_{T,\rho}(\gamma_1) = \delta_{P,\rho}(\gamma_1) + \left.\frac{\partial \gamma_1}{\partial P}\right|_{\rho} \delta_T P \qquad (5)$$

Graphically, it corresponds to comparison of γ_1 in two points in (P,ρ)-plane, shifted vertically to equalize the temperature. Of course, these two differences may not coincide.

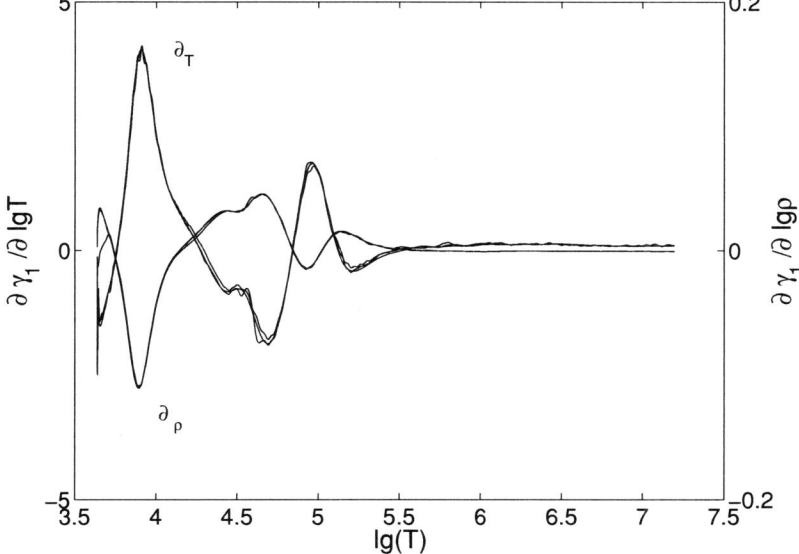

FIGURE 2. Profile of derivatives of adiabatic exponent in respect to temperature and density along points from solar model.

But there are two important moments. First, the second order difference $\delta^2_{T,\rho,P}(\gamma_1) = \delta_{T,\rho}(\gamma_1) - \delta_{P,\rho}(\gamma_1)$ depends on the derivative of adiabatic exponent, that on geometry of γ_1-surface itself, rather then on difference of EOS. In many cases the geometry of the surface is simple - adiabatic exponent is close to constant, or varying rather slowly. It means, that derivative in (5) is small. In such case the difference between two methods of comparison is negligible even for rather different type of EOS (simple illustration – two perfect gases with different specific capacities).

To check this point in solar conditions, we plot the derivatives of γ_1 along the points from a solar model on Figure 2. First observation from the Fig. 2 is a decrease of derivatives (both in respect to temperature and density) in region of high ionization that is at temperature higher $\lg T > 5.5$. In low temperature region, where ionization of hydrogen and helium take place, the geometry of γ_1-surface is complex, and shift of comparison point causes remarkable difference. Other conclusion from the figure is a slight dependence of the derivatives on type of EOS. In fact, there are four type of EOS used to plot the figure, and difference between them is small along whole range of values. It means, that for estimation of derivative any appropriate EOS can be used.

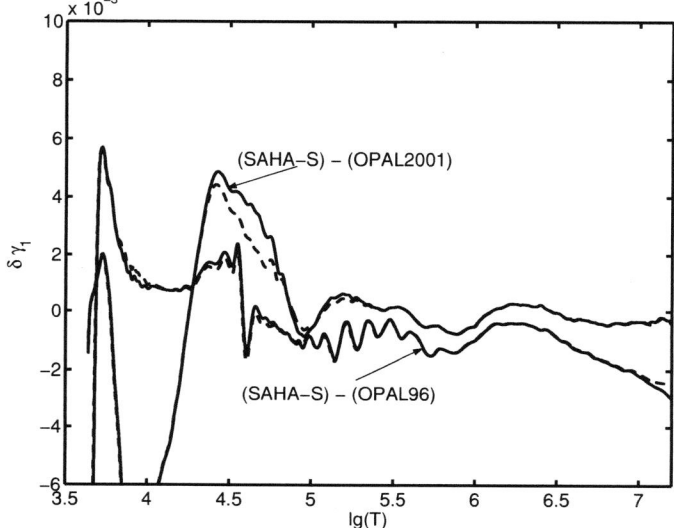

FIGURE 3. Intrinsic differences of adiabatic exponent between three EOS. Differences calculated with fixed (T,ρ) (solid) and fixed (P,ρ) (dashed).

Turn back to expression (5). Secondly, a shift difference depends on variation of pressure between EOS. This variation $\delta_T P$ comes from disturbance of EOS $P(T,\rho)$, which is not determined by incomplete $U(P,V)$. The behavior of $\delta_T P$ in solar conditions depend of type of EOS and shows more systematic character (the decrease of the shift to deep layers is less explicit, if appeared at all). Only very general expectation on amplitude of the shift may be proposed. In some cases, a difference between EOS may be described in one small parameter, like a parameter of Coulomb nonideality. In this case it may happen that both a shift and derivative are proportional to this parameter. Then the second difference will have a higher order in compare to the variation of adiabatic exponent. For example, difference in γ_1 due to including of Debye-Huckel correction to perfect gas EOS.

Generally, our study corresponding differences shows that this shift term is really small outside the ionization region. In ionization region, it may appeared to be remarkable in compare with value of $\delta\gamma_1$. So in this cases the intrinsic difference of adiabatic exponent demand specification of the conditions to be calculated.

To finalize this section, we plotted on Fig. 3 the intrinsic differences in adiabatic exponent, calculated for both pairs of variables and for three kind of EOS. First is SAHA-S rich EOS, presented in [3]. Others two are OPAL96 EOS [4] and OPAL2001 EOS [5]. It can be seen, that essential differences in adiabatic exponents are significant in the region of high ionization.

COMPARISON OF EOS ON ADIABATIC CURVES

Next step in comparison procedure is comparison of points on two adiabatic curves belonging to different EOS. Formally, one can consider situation of the same entropy of these curves, but it has a little meaning however. Entropy defined with arbitrary constant in thermodynamics, so its values in different EOS (for example, with alternated chemical composition) hardly can be compared. From view of practice, the adiabatic curve is defined by a point, which it passes through.

The case of two adiabatic curves is important for solar envelope, because of a bulk of convection zone seems to be stratified adiabatically. So comparison of solar models computed with different EOS may also be considered as a specific case of the problem.

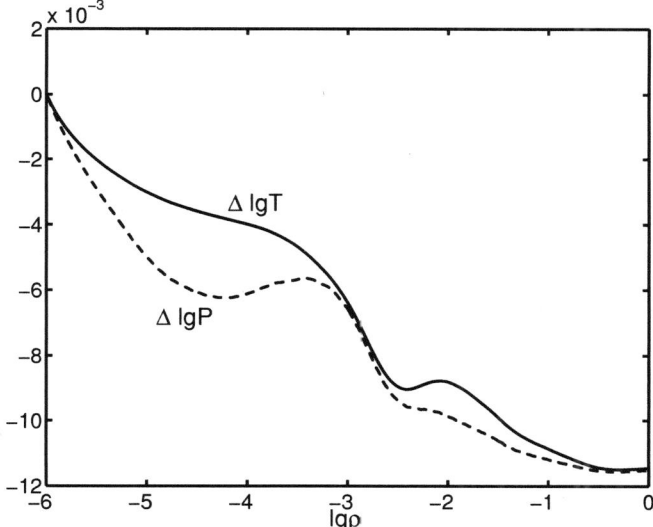

FIGURE 4. Divergence of adiabatic curves, integrated in two different EOS.

The comparison of two curves is impossible without additional restrictions. A choice of restrictions is quite arbitrary. It maybe fixed temperature, pressure, density or parameters out of thermodynamics, like a radius or a mass coordinate in the model.

From view of geometry, we have a vector of change, starting in reference point and ending in some point on the second adiabatic curve. It is not possible to describe this vector of change generally except of simple cases, when it defined in δ_P, δ_ρ. For example, for comparison on the same temperature, we already will have dependence of the vector on properties of both EOS. The inclination of isotherm $(\partial \ln P/\partial \ln \rho)_T$ is depend on EOS, but the difference of temperatures in both EOS also exists $\delta T(P,\rho) = T^{II} - T^{I}$. The situation is even more complicated, if we compare points with fixed radius of model. As rather rude approximation, we can believe that it close with comparison with fixed temperature.

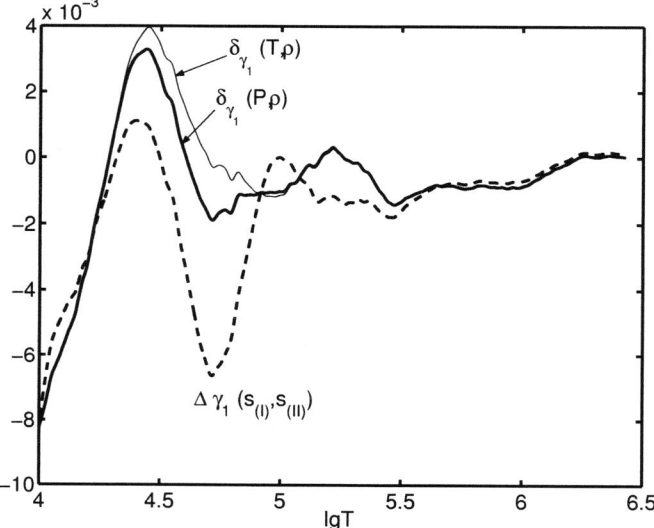

FIGURE 5. Difference of adiabatic exponent on two adiabatic curves with fixed density (dashed line). Solid lines correspond to intrinsic difference between EOS.

But the main problem is a closeness of two adiabatic curves. If compared curves are close (they intersect in proximity of compared points), one can expect, that influence of divergence between curves is small, and we have to take the intrinsic difference in γ_1 between EOS mainly. But generally the intersection point may be anywhere, and corresponding curves are far, even in case when both EOS are very similar around of comparison point.

One can introduce some qualitative measure for distance between curves, choose some vector as reference. Taking into account that we mostly interested in compare of close EOS, it may be δ_r or δ_ρ.

A practical meaning of this picture is follow. A variation of adiabatic exponent between two EOS under restrictions defined by vector $\vec{\delta}$, maybe written in linear approximation as

$$\Delta \gamma_1 = \delta_{P,\rho}(\gamma_1) + \tilde{d}\gamma_1(\vec{\delta}) \tag{6}$$

First term in right hand side of the equation is intrinsic variance of γ_1 with EOS disturbance. Second one is an intrinsic production (or convolution) 1-form $\tilde{d}\gamma_1$ (differential of adiabatic exponent) and vector of shifting $\bar{\delta}$. 1-form $\tilde{d}\gamma_1$ depend only on how speedy adiabatic exponent is changed between points, and in linear approximation can be estimated in either EOS. We also can estimate a priory where the second term is significant – in region, where adiabatic exponent is not constant. Vector $\bar{\delta}$ gives a measure of distance between compared points. In case of the situation with adiabatic curves compared under fixed density, we can easily write the expression explicitly. Using δ_P, for example

$$\Delta\gamma_1 = \delta_{P,\rho}(\gamma_1) + \left.\frac{\partial\gamma_1}{\partial P}\right|_\rho \delta_P \qquad (7)$$

In contrast to expression (5), calculation of δ_P demands specifying of adiabatic curves and depend on global properties of both EOS far from comparison point.

We present of this comparison for case of SAHA-S and OPAL2001 EOS. Two adiabatic trajectory have been computed, started at one point $\lg T_0 = 4$ $\lg\rho_0 = -6$. Integration performed up to $\lg\rho = 0$, that is slightly deeper then bottom of convection zone. On figure 4 two curves of differences in temperature and pressure at the some densities demonstrate how the adiabatic trajectories are diverging. On figure 5 the differences of adiabatic exponent are plotted. Dashed curve is differences between adiabatic curves, calculated at the same density. Marking of axis with temperature is only for purpose of illustration. Two others solid curves (thin and thick) give intrinsic differences, calculated in both senses as described in previous section. It is clear, that influence of shifting term is remarkable, especially in ionization zone. But beside of ionization, due to plane geometry of γ_1-surface, the difference in adiabatic exponent is very close to intrinsic difference despite of the large distance between adiabatic curves in this region.

CONCLUSION

Our conclusions focus on a comparison of adiabatic exponent values predicted by several modern EOS in solar conditions. Three selected descriptions of EOS – SAHA-S, OPAL96 and OPAL2001, provide elaborated description of physical states of solar plasma, and revealed to be very close to each other. So difference between them may be considered as small disturbances in EOS and treated in linear approximation. The only differences of adiabatic exponent have been analyzed and amount of information on whole EOS which can be restored is estimated.

The differences in adiabatic exponent as others differences is divided into two parts. One is intrinsic difference, which depend only on thermodynamic properties. This part is essential for correcting of EOS in future descriptions. Other part is structural difference, which depend on the conditions of comparison. Second part depends on lot of factors and hardly can be described in general way. The main finding of the study is that in deep part of the convection zone, where ionization is close to complete, the difference of adiabatic exponent can be interpreted as intrinsic one. It allows us to use the result of helioseismic inversions for the adiabatic exponent to make comparisons of available EOS formalisms, and to indicate the direction in which modern EOS should be improved.

ACKNOWLEDGMENTS

I am indebted to A. N. Starostin, I. L. Iosilievskiy, V. K. Gryaznov, W. Däppen and S. V. Ayukov for numerous helpful discussions and comments.

REFERENCES

1. Schutz, B.F., *"Geometrical Methods of Mathematical Physics"*, Cambridge University Press, 1982.
2. Kozlov, V.V., *"Teplovoe ravnovesie po Gibbsy i Puankare"*, Moscow-Izhevsk, 2002 (in Russian).
3. Ayukov, S., Baturin, V., Gryaznov, V., Iosilievskiy, I., Starostin, A., and Fortov, V., *JETF Letters*, **80**, 141 (2004).
4. Rogers, F., Swenson, F., and Iglesias, C., *Astrophys J.*, **456**, 902 (1996)
5. Rogers, F., and Nayfonov, A., *Astrophys J.*, **576**, 1064 (2002)

Heavy element settling in the Sun and equation of state

S.V. Ayukov* and V.A. Baturin*

*Sternberg Astronomical Instute, Universitetsky pr. 13, Moscow, Russia, 119992

Abstract.
We study effect of heavy element settling on the equation of state in the standard solar model. The heavy element composition is an important source of uncertainty in the solar models; the uncertainty in Z itself comes from various sources, one of them being the element diffusion/settling. Z diffusion mostly affects opacities (iron is a very important source of opacity) but also touches equation of state. The settling rate depends on the ion mass and charge therefore different chemical elements have different settling velocities; the settling rates for O and Fe differ by 10% in our estimate. We study effect of 10% change in Z diffusion rate on the equation of state.

INTRODUCTION

Chemical evolution inside the stars is a central point of the stellar physics. According modern assumptions, the outer layers of the Sun are made of hydrogen (approx. 74% by mass), helium (approx. 24% by mass) and mixture of heavy elements. These heavy elements are represented mostly by C–N–O, Ne, Fe and Si group. Total mass fraction of heavy elements is designated by Z and is assumed to be close to 0.02. Observations cannot provide us with exact value of Z but supply estimate of Z/X ratio. This ratio is typically assumed to be equal to 0.0243 ([1]).

Before 1990s chemical profiles inside the Sun were supposed to be evolved only due to nuclear burning in the core. Modern standard solar model (e.g., [2]) includes element settling and diffusion as its integral part. These processes affect distribution of elements inside the Sun. Main feature of settling is a change of helium abundance in solar convective envelope in compare to primordial value. The change in helium abundance is large enough (about 0.025 in the convection zone) to be detected with helioseismic analysis. The changes in Z profile are smaller (because Z is itself smaller) and reach at maximum 0.016 by mass fraction in the surface layers. The main effect of Z diffusion on the solar model is via opacity. We attempt to estimate effect of possible uncertainty in Z diffusion rate on the model via equation of state regardless of opacity.

DIFFUSION/SETTLING DESCRIPTION

Basic equations for element diffusion in plasma were formulated (among others) by Burgers in 1969 ([3]). Element settling in the Sun was first estimated by Noerdlinger in 1977 ([4]). Not many models included it until circa 1990, but then this effect quickly became very popular. The helioseismic reference model (Model S of [2]) published in 1996 includes helium and Z diffusion as a necessary feature of the modern standard solar model. The most pronounced effect of settling is a decrease of helium abundance in the solar convection zone by approximately 0.025 (in mass fraction of helium). The evidence for this effect on the Sun was discovered by helioseismic analysis. Heavy element settling has smaller effect on the solar structure, nevertheless it is not negligible.

Formally speaking, terms 'settling' and 'diffusion' describe different processes. 'Gravitational settling' causes heavier elements to drift down while 'diffusion' tries to compensate concentration gradients. Gravitational settling can also be correctly described as 'diffusion caused by pressure gradient'. The most significant effect in the Sun comes from settling.

Besides nuclear reactions and mixing, settling and diffusion also change element distribution inside the Sun. The physical processes behind diffusion and settling are:

- gravity: heavier elements drift towards the center; this effect can also be formulated in terms of pressure gradient (barodiffusion);
- highly charged and more massive particles drift towards the hottest part (thermodiffusion);
- concentration gradients attempt to make element distibution uniform (this effect is to be called 'diffusion');
- transfer of momentum from absorbed photons pushes ions towards the surface; the calculation of this effect requires detailed knowledge of opacity data (radiative accelerations);

In this work we use terms 'settling' or 'diffusion' to refer to cumulative result of the above processes.

The calculation of diffusion velocities is not trivial. About 10 years ago several authors formulated simplified equations suitable for direct inclusion into solar evolution codes ([5], [6], [7]). The role of radiation induced diffusion was investigated later and was found to be significant ([8]). In our computations we used formalism from [6], as in [2].

HEAVY ELEMENT DIFFUSION IN THE MODERN STANDARD SOLAR MODEL

Figure 1 displays heavy element diffusion effect in the modern standard solar model ([2]). Initially ($t = 0$; Zero-Age Main Sequence) the star is chemically uniform, and Z has constant value of 0.019628 across entire star. Nuclear reactions in the Sun only convert hydrogen to helium and do not touch heavy elements (except that C and O are partially converted into N in the CNO cycle near the center), therefore Z profile is not affected by nuclear processes. During evolution all elements except hydrogen drift towards the center, and Z increases in the center up to 0.0203 while decreasing to 0.018 in the convection zone. Thus Z change due to diffusion is most profound in the convection zone (8.0%) and is smaller in the center (3.3%).

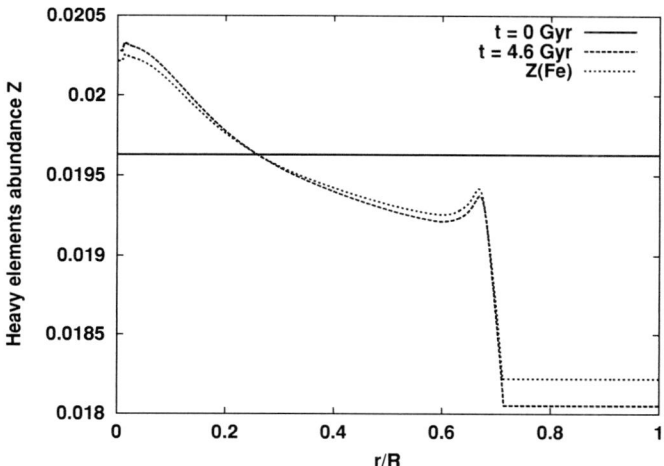

FIGURE 1. Z settling in the Sun. Based on model S of [2]. Z(Fe) curve is explained in text

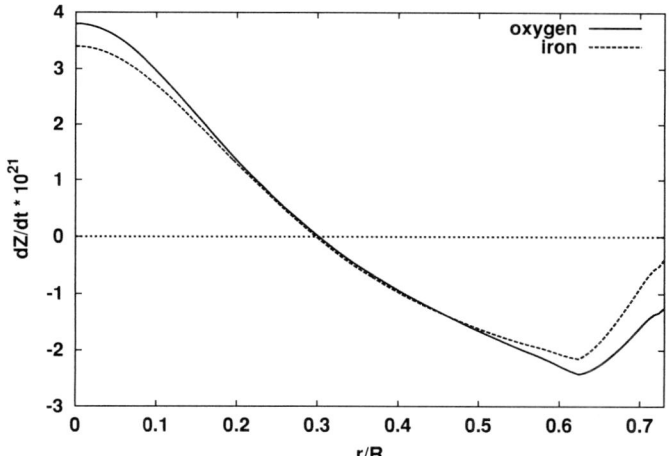

FIGURE 2. Z diffusion rate for O and Fe for Zero-Age Main Sequence model of the Sun

Z RATE EFFECT ON THE EQUATION OF STATE

In this work we present an estimation of an effect of Z diffusion rate uncertainty on the model via equation of state. Since the uncertainty isn't well known we have computed Z diffusion rate according [6] rate in Zero-Age Main Sequence model for two heavy elements, oxygen and iron (Fig. 2). These chemical elements have very different ion charges and atomic masses. The dependence on the ion charge and mass is not strong—

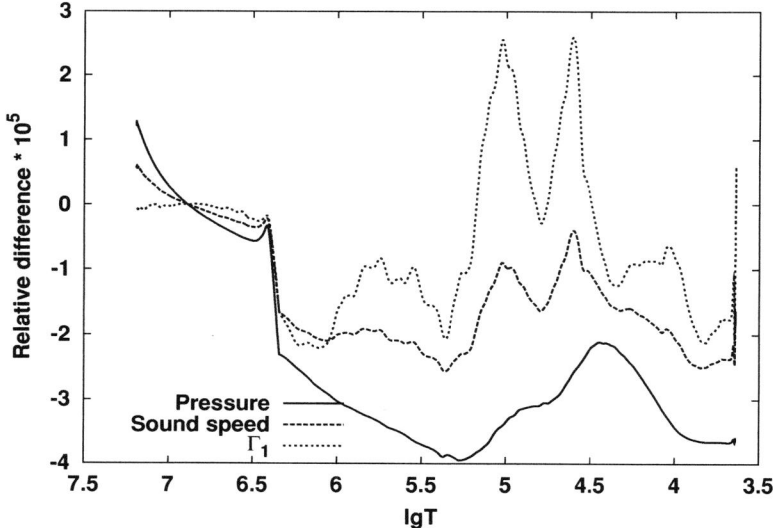

FIGURE 3. EOS difference induced by Z change

despite more than threefold difference in ion charge and mass diffusion rates differ by about 10%. The value of the element settling velocity difference between oxygen and iron (10%) was taken as an estimate of possible uncertainty in the Z settling rate.

Using obtained figure of 10% the Z settling amount in Model S was decreased by 10%; resulting Z distribution is displayed on Figure 1 as Z(Fe) curve. This small change in Z resulted in (also small) changes in thermodynamics. These changes are plotted on Figure 3 for temperature, density and hydrogen mass fraction profiles from Model S. In other words, Figure 3 shows how equation of state reacts at Z diffusion rate decrease by 10%.

It is clear that effect is small. Pressure variations reach $4 \cdot 10^{-5}$, sound speed variations are even smaller at approximately $2 \cdot 10^{-5}$. Difference look most pronounced in the convection zone ($\log_{10} T < 6.5$) due to the fact that Z change in convection zone is also maximal.

The variations shown on Figure 3 only reflect changes in thermodynamics. Thermodynamic and model comparisons ([9]) prove that actual changes in calibrated models can be several times larger, but even then Z diffusion rate uncertainty is probably too small (10^{-4} in adiabatic exponent) to have noticeable effect on solar structure.

CONCLUSIONS

This work provides estimate of Z diffusion rate uncertainty on the equation of state in the solar models. As expected this effect is rather little. It is found that changes are most significant in the convection zone; the amplitude of the effect agrees with [10].

Apparently uncertainties in the abundances themselves probably add much larger error to solar structure calculations.

REFERENCES

1. Grevesse, N., and Noels, A., *Origin and evolution of the elements*, Cambridge University Press, Cambridge, 1993, p. 15.
2. Christensen-Dalsgaard, J., Däppen, W., Ajukov, S., and et al., *Science*, **272**, 1286–1292 (1996).
3. Burgers, J. M., *Flow equations for composite gases*, Academic Press, 1969.
4. Noerdlinger, P. D., *Astron. Astrophys.*, **57**, 407–415 (1977).
5. Bahcall, J. N., and Loeb, A., *Astrophys. J.*, **360**, 267–274 (1990).
6. Michaud, G. and Proffitt, C.R., "Particle transport processes," in *Inside the stars, IAU Colloquium 137*, edited by W. W. Weiss and A. Baglin, ASP Conference Series Vol. 40, New York, 1993, p. 137.
7. Thoul, A. A., Bahcall, J. N., and Loeb, A., *Astrophys. J.*, **421**, 828–842 (1994).
8. Turcotte, S., Richer, J., Michaud, G., Iglesias, C., and Rogers, F., *Astrophys. J.*, **504**, 539–558 (1998).
9. Ayukov, S., Baturin, V., Gryaznov, V., Iosilevskiy, I., and Starostin, A., "Solar models using SAHA-S equation of state," in *Equation-of-State and Phase-Transition Issues in Models of Ordinary Astrophysical Matter*, edited by V. Celebonovic, W. Däppen, and D. Gough, these proceedings, 2004.
10. Ayukov, S., Baturin, V., Gryaznov, V., Iosilevskiy, I., Starostin, A., and Fortov, V., *JETP Letters*, **80**, 141–144 (2004).

Solar models using the SAHA-S equation of state

S.V.Ayukov*, V.A.Baturin*, V.K.Gryaznov[†], I.L.Iosilevsky** and A.N.Starostin[‡]

Sternberg Astronomical Instute, Universitetsky pr. 13, Moscow, Russia, 119992
[†]*Institute of Problems of Chemical Physics RAS, Chernogolovka, Russia*
**Moscow Institute of Physics and Techology, Dolgoprudnyi, Russia*
[‡]*Troitsk Institute for Innovation and Fusion Research, Troitsk, Russia*

Abstract.
We present solar models computed with a newly developed equation of state (EOS) of solar plasma, SAHA-S. SAHA-S is a solar version of the SAHA-* family of EOS designed for various states of matter. It incorporates modern state-of-the-art features such as arbitrary degenerated and slightly relativistic electron gas, improved formalism for Debye correction and generalized Planck-Larkin function which refines bound states contributions ([1]). We discuss standard solar models computed with SAHA-S and two versions of OPAL.

INTRODUCTION

An equation of state (EOS) is a set of functions or tables which determines all necessary thermodynamical quantities at given conditions and chemical composition of matter. Conditions are typically determined by temperature T and density ρ; chemical composition is a mass fractions of elements (see Table 1). In stars hydrogen and helium are two major elements, but role of heavier elements is not negligible.

Computing internal structure of the Sun requires knowledge of EOS in the solar interior. Solving solar structure equations requires EOS functions: pressure, internal energy and their first derivatives by temperature and density, as functions of T, ρ and chemical composition. Additionally adiabatic exponent $\Gamma_1 = \left(\frac{\partial \ln P}{\partial \ln \rho}\right)_S$ and sound speed c are required for helioseismic analysis.

The need for accurate plasma physics increased when first helioseismic inversions were performed. These inversions are able to provide high-precision profile of adiabatic exponent Γ_1 inside the Sun; the margin error related to observational noise is very small (10^{-5} through the most of lower convection zone ([2]). Thus the Sun is serving as an astrophysical laboratory allowing us to check validity of plasma physics understanding by comparing equation of state calculations with the Sun.

SAHA-S is a solar-plasma-related branch of equation of state developed by (in alphabetical order) Victor Gryaznov, Igor Iosilevsky and Andrey Starostin. In this article we study SAHA-S at conditions typical for solar interior by using new EOS to compute calibrated evolutionary model of the Sun and compare thermodynamical values between models.

TABLE 1. Heavy element mixtures used in equations of state

Chemical element	OPAL/SAHA-S	SAHA-S(rich)
C	0.1906614	0.177
N	0.0558489	0.052
O	0.5429784	0.502
Ne	0.2105114	0.097
Fe		0.075
Si		0.098

INPUT PHYSICS

Our models of the Sun are computed using the following input physics. Mass of the Sun is $1.989 \cdot 10^{33}$ grams, total luminosity is $3.846 \cdot 10^{33}$ erg/s, solar radius is $6.9599 \cdot 10^{10}$ cm. Age of the Sun is assumed be equal to 4.6 billion years. Heavy element mass fraction to hydrogen mass fraction ratio Z/X is 0.0245 ([3]).

Recent OPAL opacity tables (1995, [4]) are used. They are supplemented by Alexander, Ferguson low-temperature tables ([5]). Tables are interpolated with birational splines using G. Houdek's v9 package. Nuclear reaction rates are taken from NACRE'99 compilation ([6]). Convection is treated according Böhm-Vitenze mixing length theory.

Our models are computed with several EOS tables. The 'reference' tables are OPAL96 (original OPAL EOS tables, [7]); they were found to be very good for solar modelling: models with these EOS tables have sound speed profile close to observed (obtained by helioseismic inversion). Next tables are newer version of OPAL EOS, OPAL2001 ([8]). Comparing to OPAL96, these tables contain relativistic corrections for electrons, improved treatment of molecules and minor fixes to interpolation routines. They also have been extended for computations of low-mass stars.

The main subject of our study is newly presented SAHA-S equation of state ([9], [10]); two sets of tables with different set of heavy elements were used. Typically EOS tables are created for variable H and He abundances but fixed heavy element mixture. The total abundance of heavy elements Z can vary, but internal element distribution inside Z is fixed. For this work two sets of SAHA-S tables were calculated. One has the same heavy element mixture as OPAL96/OPAL2001 (and we refer to it as 'SAHA-S'), and second set of tables is computed using enriched composition (Fe and Si were added; we call it 'SAHA-S (rich)'); see Table 1. All these EOS tables (OPAL96, OPAL2001, SAHA-S, SAHA-S (rich)) allow interpolation by Z.

Standard solar model assumptions (spherical symmetry, no mass loss during evolution, no mixing outside convection zones) are also used. Models presented in this paper do not include helium or heavy element diffusion and settling. All models are calibrated by changing X and mixing length theory parameter (α) to have solar values of radius and luminosity at solar age. Heavy element abundance Z changes according fixed Z/X ratio. It is necessary to calibrate models with very high precision otherwise calibration errors will not allow to make proper comparisons. Calibration precision in our calculations was at 10^{-6} level; all relative differences in radius and luminosity are less than $1 \cdot 10^{-6}$.

TABLE 2. Global parameters of models. α is a mixing length theory parameter, Y_{env} is a helium mass fraction in the convection zone, T_c, ρ_c—central values for temperature and density, r_{cz}/R—bottom of convection zone position, by radius.

Model	EOS	α	Y_{env}	Z	T_c, 10^6 K	ρ_c, g/cm^3	r_{cz}/R
600-0100	OPAL1996	1.7849	0.26623	0.01754742	15.368	146.8	0.73277
600-0101	OPAL2001	1.8294	0.26675	0.01753495	15.371	146.9	0.73278
600-0108	SAHA-S	1.8434	0.26565	0.01756132	15.474	150.6	0.72760
600-0102	SAHA-S (rich)	1.8448	0.26549	0.01756514	15.468	150.6	0.72748

Most input physics data are identical (or very close) to data used to compute Model S ([11]). The most important difference is lack of helium/heavy elements diffusion in our models, and Model S was computed with older opacity tables (also from OPAL, [12]).

GLOBAL PROPERTIES OF SOLAR MODELS

Global parameters of computed solar models are listed in Table 2. Generally speaking, differences between models with different EOS are small. Enriched Z mixture has no noticeable effect on global parameters. Central temperature is close for all models (it is fixed by luminosity calibration and strong dependence of nuclear energy generation on temperature). The radius of convection zone bottom is close for models with variations of the same EOS, just like central density. This is in agreement with pressure comparison, Figure 4.

Lack of settling in these models resulted in too shallow convection zone and too high helium abundance in the envelope. The convection zone depth is known from helioseismic data and is equal to $0.713 \pm 0.003 R_\odot$ ([13]); surface helium abundance can also be obtained from frequencies.

EQUATION OF STATE AND MODEL STRUCTURE COMPARISONS

Figures 1–3 display thermodynamical values (pressure, adiabatic exponent Γ_1, sound speed) computed for all EOS tables considered and plotted as differences between specific EOS and 'reference' EOS tables, OPAL96. Conditions at points of comparison (temperature, density, X and Z) are taken from Model S.

Figures 4–6 display differences between values from model profiles. Differences are computed at fixed radius points and plotted on temperature scale for presentation purposes. Comparison of quickly varied values such as pressure or sound speed give sophisticated picture due to difficulties with synchronization of comparison points. In other words, we subtract pressure in two models at the same radius, but due to overall structure change, radius scale has also changed and we get huge differences just because pressure varies quickly with radius.

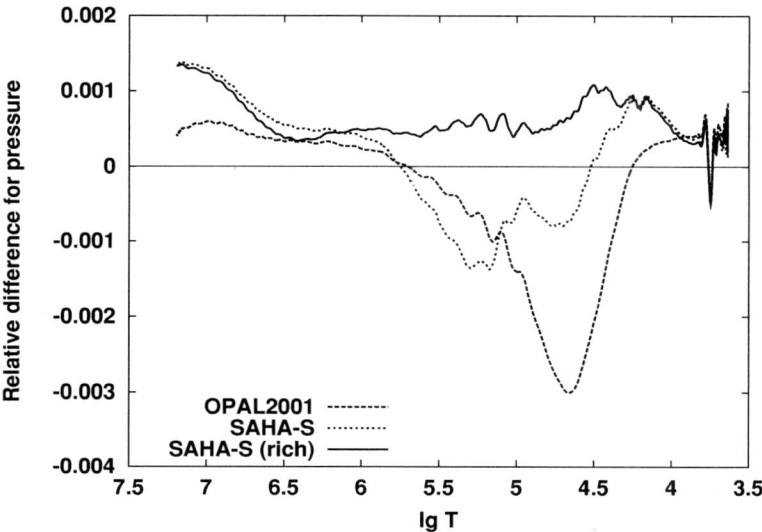

FIGURE 1. Equation of state comparison. Pressure. Zero line at this and next figures corresponds to OPAL96 equation of state.

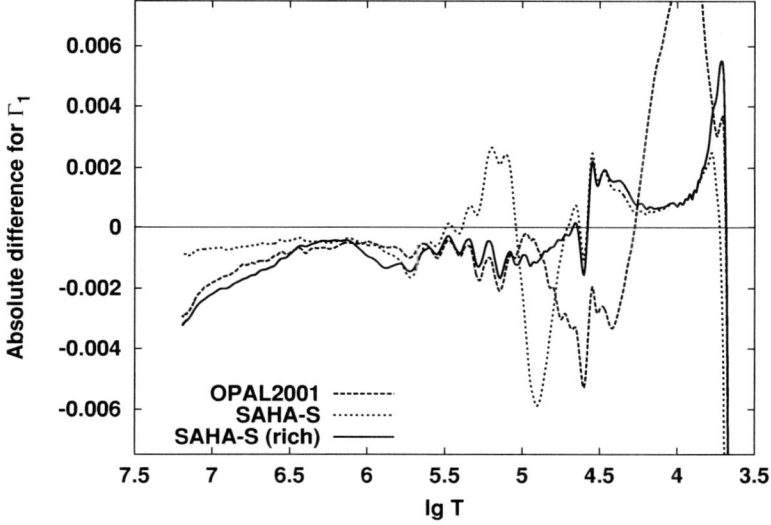

FIGURE 2. Equation of state comparison. Adiabatic exponent Γ_1

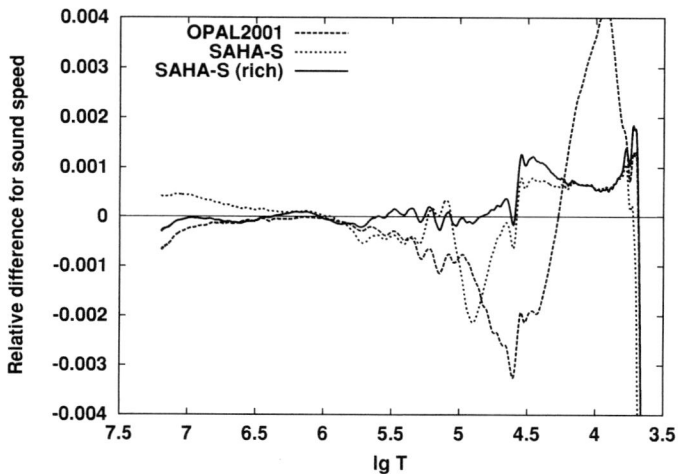

FIGURE 3. Equation of state comparison. Sound speed

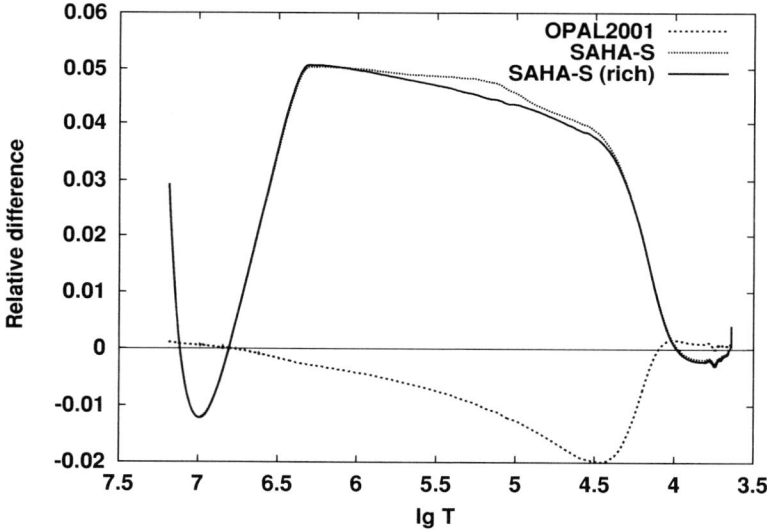

FIGURE 4. Model comparison. Pressure

Let's select two distinct areas in the star interior for comparison of thermodynamical and structural values: lower (adiabatic) part of the convection zone and radiative core (everything below convection zone).

a) From $\Gamma_1 - \Gamma_{1OPAL96}$ EOS comparison (Figure 2) one can find that OPAL2001 is close to SAHA-S (rich); these two EOS are much closer to each other in terms of Γ_1

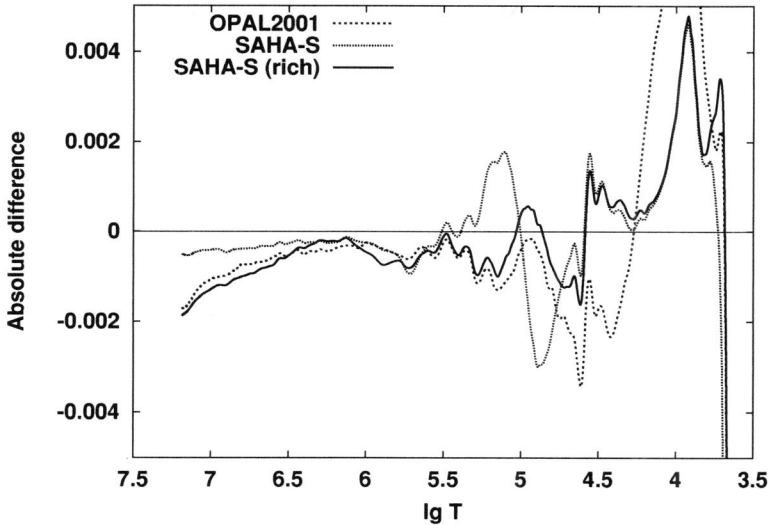

FIGURE 5. Model comparison. Adiabatic exponent Γ_1

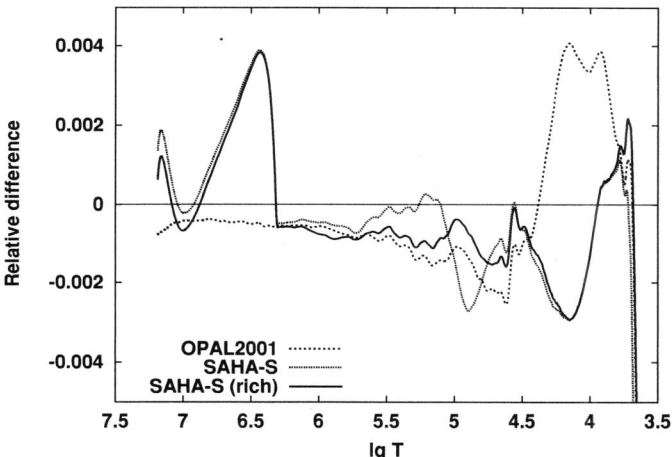

FIGURE 6. Model comparison. Sound speed

than pair of SAHA-S and SAHA-S (rich) or even pair of OPAL96 and OPAL2001. This is true both in the core and lower part of convection zone. At region of partial ionization ($\log_{10} T < 4.5$) SAHA-S is close to SAHA-S (rich). In this part of the Sun OPAL2001 also shows surprising difference from other EOS including OPAL96 and SAHA-S family.

b) Difference of Γ_1 in models (Figure 5) exhibits similar behaviour to the picture described in a). OPAL2001 and SAHA-S (rich) are rather close in the core and lower part of the convection zone. The thermodynamical difference is about 1.5 times larger than difference in model profiles. Closeness of SAHA-S (rich) and OPAL2001 below $T = 10^5$ K is most clearly seen in thermodynamic comparison and also found in model comparison. In helium ionization zone (peaks near $\log_{10} T \approx 4.7$) model profiles have similar shape but different amplitude in OPAL2001 and SAHA-S (rich).

c) Sound speed plots show vastly different picture reflecting structure differences between models. The most profound feature on Figure 6 is a large peak just below the convection zone. This is in accordance with Table 2, which confirms identical convection zone depths between OPAL96–OPAL2001 models and SAHA-S–SAHA-S (rich) models, whereas difference in $r_{cz}/R_\odot$ between EOS groups reaches 0.5% $R_\odot$. The reason of this difference is discussed below.

It is also worth noting that sound speed c in the lower part of the convection zone is smaller in all models than in reference model (OPAL96). Similarly to the Γ_1 comparison OPAL2001 model and SAHA-S (rich) model again have close sound speed profiles, but this feature in c is less explicit than in Γ_1 comparison.

Thermodynamical comparisons of sound speed differences (3) largely reminds of Γ_1 analysis. Except SAHA-S all equations of state show similar behaviour in the radiative core in the thermodynamical comparison. Lower part of the convection zone shows surprising closeness of SAHA-S (rich) to OPAL96. We don't have further comments for this fact at the moment but have to note that difference between SAHA-S and SAHA-S (rich) (i.e. effect of different Z mixture) is larger than (surprisingly small) difference between SAHA-S (rich) and OPAL96.

d) Let's return to the convection zone depth in the models with different equations of state. Comparison of global model parameters indicates that helium abundance variations (0.001 in mass fraction, Table 2) due to EOS change are not significant at our level of accuracy. Given identical opacity tables in all models we can conclude than the difference in convection zone depth is caused only by EOS change itself. Specifically the adiabatic part of the convection zone may change either 1) due to variation of entropy in different models or 2) due to change of shape of the adiabate which is determined by EOS. The first reason can be verified by checking pressure comparison plot (Figure 4). The pressure is compared here at given points on radius $r/R_\odot$ but additional study confirmed that comparison at fixed temperature T does not change general behaviour and conclusions. We can draw the following conclusions from these plots:

1. The difference in convection zone depths between models is caused by different entropy of the adiabatic part of the convection zone. SAHA-S EOS models give higher pressure than in OPAL models, so specific entropy is smaller in models with SAHA-S and SAHA-S (rich).
2. SAHA-S, SAHA-S (rich) and OPAL2001 have similar shapes of adiabates; OPAL96's adiabate has slightly different shape. This conclusion can be derived from Γ_1 comparison (see a) and b)), but direct comparison of Figures 4 and 5 cannot be performed.
3. The same convection zone depth in OPAL96 and OPAL2001 models is a coincidence. Despite the large difference in pressure in the upper layers of the convection

zone this difference is going down towards the bottom (Γ_1 in OPAL2001 is smaller) and vanishes near the bottom of the convection zone. Given that convection zone depth is determined by P and T when opacities and chemical composition are fixed, it is clear that convection zone depth must be identical in these models. On the contrary SAHA-S models have larger pressure in outer layers which further increases downwards which leads to significant change of convection zone depth relatively to OPAL96 model.

General preliminary conclusion here is that OPAL2001 and both SAHA-S and SAHA-S (rich) EOS have more similarity in adiabatic part of the convection zone than OPAL96 model.

e) Analysis of thermodynamical pressure comparison (Figure 1) shows that SAHA-S has an excess in pressure in all parts of the star comparing to OPAL96. SAHA-S also demonstrates that Z mixture has significant effect on the pressure. OPAL2001 has a pressure deficit in the ionization region (probably due to ionization at higher temperatures) and smaller pressure in the center respectively to SAHA-S. It must be noted that pressure differences in the center are very different from adiabatic thermodynamical values behaviour.

CONCLUSIONS

Comparison of the same values (e.g., pressure) in purely thermodynamical sense and in calibrated models shows that these two ways of analysis give results which are far from identical. Thermodynamical comparisons (at fixed temperature, density and chemical composition) reveal features of the equation of state itself. Model comparisons (at fixed points by radius) are greatly affected by model structure changes; they reflect many EOS features mixed via model calbration to modern values of radius and luminosity. These comparisons are harder to analyze at physical level but they show real model (i.e. potentially observable) response to the EOS changes. We find that thermodynamical and model comparisons can differ very significantly (50% and more for Γ_1).

Two realizations of new equation of state, SAHA-S, with different metal compositions appear to be close to OPAL2001. Original version of OPAL (OPAL96) was found to be less similar to SAHA-S and OPAL2001 results. This is unclear because the physical differences between OPAL2001 and OPAL96 do not seem to be large.

The big difference in pressure (up to 0.15%) in the central area of the Sun between OPAL and SAHA-S causes significant structure differences (convection zone depth differs by $0.005\ R_\odot$).

SAHA-S and SAHA-S (rich) EOS tables differ by chemical composition, namely Fe and Si are combined with Ne in SAHA-S (see Table 1); the total mass fraction of heavy elements remains the same. This change in composition might seem to look minor, but calculations show that it may well cause differences of the same order as differences between equations of state themselves.

Concluding on the attempts to estimate quality of the newly developed EOS, SAHA-S, one have to say that the differences between plasma EOS (SAHA-S and OPAL) do exist. Nevertheless given small values of these differences it is clear that agreement between SAHA-S and OPAL is impressive. Taking into account that these equations of state use different approach to plasma description it could be said that both EOS do not contain significant errors in underlying physics at $(3-5)\cdot 10^{-3}$ level (in adiabatic exponent). Heavy element mixture modifications affect EOS very noticeably at this level of accuracy.

ACKNOWLEDGMENTS

The authors are grateful to S.V. Vorontsov, W. Däppen and V.E. Fortov for numerous helpful discussions.

REFERENCES

1. Starostin, A., Roerich, V., and More, R., *Contrib. Plasma Phys.*, **43**, 369 (2003).
2. Vorontsov, S., "Helioseismic structural inversion with SOHO MIDI data," in *From Solar Min to Max: Half a Solar Cycle with SOHO*, edited by A. Wilson, Proc. SOHO 11 Symposium, ESA SP-508, Noordwijk, The Netherlands, 2002, pp. 107–110.
3. Grevesse, N., and Noels, A., *Origin and evolution of the elements*, Cambridge University Press, Cambridge, 1993, p. 15.
4. Iglesias, C., and Rogers, F., *Astrophys. J.*, **464**, 943 (1996).
5. Alexander, D., and Ferguson, J. W., *Astrophys. J.*, **437**, 879 (1994).
6. Angulo, C., Arnould, M., Rayet, M., and *et al.*, *Nuclear Physics A*, **656**, 3–183 (1999), URL http://pntpm.ulb.ac.be/Nacre/nacre.htm.
7. Rogers, F., Swenson, F., and Iglesias, C., *Astrophys J.*, **456**, 902 (1996).
8. Rogers, F., and Nayfonov, A., *Astrophys. J.*, **576**, 1064 (2002).
9. Gryaznov, V., Ayukov, S., Baturin, V., Iosilevskiy, I., Starostin, A., and Fortov, V., "SAHA-S model: Equation of state and thermodynamic functions of solar plasma," in *Equation-of-State and Phase-Transition Issues in Models of Ordinary Astrophysical Matter*, edited by V. Celebonovic, W. Däppen, and D. Gough, these proceedings, 2004.
10. Ayukov, S., Baturin, V., Gryaznov, V., Iosilevskiy, I., Starostin, A., and Fortov, V., *JETP Letters*, **80**, 141–144 (2004).
11. Christensen-Dalsgaard, J., Däppen, W., Ajukov, S., and et al., *Science*, **272**, 1286–1292 (1996).
12. Rogers, F., and Iglesias, C., *Astrophys. J.*, **401**, 361 (1992).
13. Christensen-Dalsgaard, J., Gough, D., and Thompson, M., *Astrophys. J.*, **378**, 413–437 (1991).

The effect of using different EOS in modelling the α *Centauri* binary system

A. Miglio

Institut d'Astrophysique et Géophysique de l'Université de Liège, Belgium

Abstract. In this preliminary study we investigate the effects of using different equations of state (CEFF and OPAL) in the calibration of the binary system α *Centauri*. Constraints coming from the detection of acoustic oscillations in α *Centauri* A and B are included in the modelling.

INTRODUCTION

The visual binary system α *Centauri* represents a promising target to test our understanding of stellar structure and evolution due to its numerous and stringent observational constraints, including the recent detection of acoustic oscillations frequencies in both components of the system [4], [6].

MODELLING α CENTAURI

The calibration of the system consists in defining a goodness-of-fit measurement (χ^2), that includes both seismic and non-seismic constraints weighted with their observational uncertainties, and then minimizing the χ^2 using a Levenberg-Marquardt optimization procedure. Models for components A and B, computed with the same initial chemical composition, are fitted to their observational constraints at the same age. We take as observational constraints the small ($\delta\nu$) and large ($\Delta\nu$) frequency separations (see e.g. [8]) derived from the observed oscillation frequencies by [4] and [6] as well as the effective temperature, luminosity and metallicity (Table 1). A more detailed description of the calibration procedure and of the choice of observational constraints will be reported in [16].

We take as free parameters the age, the initial chemical composition (X,Z) and the mixing length parameter for each component: α_A and α_B. The masses are fixed and taken equal to the values given by [18] (M_A= 1.105 $M_\odot$ and M_B=0.934 $M_\odot$). We have also allowed a variation of the masses within their error bars [16], but its effect is not relevant while studying the effect of different EOS on the calibration of the system.

The stellar models are computed using CLES (Code Liégeois d'Evolution Stellaire). The opacity tables are those of OPAL96 [13] complemented at $T < 6000$ with Alexander

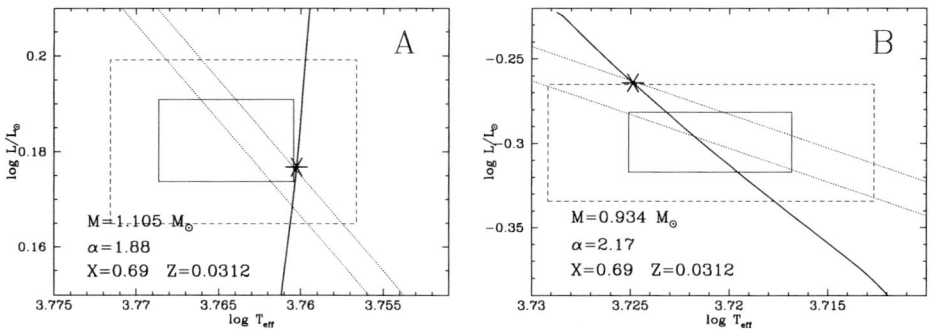

FIGURE 1. Position of models A0 and B0 in the HR diagram with 1σ and 2σ error boxes in luminosity, effective temperature and radius (2σ, assuming the values determined by [14]).

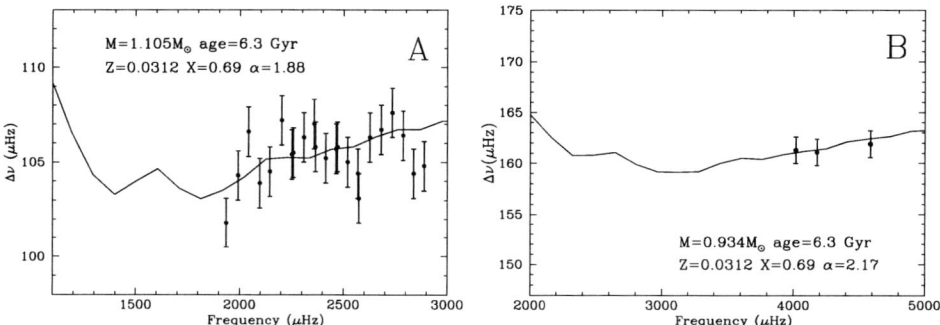

FIGURE 2. Comparison between observed and predicted (model A0 and B0) large frequency separation $\Delta\nu$ for α Cen A and B.

and Ferguson opacities [1]. The solar Z-distribution from [12] is adopted in the opacities and in the equation of state. Nuclear reaction rates are taken from [7] and the screening factor from [20]. Convective zones are treated with the classical mixing length theory [3] with the formulation by [10]; atmospheric boundary conditions [15] are applied at $T = T_{\text{eff}}$. The code includes microscopic diffusion of H,He and Z using the subroutine by [21]. The equation of state can be chosen among CEFF [9] and OPAL [19].

Fig. 1 shows the HR diagram location of the α Cen A and B models that best fit the classical (T_{eff}, L) and seismic observables (hereafter A0 and B0, see Table 2); in Figure 2 and 3 we plot the corresponding large and small separations.

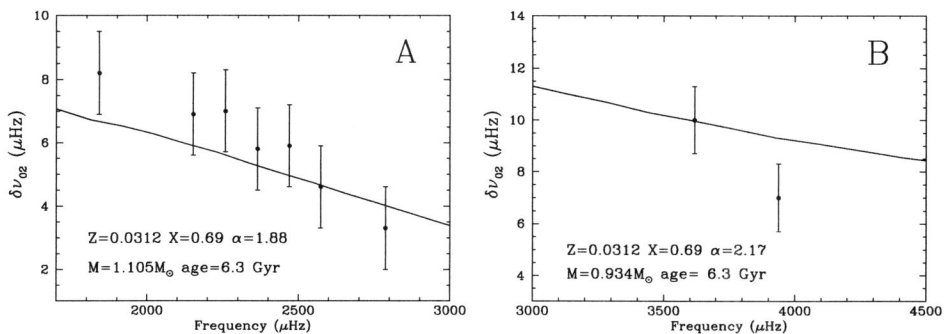

FIGURE 3. Observed versus predicted small frequency separation (δv).

TABLE 1. Non-asteroseismic constraints assumed in the calibration

	A	B	Ref
T_{eff} (K)	5810 ± 50	5260 ± 50	[17]
$L/L_\odot$	1.522±0.030	0.503±0.020	[11]
$(Z/X)_S$	0.039±0.006	0.039±0.006	[22]

INTRINSIC DIFFERENCE

In order to analyze the differences between CEFF and OPAL equations of state, we take the internal structure ($T - \rho$) of the best models (A0 and B0) calibrated using OPAL and we estimate c^2 and Γ_1 using CEFF. In Fig. 4 we show the internal structure of the models A0 and B0 as well as the structure of a solar model for comparison.

The differences between sound speed and first adiabatic exponent, due solely to the use of a different EOS, are shown in Fig. 5 and 6. These differences are of the same order of those predicted in solar models [2], they appear to be larger in the lower-mass model B0 than in model A0 and are mainly located in the hydrogen and helium ionization regions. Such a small difference in the internal structure of a model propagates in a variation of the observables of each model (e.g. effective temperature, luminosity) that could be easily compensated by a re-adjustment of the free parameters in our modelling. Slightly changing the initial chemical composition, the mixing length parameter or the age, would easily let the models computed with different equations of state satisfy the observational constraints.

TABLE 2. Model parameters for reference models A0 and B0.

Model	$M/M_\odot$	α	Age (Gyr)	X	Z
A0	1.105	1.88	6.3	0.690	0.0312
B0	0.934	2.17	6.3	0.690	0.0312

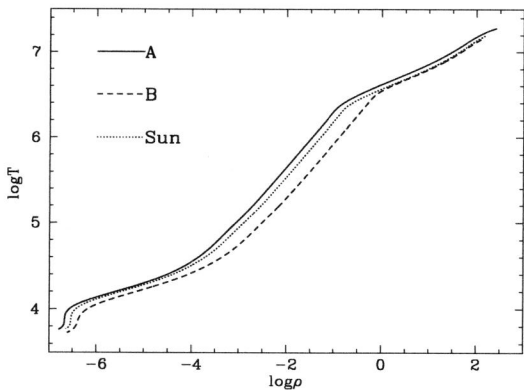

FIGURE 4. Internal temperature-density profile of reference models A0 and B0 compared to the profile of a solar model (dotted line).

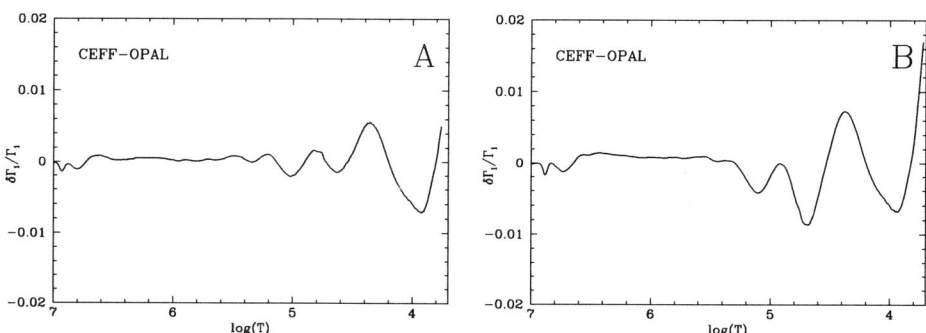

FIGURE 5. Intrinsic difference in Γ_1 in model A0 (left panel) and B0 (right panel).

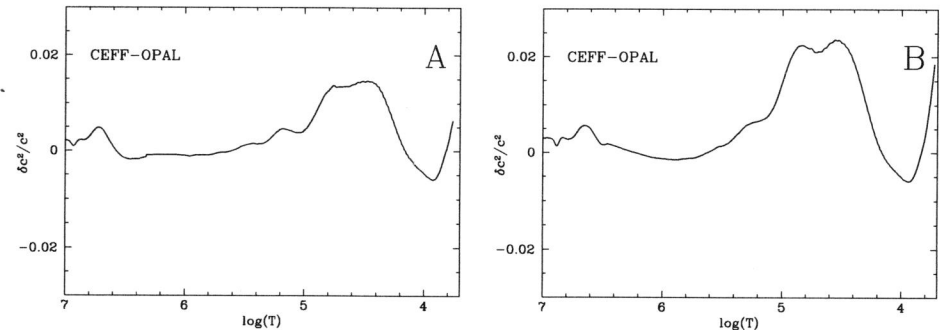

FIGURE 6. Intrinsic difference in squared adiabatic sound speed c^2.

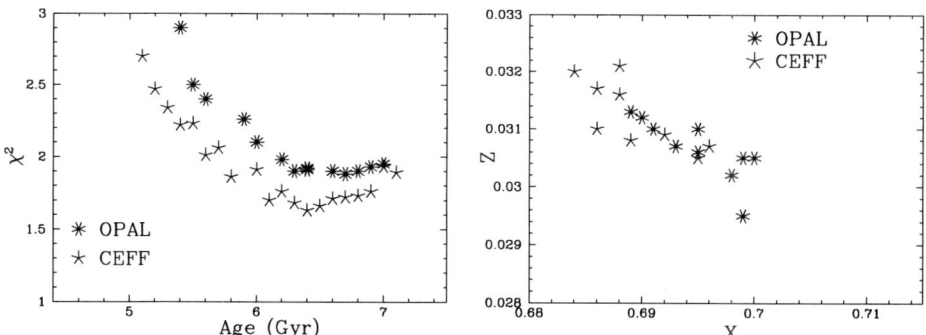

FIGURE 7. χ^2 of the best model as a function of age. Initial chemical composition of models calibrated with CEFF and OPAL that have a similar χ^2.

CALIBRATION USING OPAL AND CEFF

Both CEFF and OPAL lead to a quantitatively (χ^2) similar calibration, nonetheless the difference between model parameters (in particular X,Z and Age) corresponding to good-fit models computed with CEFF and OPAL could represent a useful estimate of systematic uncertainties in the parameters due to the use of a different EOS in the modelling. For this purpose we have run our calibration algorithm with both EOS; the results are shown in Fig. 7. The initial helium content and metallicity of models that have a similar fit to the observational constraints slightly differ (right panel) and no significant difference in the age estimated using CEFF and OPAL calibrations is found (left panel).

CONCLUSIONS

In agreement with previous works (e.g. [5]) we find that present day accuracy of observed p-mode oscillations in solar-type stars is not sufficient to isolate and detect any "equation of state effects" in the calibration of the binary system α Centauri. We show nonetheless that the comparison between calibrations performed using different equations of state could provide an additional source of systematic uncertainty on the model parameters.

ACKNOWLEDGMENTS

The author acknowledges financial support from the Prodex-ESA Contract 15448/01/NL/Sfe(IC) and is thankful to J. Montalbán, A. Noels and R. Scuflaire for very useful discussions and help in the redaction of the paper. Thanks are also due to M.P. Di Mauro and to the organizers of the meeting.

REFERENCES

1. Alexander, D. R., Ferguson, J. W. 1994 ApJ 437, 879
2. Basu S. and Christensen-Dalsgaard, J. 1997 A&A 322, L5
3. Böhm-Vitense, E. 1958, Z. Astrophys., 46, 108
4. Bouchy, F., Carrier, F. 2002 A&A 390, 205
5. Brown, T. M., Christensen-Dalsgaard, J., Weibel-Mihalas et al. 1994 ApJ 427, 1013
6. Carrier, F. and Bourban, G. 2003 A&A 406L, 23
7. Caughlan, G. R., Fowler, W. A. 1988, *Atomic data and nuclear data tables*, 40, 283
8. Christensen-Dalsgaard, J. 1988, in *Advances in Helio- and Asteroseismology*, eds. J. Christensen-Dalsgaard, & S. Frandsen (Dordrecht: Reidel), IAU Symp., 123, 295
9. Christensen-Dalsgaard, J. and Däppen W. 1992 Astronomy and Astrophysics Review 4, 267
10. Cox, J. P., & Giuli, R. T. 1968, in *Principles of Stellar Structure*, eds. Gordon and Breach, Chap. 14
11. Eggenberger, P., Charbonnel, C., Talon, S. et al. 2004 A&A 417, 235
12. Grevesse N., Noels, A. 1993 *La formation des éléments chimiques*, eds. B. Hauck, S. Paltani, D. Raboud, AVCP, 205
13. Iglesias, C.A., Rogers, F.J. 1996 ApJ 464, 943
14. Kervella, P., Thévenin, F., Ségransan et al. 2003 A&A 404, 1087
15. Kurucz, R. L., 1998, http://kurucz.harvard.edu/grids.html
16. Miglio et al., in preparation
17. Neuforge-Verheecke, C. and Magain, P. 1997 A&A 328, 261
18. Pourbaix, D., Nidever, D., McCarthy, C. et al. 2002 A&A 386, 280
19. Rogers, F. J., Nayfonov, A. 2002 ApJ 576, 1064
20. Salpeter, E. E. 1954 Austral. J. Phys., 7, 373
21. Thoul, A., Bahcall, J. N., Loeb, A. 1994 ApJ 421, 828
22. Thoul, A., Scuflaire, R., Noels, A. et al. 2003 A&A 402, 293

Asteroseismic helium abundance determination

G. Houdek

Institute of Astronomy, University of Cambridge, Cambridge CB30HA, UK

Abstract. We investigate the influence of rapid variations in the stratification of solar-type stars on the low-degree acoustic oscillation frequencies. The signature in the oscillation frequencies that is produced by these rapid variations is approximated by periodic functions with periods that are twice the acoustic depths of the centres of the rapid variations. We show how an approximation to these variations leads to a simple representation of the oscillatory contributions to the second frequency differences, which can be used as a diagnostic of the gross structure of the star, such as the helium abundance and the extent of the outer convection zone.

INTRODUCTION

Abrupt variation in the stratification of a star (relative to the scale of the inverse radial wavenumber), such as that produced by the ionization of helium or by the sharp transition from radiative to convective heat transport at the base of the convection zone, induces small oscillatory components (with respect to frequency) in the cyclic eigenfrequencies $v_{n,l}$ of the oscillation. One might therefore wonder whether the variation of the sound speed induced by helium ionization enables one to determine the helium abundance from the low-degree eigenfrequencies. A convenient and easily evaluated measure of the oscillatory component is the second difference

$$\Delta_2 v_{n,l} \equiv v_{n-1,l} - 2v_{n,l} + v_{n+1,l}, \tag{1}$$

with respect to radial order n for modes with the same degree l [2]. Any localized region of rapid variation of either the sound speed c or a spatial derivative of it, which here we call an acoustic glitch, induces an oscillatory component in $\Delta_2 v$ with a 'cyclic frequency' approximately equal to twice the acoustic depth

$$\tau = \int_{r_{\text{glitch}}}^{R} c^{-1} \, dr \tag{2}$$

of the glitch, where r is a radial coordinate, and R is the stellar radius. The amplitude of the oscillatory component depends on the amplitude of the glitch, and decays with v once the inverse radial wavenumber of the mode becomes comparable with or greater than the radial extent of the glitch. By calibrating a theoretical representation of the effect of glitches against the observations one can learn, in principle, about the characteristics of those glitches. Because only low-degree modes are used, the l-dependence can safely be ignored, even down to the base of the convection zone.

THE VARIATIONAL PRINCIPLE

We seek a continuous function $y(x)$, satisfying the boundary conditions $y(a) = y_a$ and $y(b) = y_b$, that minimizes the expression

$$\mathcal{H} = \int_a^b F[x, y(x), y'(x)]\,dx = \mathcal{H}[y(x)] \to \text{minimum}, \tag{3}$$

where here a prime (') denotes a derivative with respect to x. The necessary condition for the minimum is the Euler-Lagrange equation [e.g., 1]

$$\frac{\partial F}{\partial y} - \frac{d}{dx}\frac{\partial F}{\partial y'} = 0. \tag{4}$$

If the problem has also to obey the following constraint (e.g., isoperimetric problems)

$$I = \int_a^b G[x, y(x), y'(x)]\,dx = \text{constant}, \tag{5}$$

the necessary condition to minimize $\mathcal{H}[y(x)]$ is then

$$\frac{\partial h}{\partial y} - \frac{d}{dx}\frac{\partial h}{\partial y'} = 0, \qquad h = F + \Lambda G, \tag{6}$$

where Λ is a Lagrangian multiplier.

Sturm-Liouville equations as Euler-Lagrange equations for an isoperimetric problem

We consider two problems. The first problem is to find the solutions to the Sturm-Liouville equations:

$$L[y] + \lambda \rho(x) y(x) = 0, \tag{7}$$

$$L = \frac{d}{dx}\left[p(x)\frac{d}{dx}\cdot\right] - q(x), \tag{8}$$

$$y(a) = 0, \qquad y(b) = 0, \tag{9}$$

where λ is an eigenvalue and the functions $p(x)$, $p'(x)$, $q(x)$ and $\rho(x)$ are continuous, and $p(x) > 0$ and $r(x) > 0$ on the interval $a \le x \le b$.
The second problem is to minimize the isoperimetric problem:

$$\mathcal{H}[y] = \int_a^b (py'^2 + qy^2)\,dx \to \text{minimum}, \tag{10}$$

$$I[y] = \int_a^b \rho y^2\,dx = \text{constant}. \tag{11}$$

The two problems are identical, i.e., both have the same solution, if and only if the Lagrangian multiplier

$$\Lambda = \lambda. \tag{12}$$

It can also be shown that the solution of the Sturm-Liouville equations (7) is the solution to the isoperimetric problem

$$\mathscr{R}[y] = \frac{\mathscr{K}[y]}{I[y]} \rightarrow \text{minimum (stationary)}, \tag{13}$$

if $\lambda = \lambda_0$, in which

$$\lambda_0 = \min \mathscr{R}[y(x)] = \frac{\int_a^b (py'^2 + qy^2)\,dx}{\int_a^b py^2\,dx} = \frac{\mathscr{K}}{I} \tag{14}$$

is the lowest eigenvalue of the Sturm-Liouville problem (7). Equation (14) is a variational principle for the eigenvalue λ.

Perturbation theory applied to eigenvalue problems

If we expand equation (7) to second order and take into account that the linear operator (8) is self-adjoint (hermitian), it can be demonstrated that a 'trial function' $y(x)$ for (7) and consequently also for (14), which is good to 'first order', yields an approximate eigenvalue λ, which is good to 'second order'.

Let $L = L^0 + Q$, where L^0 is the unperturbed (zero-order) linear operator and Q is small, and expand the eigenvectors y_j and eigenvalues λ_j, $j = 1, 2, 3, \ldots$, in powers of Q [e.g. 3, Chapter 31.9]:

$$\lambda_j = \lambda_j^0 + \lambda_j^{(1)} + \lambda_j^{(2)} \ldots, \tag{15}$$

$$y_j = y_j^0 + \sum_i a_{ij}^{(1)} y_i^0 + \sum_i a_{ij}^{(2)} y_i^0 + \ldots, \tag{16}$$

where we assumed that the unperturbed eigenvectors y_i^0 form a complete orthonormal set and where the coefficients $a_{ij}^{(k)}$ are in general complex. Comparing the coefficients of the first-order terms leads to the important solution

$$\lambda_j^{(1)} = \int_a^b y_j^0 Q[y_j^0]\,dx, \tag{17}$$

i.e., the $\lambda_j^{(1)}$ depend only on the unperturbed (zero-order) eigenvectors y_j^0. If we assume that the eigenvectors y_j are not normalized, equation (17) can be written in view of equation (14) as

$$\lambda_j^{(1)} = \frac{\int_a^b y_j^0 Q[y_j^0]\,dx}{\int_a^b y_j^0 y_j^0\,dx} := \delta\omega^2, \tag{18}$$

where $\delta\omega^2$ is the change (perturbation) in the squared angular eigenfrequency ω^2, and we further assumed that the weight function $\rho(x) = 1$ for all points in $a \leq x \leq b$.

Variational principle for nonrotating stars

The linearized conservation equations for adiabatic, nonradial stellar oscillations can be written as (see Christensen-Dalsgaard, these proceedings)

$$\rho' + \nabla \cdot (\rho_0 \boldsymbol{\xi}) = 0, \tag{19}$$

$$\omega^2 \rho_0 \boldsymbol{\xi} - \nabla p' + \rho_0 \nabla \Phi' + \rho' \nabla \Phi_0 = 0, \tag{20}$$

$$\nabla^2 \Phi' = -4\pi G \rho', \tag{21}$$

$$p' - c_0^2 \rho' = (c_0^2 \nabla \rho_0 - \nabla p_0) \cdot \boldsymbol{\xi}, \tag{22}$$

where ρ is density, p is pressure, Φ is the gravitational potential, V is the volume and $\boldsymbol{\xi}$ is the displacement eigenfunction; a subscript '0' indicates an unperturbed quantity and a prime (') indicates here an Eulerian perturbation. Eliminating the density perturbation ρ' with equation (19) and assuming the Cowling approximation, i.e., $\Phi' = 0$, the pulsation equations are reduced to a Sturm-Liouville-type differential equation:

$$\omega^2 \boldsymbol{\xi} = \mathscr{L}(\boldsymbol{\xi}), \tag{23}$$

$$\mathscr{L}(\boldsymbol{\xi}) = -\rho_0^{-1} \nabla (c_0^2 \rho_0 \nabla \cdot \boldsymbol{\xi} + \boldsymbol{\xi} \cdot \nabla p_0) + \rho_0^{-2} \nabla \cdot (\rho_0 \boldsymbol{\xi} \nabla p_0). \tag{24}$$

The operator $\rho_0 \mathscr{L}$ is hermitian and if $\nabla p_0 = 0$ at the stellar surface, the angular eigenfrequency ω obeys a variational principle [e.g., 4]

$$\omega^2 = \frac{\int_V \rho_0 \boldsymbol{\xi}^* \cdot \mathscr{L}(\boldsymbol{\xi}) \, dV}{\int_V \rho_0 \boldsymbol{\xi}^* \cdot \boldsymbol{\xi} \, dV} = \frac{\mathscr{K}}{I}, \tag{25}$$

in which

$$\mathscr{K} \simeq \int_V (\mathscr{K}_1 + \mathscr{K}_2 + \mathscr{K}_3 + \mathscr{B}) \, dV. \tag{26}$$

The $\mathscr{K}_i$ and $\mathscr{B}$ are [cf. 5]

$$\mathscr{K}_1 = \gamma p (\mathrm{div}\,\boldsymbol{\xi})^2, \tag{27}$$

$$\mathscr{K}_2 = 2(\boldsymbol{\xi} \cdot \nabla p)(\mathrm{div}\,\boldsymbol{\xi}), \tag{28}$$

$$\mathscr{K}_3 = (\boldsymbol{\xi} \cdot \nabla p)\, \boldsymbol{\xi} \cdot \nabla \ln \rho, \tag{29}$$

$$\mathscr{B} = \text{surface term}, \tag{30}$$

where $\gamma = (\partial \ln p / \partial \ln \rho)_s$ is the first adiabatic exponent (s is entropy). The change $\delta\omega$ in the angular eigenfrequency ω due to abrupt variations in the stratification of the star can be written as (see also equation (18))

$$\delta\omega = \frac{\delta\mathscr{K} - \delta\mathscr{B} - \omega \delta I}{2I\omega}. \tag{31}$$

THE SEISMIC DIAGNOSTIC MODEL

We separate $\delta\omega$ into a smoothly varying $\delta_{\text{smooth}}\omega$ and an oscillatory $\delta_{\text{osc}}\omega$ component:

$$\delta\omega = \delta_{\text{smooth}}\omega + \delta_{\text{osc}}\omega, \tag{32}$$

and approximate, with a suitable eigenfunction normalization, the oscillatory component, associated with helium ionization, as follows:

$$\delta_{\text{osc}}\omega \simeq \delta_\gamma\omega := \frac{\delta_\gamma \mathcal{K}_1}{2\omega I}, \tag{33}$$

with

$$\delta_\gamma \mathcal{K}_1 \simeq -\frac{1}{2}\omega \int \frac{\delta\gamma}{\gamma} \cos 2(\omega\tau + \varepsilon)\,d\tau, \tag{34}$$

and

$$I = \int \rho\,\boldsymbol{\xi}\cdot\boldsymbol{\xi}\,r^2\,dr \simeq \frac{1}{\omega}\int \cos^2(\omega\tau+\varepsilon)\,d\tau \simeq \frac{1}{2\omega}T, \tag{35}$$

where T is the sound travel time from the surface to the centre of the star, and ε is a phase constant. The acoustic glitch induced by helium ionization is first represented by a Gaussian function about the acoustic depth $\tau = \tau_{\text{II}}$ of the centre of the HeII ionization region:

$$\frac{\delta\gamma}{\gamma} \simeq -\frac{1}{\sqrt{2\pi}}\frac{\Gamma_{\text{II}}}{\Delta_{\text{II}}}e^{-(\tau-\tau_{\text{II}})^2/2\Delta_{\text{II}}^2}. \tag{36}$$

Thus the effect of this glitch is to impart on the eigenfrequencies an additional, oscillatory component, $\delta_\gamma\omega$, which can be shown to be given approximately by

$$\delta_\gamma\omega \simeq A_{\text{II}}\omega\,e^{-2\Delta_{\text{II}}^2\omega^2}\cos 2(\omega\tau_{\text{II}} + \varepsilon_{\text{II}}), \tag{37}$$

with

$$A_{\text{II}} = -\frac{1}{2}\Gamma_{\text{II}}T^{-1}. \tag{38}$$

The acoustic glitch at the base of the convection zone ($r = r_c$) is essentially a discontinuity in the second derivative of density (a local mixing-length model – as we used in constructing our theoretical test models – leads to a true discontinuity) and is to impart a further oscillatory component, $\delta_c\omega$, which is given approximately by

$$\delta_c\omega \simeq A_c\omega^{-1}\cos 2(\omega\tau_c + \varepsilon_c), \tag{39}$$

with

$$A_c = -\frac{1}{8}c_c^2\tau_c T^{-1}\left[\frac{d^2\ln\rho}{dr^2}\right]_{r_c-}^{r_c+}, \tag{40}$$

where $c_c = c(r_c)$. The amplitude A_c is a measure of the discontinuity in the second derivative of ρ at the base of the convection zone. Smooth contributions to the eigenfrequencies (arising, in part, from refraction in the stellar core and from hydrogen ionization

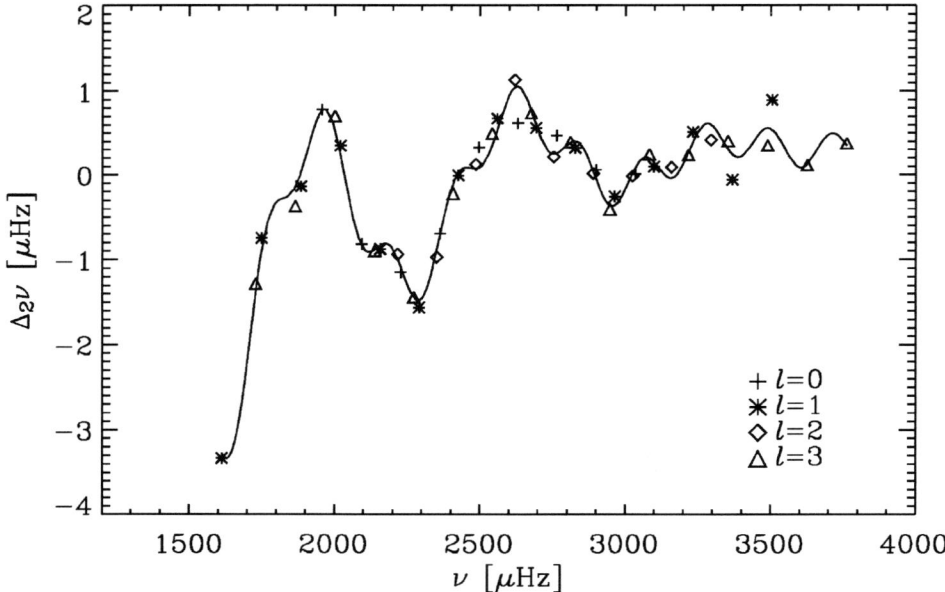

FIGURE 1. The symbols are second differences Δ_2, defined by equation (1), of low-degree solar frequencies obtained from MDI. The curve is an approximation to the second derivative with respect to order n (see equation (43), in which we have approximated $\partial \nu/\partial n$ by a constant) for $\delta \nu$ (44) whose twelve parameters have been adjusted to fit the curve to the data by least squares.

and the superadiabaticity in the upper boundary layer of the convection zone lying in the upper evanescent region of the acoustic mode) are approximated by a fourth-degree polynomial in ω^{-1}. The complete fitting function is then given by

$$\Delta_2 \nu \simeq \Delta_2 \left(\delta_\gamma \nu + \delta_c \nu \right) + \sum_{i=0}^{4} a_i/\nu^i \qquad (41)$$

$$\simeq \frac{\partial^2}{\partial n^2} \left(\delta_\gamma \nu + \delta_c \nu \right) + \sum_{i=0}^{4} a_i/\nu^i, \qquad (42)$$

where we have now converted to cyclic frequency ν, and in which we have now regarded n as a continuous variable. The total seismic diagnostic becomes:

$$\Delta_2 \nu \simeq \left(\frac{\partial \nu}{\partial n} \right)_l^2 \frac{d^2 \delta \nu_{\rm osc}}{d\nu^2} + \sum_{i=0}^{4} \frac{a_i}{\nu^i}, \qquad (43)$$

$$\delta \nu_{\rm osc} \simeq A_{\rm II} \nu \, e^{-8\pi^2 \Delta_{\rm II}^2 \nu^2} \cos(4\pi \nu \tau_{\rm II} + 2\varepsilon_{\rm II})$$
$$+ A_{\rm c} (4\pi^2 \nu)^{-1} \cos(4\pi \nu \tau_{\rm c} + 2\varepsilon_{\rm c}). \qquad (44)$$

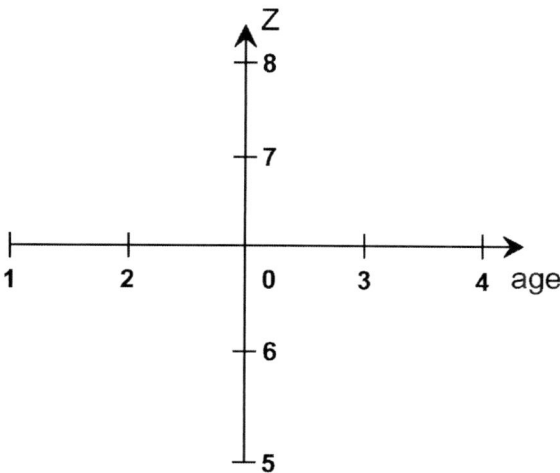

FIGURE 2. Denotation of the nine calibrated solar models which are used for testing the formulation of the second frequency differences. The 'central model' is model 0; the set of the four models (1-4) have a constant value of Z but varying age; the second set of the models (5-8) have constant age but varying Z.

In Fig. 1 second differences Δ_2, defined by equation (1), are plotted as symbols for low-degree solar oscillations, obtained from the Michelson Doppler Imager (MDI) instrument on the SOlar and Heliospheric Observatory (SOHO) satellite. The curve is the result of the seismic diagnostic (43)–(44), in which $\partial v/\partial n$ is approximated by a constant, and for which the twelve parameters A_{II}, A_c, Δ_{II}, τ_{II}, τ_c, ε_{II}, ε_c and a_i are adjusted to fit by least squares the theoretical curve to the data.

One should expect such an expression to represent the second differences of the actual frequencies more faithfully than the more arbitrary expressions that have been used before [e.g., 6].

TESTING THE FORMULATION ON THEORETICAL MODELS

Two sets of calibrated evolutionary models for the Sun are used. The models were computed by Gough & Novotny [7] and were carefully examined and adjusted to satisfy hydrostatic equilibrium to high precision. One set of models has a constant value for the heavy-element abundance Z but varying age, the other has constant age but varying Z (see Fig. 2). For all nine test models we determined the properties of the HeII bump in γ by fitting a Gaussian (plus a constant) to γ. Details of this fitting process are illustrated in Fig. 3 for the central model 0.

The three parameters, Γ, τ_γ and Δ so determined are plotted in Fig. 4 for the nine test models as a function of the helium abundance Y. Also shown in Fig. 4 (upper left panel) is the standard measure $\tilde{\chi}^2$ for the goodness of the fit.

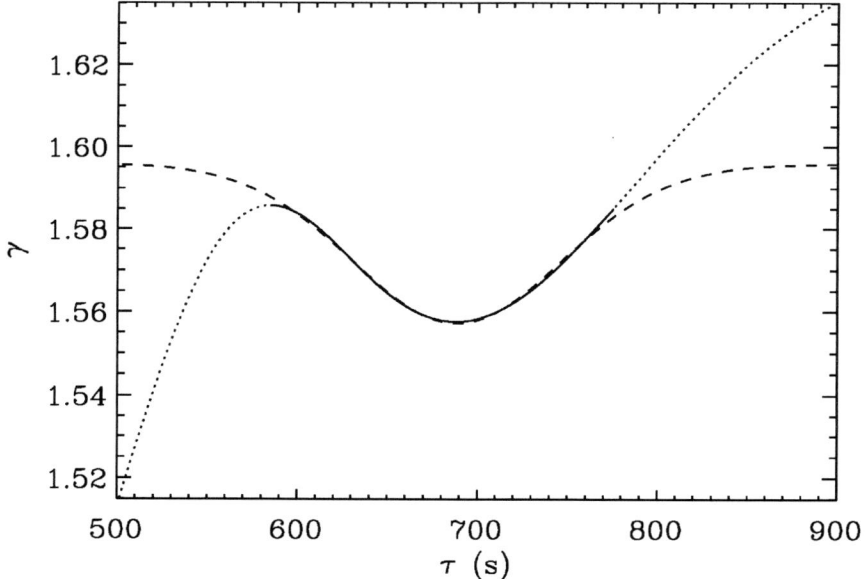

FIGURE 3. Adiabatic exponent γ (solid and dotted curve) for the central stellar model 0 as a function of the acoustic radius τ. The dashed curve is a Gaussian $\Gamma \exp[-(\tau - \tau_\gamma)^2/2\Delta^2]$ plus a constant whose parameters (Γ, τ_γ, Δ) have been adjusted to fit γ in the region in which the curve is solid.

The contribution to $\Delta_2 \nu$, associated with the acoustic glitch of the HeII ionization, may be written as

$$\Delta_2 \nu_\gamma \simeq A \cos\left[4\pi \tau_{II} \nu - \tan^{-1}\left(\frac{b}{a}\right) + 2\varepsilon_{II}\right], \tag{45}$$

where the amplitude

$$A(\nu) = A_{II}\sqrt{a^2 + b^2}\, e^{-8\pi^2 \Delta_{II}^2 \nu^2} \tag{46}$$

is evaluated at $\nu = 2\,\mathrm{mHz}$, and in which

$$a(\nu) = 8\pi(16\pi^2 \Delta_{II}^4 \nu^2 - 3\Delta_{II}^2 - \tau_{II}^2)\nu, \tag{47}$$
$$b(\nu) = 4(16\pi^2 \Delta_{II}^2 \nu^2 - 1)\tau_{II}. \tag{48}$$

Using Christensen-Dalsgaard's adiabatic pulsation code we computed low-degree eigenfrequencies for all nine models to which we calibrated the twelve parameters of the seismic diagnostic (43)–(44). The results for the calibrated parameters A, τ_{II} and Δ_{II} are shown in Fig. 5 as a function of Y of the nine test stellar models. They can be compared with the fitted HeII parameters Γ, τ_γ and Δ of the first adiabatic exponent for the nine test models in Fig. 4.

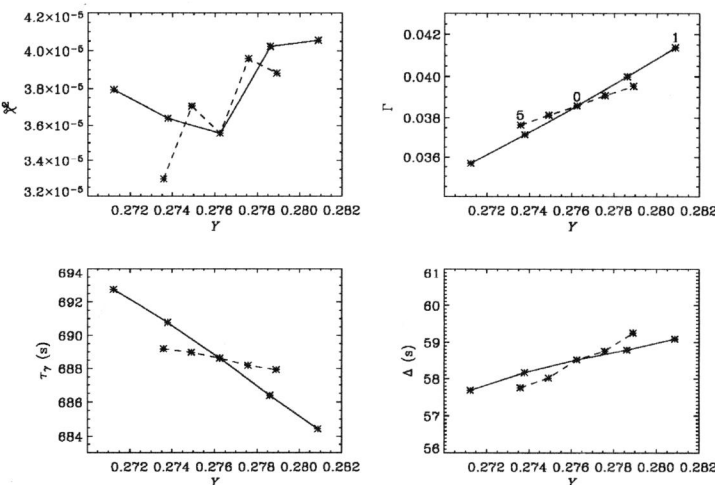

FIGURE 4. Parameters of the fitted Gaussian to the HeII bump in γ for all nine test models as a function of Y. The central model 0 and the models 1 and 5 are indicated in the upper right panel. The upper left panel is the standard measure $\tilde{\chi}^2$ for the goodness of the fit between γ and the calibrated Gaussian.

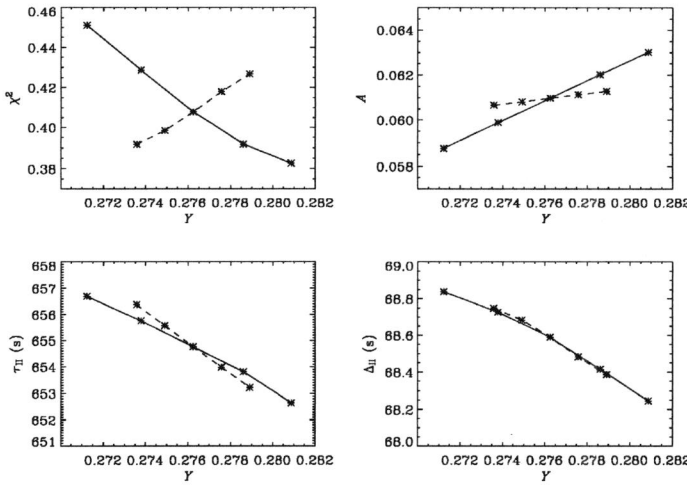

FIGURE 5. Calibrated parameters of the seismic diagnostic (43)–(44) for the nine test models. The parameters A, τ_{II} and Δ_{II} have been adjusted to fit the second differences Δ_2, defined by equation (1), of the low-degree (l=0,1,2), adiabatically computed, model frequencies. The smallest frequency value used in the least-squares fitting is $\nu \simeq 1321\,\mu\text{Hz}$. The upper left panel is the standard measure χ^2 of the goodness of the fit between Δ_2 of the modelled frequencies and the calibrated seismic diagnostic. The line styles are as in Fig. 4.

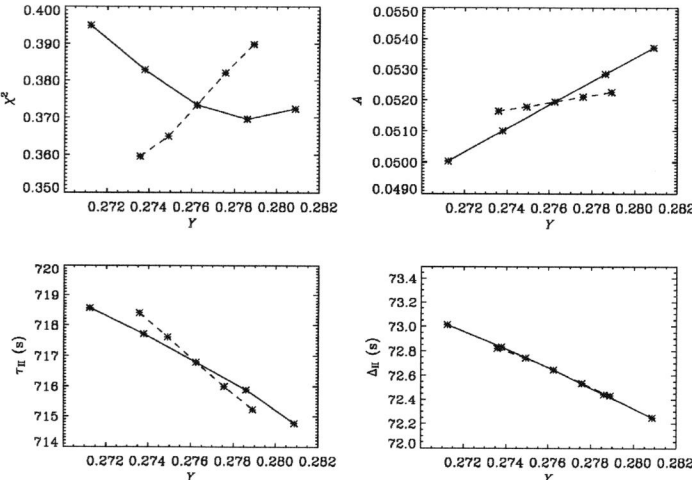

FIGURE 6. Calibrated parameters of the seismic diagnostic (56)–(57) which includes an improved treatment of the phase function ϕ (53)–(54). Results are shown for the nine test models. The parameters A, τ_{II} and Δ_{II} have been adjusted to fit the second differences Δ_2, defined by equation (1), of the low-degree ($l=0,1,2$), adiabatically computed, model frequencies. The smallest frequency value used in the least-squares fitting is $\nu \simeq 1321\,\mu$Hz. The upper left panel is the standard measure χ^2 of the goodness of the fit between Δ_2 of the modelled frequencies and the calibrated seismic diagnostic. The line styles are as in Fig. 4.

FURTHER IMPROVEMENTS

Improved treatment of the phase function ϕ

The phase function is given by [e.g., 5]

$$\phi(r) = \int_{r_t}^{r} K(r)\,dr, \tag{49}$$

where r_t is the radius at which the mode is refracted. K is the wavenumber which becomes in the limit of low-degree modes:

$$K(r) \simeq \frac{\omega}{c}\left(1 - \frac{\omega_a^2}{\omega^2}\right)^{1/2}. \tag{50}$$

We assume that the outer stellar layers can be approximated by a polytrope of index m, for which the acoustic cut-off frequency is

$$\omega_a^2 = \frac{(m+1)^2}{\tau^2}. \tag{51}$$

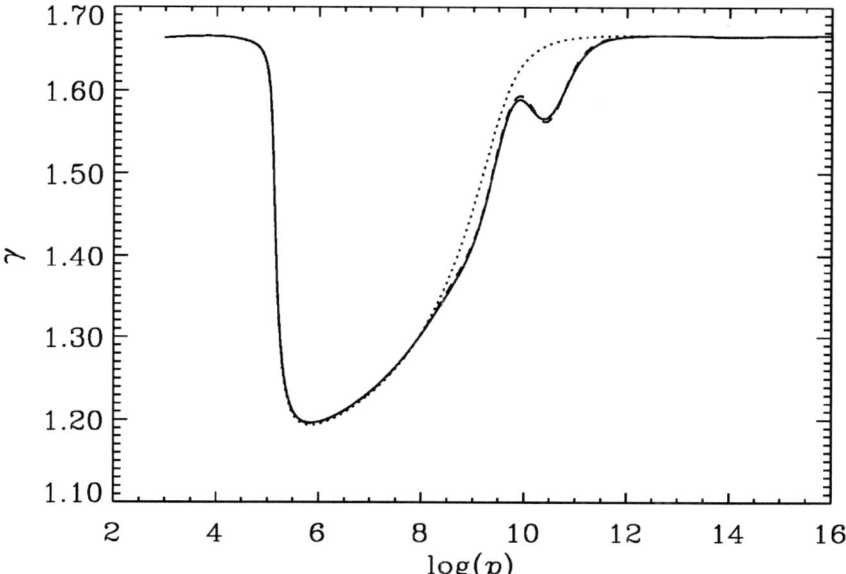

FIGURE 7. The solid curve is the adiabatic exponent $\gamma(\rho,T,Y_\odot)$ in model S [8]; the dotted curve is the adiabatic exponent $\gamma(\rho,T,0)$. The dashed curve is the first two terms of the Taylor expansion of γ about $Y = 0$ (see equation (58)).

Because the acoustic glitch associated with the helium ionization is confined to a small region in acoustic radius we can expand the phase function $\phi(\tau)$ about the centre of the acoustic glitch:

$$\phi(\tau) \simeq \phi(\tau_{II}) + \omega \left[1 - \left(\frac{m+1}{\omega\tau}\right)^2\right]^{1/2}(\tau - \tau_{II}) + O\left[(\tau - \tau_{II})^2\right]. \tag{52}$$

The zero-order term in the expansion (52) is given by

$$\phi(\tau_{II}) \simeq \omega \int_{\alpha/\omega}^{\tau_{II}} \left[1 - \left(\frac{\alpha}{\omega\tau}\right)^{1/2}\right]^{1/2} d\tau = \kappa\omega\tau_{II} + \alpha\left[\tan^{-1}\left(\frac{1}{\kappa}\frac{\alpha}{\omega\tau_{II}}\right) - \frac{\pi}{2}\right], \tag{53}$$

with

$$\alpha \equiv 1+m, \qquad \kappa \equiv \sqrt{1 - \left(\frac{\alpha}{\omega\tau_{II}}\right)^2}. \tag{54}$$

The improved function for representing the change in frequency associated with the helium ionization is then given by

$$\delta_\gamma \nu \simeq A_{II}\kappa^{-1}\nu e^{-8\kappa^2\pi^2\Delta_{II}^2\nu^2}\cos 2[\phi(\tau_{II}) + \varepsilon_{II}], \tag{55}$$

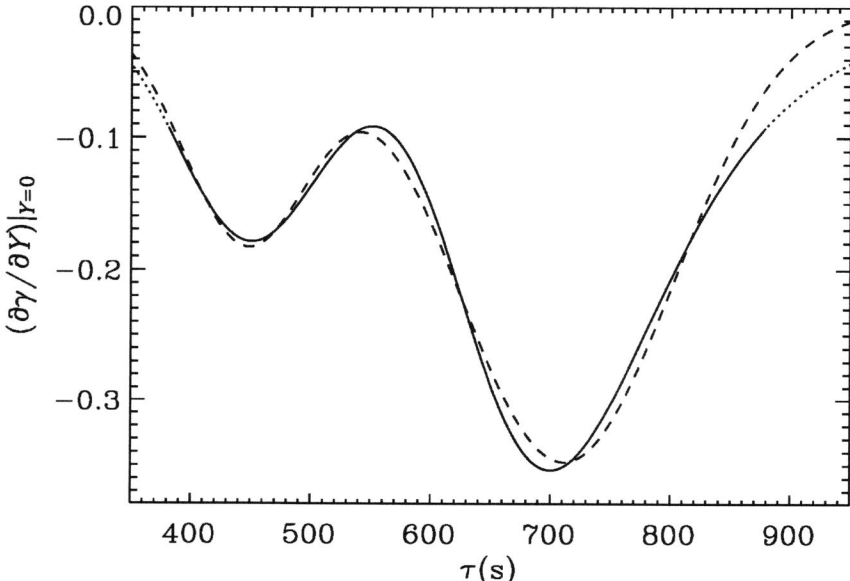

FIGURE 8. The solid and dotted curve is $(\partial \gamma/\partial Y)|_{Y=0}$ for the central model 0. The dashed curve is the sum of the two Gaussian functions used in the construction of the seismic diagnostic (62)–(63) whose parameters have been adjusted to fit $(\partial \gamma/\partial Y)|_{Y=0}$ in the region in which the curve is solid.

and the complete seismic diagnostic becomes:

$$\Delta_2 \nu \simeq \left(\frac{\partial \nu}{\partial n}\right)_l^2 \frac{d^2 \delta \nu_{\rm osc}}{d\nu^2} + \sum_{i=0}^{4} \frac{a_i}{\nu^i}, \qquad (56)$$

$$\delta \nu_{\rm osc} \simeq A_{\rm II} \kappa^{-1} \nu e^{-8\kappa^2 \pi^2 \Delta_{\rm II}^2 \nu^2} \cos 2[\phi(\tau_{\rm II}) + \varepsilon_{\rm II}]$$
$$+ A_{\rm c}(4\pi^2 \nu)^{-1} \cos(4\pi\nu\tau_{\rm c} + 2\varepsilon_{\rm c}). \qquad (57)$$

Results for the seismic diagnostic (56)–(57) are shown in Fig. 6. It shows that the seismic diagnostic (56)–(57) results in a better fit to the adiabatically computed eigenfrequencies than the seismic diagnostic (43)–(44) shown in Fig. 5, illustrated by a smaller χ^2 (upper left panel).

Including HeI ionization

Instead of approximating γ by a Gaussian we expand γ about the helium abundance $Y = 0$:

$$\gamma \simeq \gamma|_{Y=0} + (\partial \gamma/\partial Y)|_{Y=0} Y. \qquad (58)$$

The variation $\delta\gamma$ in the adiabatic exponent γ induced by helium ionization then becomes:

$$\delta\gamma \simeq (\partial\gamma/\partial Y)|_{Y=0}\, Y. \tag{59}$$

The first-order expansion (58) is illustrated in Fig. 7 for a solar model; the solid curve is the adiabatic exponent $\gamma(\rho,T,Y_\odot)$, the dotted curve is $\gamma(\rho,T,0)$ and the dashed curve shows the right-hand-side of equation (58). A plot of $(\partial\gamma/\partial Y)|_{Y=0}$ is shown in Fig. 8 for the central model 0 (solid and dotted curve), which shows the two glitches induced by HeI and HeII ionization, and which we represent by the sum of two Gaussian functions about the acoustic depths $\tau = \tau_\mathrm{I}$ (HeI) and $\tau = \tau_\mathrm{II}$ (HeII):

$$\frac{\delta\gamma}{\gamma} \simeq \frac{Y}{\sqrt{2\pi}}\left[\frac{\Gamma_\mathrm{I}}{\Delta_\mathrm{I}}e^{-(\tau-\tau_\mathrm{I})^2/2\Delta_\mathrm{I}^2} + \frac{\Gamma_\mathrm{II}}{\Delta_\mathrm{II}}e^{-(\tau-\tau_\mathrm{II})^2/2\Delta_\mathrm{II}^2}\right]. \tag{60}$$

The dashed curve in Fig. 8 is a least-squares fit of the two Gaussian functions to $(d\gamma/dY)|_{Y=0}$ of the central stellar model 0.

Following the same procedure as discussed before, the extended fitting formula, which takes into account the acoustic glitches from both the HeI and the HeII ionization becomes:

$$\delta_\gamma v \simeq \tilde{A}_\mathrm{II} v \left[\mu\beta\, e^{-8\pi^2\mu^2\Delta_\mathrm{II}^2 v^2}\cos 2(2\pi v\eta\,\tau_\mathrm{II}+\varepsilon_\mathrm{II}) + e^{-8\pi^2\Delta_\mathrm{II}^2 v^2}\cos 2(2\pi v\tau_\mathrm{II}+\varepsilon_\mathrm{II})\right], \tag{61}$$

where $\tilde{A}_\mathrm{II} = A_\mathrm{II} Y$ is now used amongst the fitting parameters instead of A_II, and $\beta = \Gamma_\mathrm{I}\Delta_\mathrm{II}/\Gamma_\mathrm{II}\Delta_\mathrm{I}$, $\eta = \tau_\mathrm{I}/\tau_\mathrm{II}$ and $\mu = \Delta_\mathrm{I}/\Delta_\mathrm{II}$ are assumed to be constants with values determined from fitting the two Gaussian functions (see equation (60)) to $(d\gamma/dY)|_{Y=0}$ for the central stellar model 0 (see Fig. 8). Results of adiabatically stratified stellar models whose radii and masses vary by a factor of about five, indicate that the constants β, η and μ vary at most by about 35%. These results support the assumption that β, η and μ can be treated as constants. The complete seismic diagnostic is then given by

$$\Delta_2 v \simeq \left(\frac{\partial v}{\partial n}\right)_1^2 \frac{d^2\delta v_\mathrm{osc}}{dv^2} + \sum_{i=0}^{4}\frac{a_i}{v^i}, \tag{62}$$

$$\delta v_\mathrm{osc} \simeq \tilde{A}_\mathrm{II} v\left[\mu\beta\, e^{-8\pi^2\mu^2\Delta_\mathrm{II}^2 v^2}\cos 2(2\pi v\eta\,\tau_\mathrm{II}+\varepsilon_\mathrm{II})+e^{-8\pi^2\Delta_\mathrm{II}^2 v^2}\cos 2(2\pi v\tau_\mathrm{II}+\varepsilon_\mathrm{II})\right]$$
$$+A_\mathrm{c}(4\pi^2 v)^{-1}\cos 2(2\pi v\tau_\mathrm{c}+\varepsilon_\mathrm{c}). \tag{63}$$

We continue to use A, defined by equations (46)–(48) (but with A_II replaced by $\tilde{A}_\mathrm{II}$), as a measure of the amplitude of the helium-induced oscillatory component. The values $\beta = 0.20$, $\eta = 0.68$ and $\mu = 0.63$ were adopted for all the fits. In Fig. 9 the calibrated values of A, τ_II and Δ_II are plotted for the nine test models using the extended seismic diagnostic (62)–(63), which includes an approximate description of the acoustic glitches of both the HeI and HeII ionizations. A comparison of χ^2 between Figures 5, 6 and 9 shows that the extended seismic diagnostic (62)–(63) results in an even better fit to the adiabatically computed eigenfrequencies than the seismic diagnostic (43)–(44), which includes a description of only the HeII ionization, and the seismic diagnostic (56)–(57) with the improved treatment of the phase function.

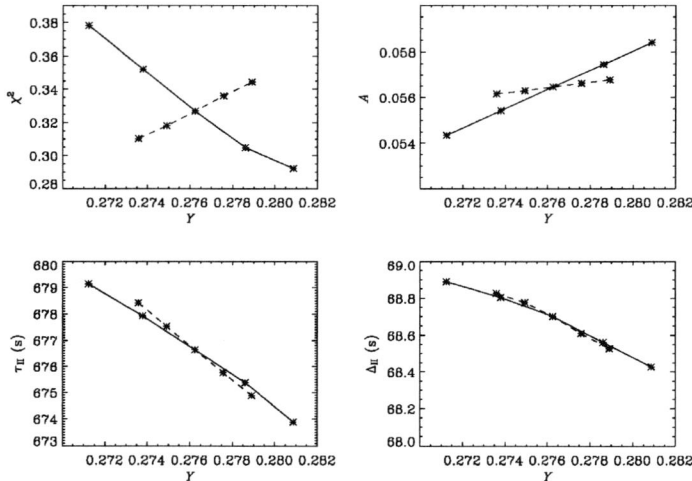

FIGURE 9. Calibrated parameters of the seismic diagnostic (62)–(63), which includes a description for both the HeI and the HeII ionization zones. Results are displayed for the nine test models. The parameters A, τ_{II} and Δ_{II} have been adjusted to fit the second differences Δ_2, defined by equation (1), of the low-degree ($l=0,1,2$), adiabatically computed, model frequencies. The smallest frequency value used in the least-squares fitting is $\nu \simeq 1321\,\mu\text{Hz}$. The upper left panel is the standard measure χ^2 of the goodness of the fit between Δ_2 of the modelled frequencies and the calibrated seismic diagnostic. The line styles are as in Fig. 4.

A least-squares fit of the seismic diagnostic (62)–(63) to Δ_2 of the adiabatically computed eigenfrequencies of the central stellar model 0 is shown in Fig. 10. It demonstrates how well the oscillatory contributions to the second frequency differences are approximated by our seismic diagnostic, using only low-degree acoustic modes.

The results of this investigation indicate that we can calibrate Y from stellar second frequency differences. However, we have yet carried out only a rough assessment of the likely precision with which the calibration can be accomplished. That assessment suggests that the calibration is superior to the more naive procedures of the kind carried out by e.g., Basu et al. [6].

SUMMARY AND CONCLUSION

- The amplitude measure of the oscillatory fitting function is approximately proportional to Y.
- With plausible data errors we expect the seismic diagnostic to be able to distinguish between stellar models with helium abundances differing by $Y < 0.005$.

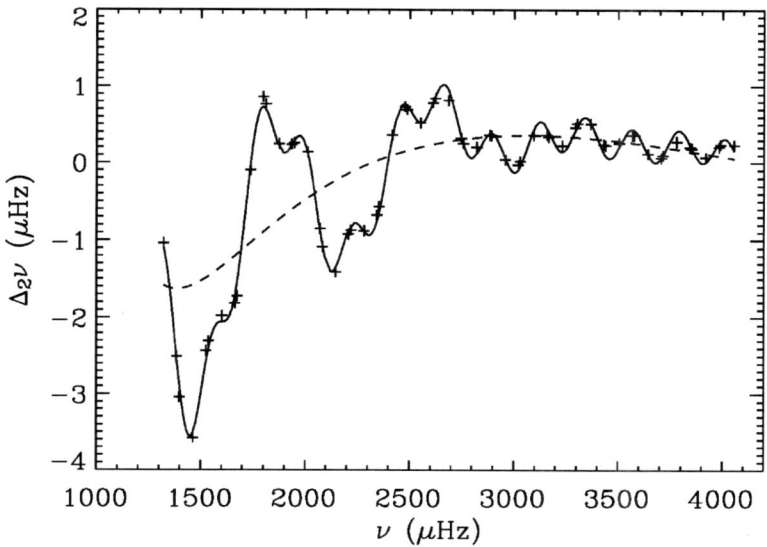

FIGURE 10. The symbols are second differences Δ_2, defined by equation (1), of low-degree ($l=0,1,2$) frequencies obtained from adiabatic pulsation calculations of the central model 0. The solid curve is an approximation to the second derivative with respect to order n (see equation (62), in which we have again approximated $\partial \nu/\partial n$ by a constant) for $\delta\nu_{osc}$ (see equation (63) which includes a description for both the HeI and HeII ionization zones) plus a fourth-degree polynomial whose twelve parameters have been adjusted to fit the theoretical (solid) curve to the data by least squares. The dashed curve represents the smooth contributions (last term in equation (62)).

ACKNOWLEDGMENTS

I am very grateful to Douglas Gough for many helpful discussions and for the support by the workshop organizers. Support by the Particle Physics and Astronomy Research Council of the United Kingdom is acknowledged.

REFERENCES

1. Sagan H., 1989, Boundary and Eigenvalue Problems in Mathematical Physics, Dover Publications, New York.
2. Gough D. O., 1990, in: Progress of Seismology of the Sun and Stars, Lecture Notes in Physics, Vol. 367, Y. Osaki, H. Shibahashi (eds.), Springer Verlag, p. 283.
3. Schiff L. I., 1949, Quantum Mechanics, McGraw Hill, New York.
4. Chandrasekhar S., 1963, ApJ 138, 896.
5. Gough D. O., 1993, in: Astrophysical fluid dynamics, Zahn J-P., Zinn-Justin J. (eds.), Amsterdam, Elsevier, p. 399.
6. Basu S., Mazumdar A., Antia H. M., Demarque P., 2004, MNRAS 350, 277.
7. Gough D. O., Novotny E., 1990, Solar Physics 128, 143.
8. Christensen-Dalsgaard J., et. al., 1996, Science 272, 1286.

Quantum statistical corrections on the enhancement factor in solar fusion reactions

A. Perez [1,2], G. Chabrier [3]

[1] *Laboratoire de Physique Théorique (UMR CNRS/ULP 7085), Université Louis Pasteur de Strasbourg, 67084 Strasbourg Cedex, France*
[2] *Department of Applied Physics, Jerusalem College of Technology, 92221 Jerusalem, Israel*
[3] *Centre de Recherche Astrophysique de Lyon (UMR CNRS 5574), Ecole Normale Superieure, 69364 Lyon Cedex 07, France*

Abstract. The enhancement factor to the thermonuclear reaction rate in the center of the Sun is calculated using the Feynman-Kac path integral of the grand-canonical partition function. Using the generalized virial expansion for a multicomponent quantum plasma, we have calculated the truncated pair distribution function for solar conditions up to the order $\rho^{5/2}$ in density. The enhancement factor is calculated for the $p-p$ channel. This result extends the well-known Salpeter calculation of the enhancement factor in the classical limit and exhibits the various classical and quantum corrections beyond the Debye-Hückel screening correction. In this issue, it comes out that quantum corrective terms related to the statistics are the leading corrections beyond the Salpeter's result. Indeed, this result can induce significant changes in the prediction of solar neutrino deficit.

INTRODUCTION.

In the core of ordinary stars like the Sun, the plasma is dense enough for non-ideal corrections to the Coulomb gas to be taken into account. These corrections include both classical Debye-Hückel screening mechanism and quantum corrections such as long ranged diffraction effect, scattering, formation of bound states, effects related to the statistics of particles and dynamical corrections. This influences naturally results on solar models and on the estimation of the thermonuclear reaction rate.

In the case of the Solar plasma, these corrections have to be precisely taken into account in order to ensure proper calculation of screening effect to the thermonuclear rate. This affects the theoretical prediction of the solar neutrino production rate and - as a consequence - the counting rate in the different experiments related to the detection of solar neutrinos. This question has been first investigated by Schatzman[1] and Salpeter [2] who have shown that surrounding particles in a plasma at equilibrium lower the repulsive barrier between two interacting particles (typically in a region smaller than the Debye sphere) as they approach to each other. Consequently, when fusion of two particles occurs, this additive energy influences the thermonuclear reaction and enhances the rate normally obtained in the case of particles in vacuum (i.e. infinite dilute limit case). In his paper of 1954, Salpeter has established the fundamental theory of the screening correction to the thermonuclear rate and has shown rigorously that an effective interaction energy is gained by any two screened ions separated by a distance a in the dilute plasma. One of the fusing ions with a charge e_α creates at a distance r a Debye-Hückel electrostatic potential $\varphi_\alpha = e_\alpha \exp(-\kappa r)/r$ where κ defines the inverse Debye radius $(1/\kappa = (4\pi \sum_\alpha \beta e_\alpha^2 \rho_\alpha)^{1/2})$. As a consequence, the concentration of the second ion e_β is increased in the neighborhood of e_α by a statistical Boltzmann weight proportional to $\exp(-\beta e_\beta \varphi_\alpha(a))$. The screening factor f reads:

$$f = \lim_{a \to 0} \frac{\exp(-\beta e_\beta \varphi_\alpha(r))}{\exp(-\beta e_\alpha e_\beta / r)} \qquad (1)$$

where the denominator term defines the statistical contribution to the concentration of ions of species e_β in the case of an infinite dilute gas. At a distance close to the nuclear fusion radius, the Debye-Hückel potential can be linearized, leading to the Salpeter's static screening factor $f_S = \exp(\beta e_\alpha e_\beta \kappa)$ which is scaled only by the Debye length $1/\kappa$. Indeed, the Salpeter formulation considers that, inside the Debye sphere, every particle has reached its thermodynamical equilibrium leading to the enhancement of the fusion rate through the drop of the electrostatic potential barrier. In this formulation, spherical symmetry is kept and quantum corrective terms (e.g. scattering, statistical effects) are neglected.

Different models correcting the Salpeter static screening formula have been investigated during the last decade. An extensive review of models based on a mean field approximation of the screening potential can be found in Dzitko *et al.* [3]. Using WKB theory, Bahcall *et al.* [4] have shown that the regular Gamow factor must be replaced by an effective penetration peak taking into account the Debye-Hückel screening factor. The derivation of the screening factor to the thermonuclear rate retrieves indeed at moderate density the Salpeter factor. Using the free energy excess of two fusing ions in a moderate dense plasma, DeWitt *et al.* [5] gave an analytical derivation of the screening factor to the thermonuclear rate which converges again at low screening to the Salpeter formulation. Brüggen & Gough [6] have shown either that under thermal equilibrium conditions, the effective potential between the two interacting ions contributing to the fusion rate is the free energy difference between the bound entity (when the fusion has occurred) and initial ions entities. These authors [7] have derived an expression of the screening factor by solving the generalized Poisson-Boltzmann equation. Brown & Sawyer have developed a quantum perturbation scheme formalism going beyond the Salpeter formulation. The leading correction to this screening factor is coming from the Fermi-Dirac statistic. Other quantum corrections exists but they are negligible in regards to the quantum statistic correction. An extensive review of these formalisms and their coincidence with Salpeter's formulation can be found in the paper of Bahcall *et al.* [8].

Claims on the lack of consistency of the static Salpeter formulation have raised controversy during the last few years [9, 10, 11]. First attempts for describing dynamical effects have been mentioned by Mitler [12] and Carraro [13] claiming that themodynamical equilibrium cannot be ensured at distance far smaller than the Debye length. In their papers, Shaviv & Shaviv [11, 14] have shown that dynamic effects on the screening have to be taken into account. Spherical symmetry is no longer conserved and inside the Debye radius, the number of electrons cannot justify a thermal equilibrium treatment of the particles. Their method has been performed using molecular dynamics and the results show quantitative changes in the predicted nuclear rate. These changes affect significantly the predicted neutrino fluxes. The spherical invariance symmetry is broken and the scattering particles feel a fluctuating potential due to the local potential that can no longer be considered as a smooth one.

Along these different attempts for evaluating corrections to the screening factor, we propose in this paper a rigorous formalism based on the Feynman-Kac (FK) path-integral representation of the grand-partition function. This formalism provides systematic prescriptions for calculating the virial expansion of the thermodynamic quantities characterizing the solar plasma. The FK representation [15, 16] allows to treat exactly on the same level the different physical effects occurring in the solar plasma such as: the Debye screening, long ranged diffraction corrections, scattering, formation of bound states in the plasma and corrections to the ideal gas related to the fermionic or bosonic nature of the particles. The virial expansion has been carried out up to the order $\rho^{5/2}$ in the density ρ. In this paper, we first derive the expression of the pair distribution function $g(r)$ of two particles moving towards each other and immersed in a moderately dense plasma. The screening factor for solar fusion reactions at very short distances is then calculated according to:

$$f = \lim_{a \to 0} \frac{g(r)}{g_0(r)} \qquad (2)$$

where $g(r)$ is calculated at a distance ε of nuclear radius dimension. According to [17], since in coulombic matter, ε is much smaller than the Debye length $1/\kappa$, one can evaluate the pair distribution at $\varepsilon \to 0$. In the latter expression, $g_0(r)$ normalizes the fusion reactions rate and represents the pair distribution of two fusing ions in a infinitely dilute plasma.

The paper is structured as follows: in section 2, we describe the model and summarize the formalism of the Feynman-Kac representation of the path integral at finite temperature T. In section 3, the virial expansion of the pair distribution function up to order $\rho^{5/2}$ is detailed. In section 4, we derive the expression of the screening factor for the $p-p$ channel up to the order $\rho^{5/2}$ in solar conditions taking into account classical corrections beyond the debye factor, short ranged, long ranged and statistical quantum corrections. We show that the leading correction to the Salpeter's formulation comes from two sources. The first one comes from the exact determination of the overall density of the quantum plasma which needs to be corrected by statistical effects due to the exchange of two identical particles. The second contribution comes from the evaluation of the screening factor along the $p-p$ channel: exchange of the two ions at a nuclear distance a brings an additive term proportional to ρ^2.

THE FEYNMAN-KAC REPRESENTATION OF THE GRAND PARTITION FUNCTION

We consider a multicomponent system made of point charges particles e_α with mass m_α and spin σ_α; the species index α specifies the nature of the particle. Species carrying positive and negative charges are present in the plasma in order to ensure the electroneutrality of the whole system. Two charges e_α and e_β separated by a distance r are interacting through the Coulomb potential $e_\alpha e_\beta/r$. For conciseness, in the following, a symbol i will stand for $i = [k; \alpha]$ with k running from 1 to the number N_α of charges of species α and α running from 1 to the number n_s of species (the total number of particles is therefore $N = \sum_\alpha N_\alpha$).

The hamiltonian corresponding to the system of N charges enclosed in a box with volume Λ reads $H_{N,\Lambda} = -\sum_i (\hbar^2/2m_i)\Delta_i (1/2)\sum_{i \neq j} e_i e_j/|\vec{r}_i - \vec{r}_j|$. We impose Dirichlet type boundary conditions in order to ensure that the eigenfunctions of $H_{N,\Lambda}$ vanish at the surface of the box.

The grand-canonical partition function

When the system reaches the thermodynamic limit, the grand-canonical partition function of the system at the temperature T ($\beta = 1/k_B T$) reads

$$\Xi_\Lambda = Tr_\Lambda \exp[-\beta(H_{N,\Lambda} - \sum_\alpha \mu_\alpha N_\alpha)] \qquad (3)$$

Here, the trace Tr_Λ runs over all the states satisfying the above boundary conditions and is symmetrized or anti-symmetrized according to the statistics of each species.

In the thermodynamic limit, the local density of any species α becomes uniform and reduces to $\rho_\alpha = z_\alpha \partial(\ln\Xi_\Lambda/\Lambda)/\partial z_\alpha$ where $z_\alpha = \exp(\beta\mu_\alpha)$ is the fugacity associated to the particle α. According to Ref. [16], the grand-canonical partition function Ξ_Λ can be rewritten in the configuration space and the spin space as

$$\Xi_\Lambda = \Xi_\Lambda^{MB} \sum_{n=2}^{\infty} \Xi_\Lambda^{(n)} \qquad (4)$$

Where the first contribution to the r.h.s. of this equation ($n = 0$) corresponds to the Maxwell-Boltzmann (MB) statistics and reads

$$\Xi_\Lambda^{MB} = \sum_{N_\alpha=0}^{\infty} \prod_\alpha \frac{z_\alpha^{N_\alpha}}{N_\alpha!} (2\sigma_\alpha)^{N_\alpha} \int_{\Lambda^N} \prod_i d\vec{r}_i < \vec{R}_N | \exp(-\beta H_{N,\Lambda}) | \vec{R}_N > \qquad (5)$$

In the latter equation, the ket $|\vec{R}_N >$ denotes the tensorial product $\otimes |\vec{r}_i >$ over the $N = \sum_\alpha N_\alpha$ particles of the plasma.

The second term in the r.h.s. of Equ. (5) (e.g. ($n = 2$)) describes the exchange of two identical particles and is obtained by collecting all the sets of permutations which exchange two indices associated to a given species α and leave the remaining ones unchanged. This term reads:

$$\Xi_\Lambda^{(2)} = \sum_\alpha (-1)^{2\sigma_\alpha} \frac{z_\alpha^2}{2!} (2\sigma_\alpha)^2 \sum_{N_\gamma=2}^{\infty} \frac{z_\gamma^{N_\gamma-2}}{(N_\gamma-2)!} (2\sigma_\gamma)^{N_\gamma-2}$$
$$\times \int_{\Lambda^N} d\vec{r}_1 d\vec{r}_2 d\vec{R}_{N-2} < \vec{r}_2 \vec{r}_1 \vec{R}_{N-2} | \exp(-\beta H_N) | \vec{r}_1 \vec{r}_2 \vec{R}_{N-2} > \qquad (6)$$

Where the summation over the index γ runs over all particles of species which have not been exchanged.

Since the virial expansion is truncated at the order $\rho^{5/2}$ in this fugacity development, we have to retain in Equ. (4) only the first two terms Ξ_Λ^{MB} and $\Xi_\Lambda^{(2)}$ (this last term contributes as a ρ^2 term because of its short-range character).

The Feynman-Kac representation of the partition function

Equs. (5) and (6) can be reformulated using the Feynman-Kac (FK) path-integral representation. This formulation allows to build a formal equivalence between the quantum system made of point particles (for which no perturbative treatment of Ξ_Λ can be made by construction) and a classical system made of extended filaments interacting via two-body forces. In this context, the virial expansion of the equivalent grand-partition function Ξ_Λ^* becomes possible since all its constituting terms are classical. The main steps of this formulation for the coulombic hamiltonian are summarized hereafter.

The FK representation of the N-body matrix element $|\vec{r}_N>$ of the equilibrium density matrix reads [19]

$$< \vec{R}'_N | \exp(-\beta H_{N,\Lambda}) | \vec{R}_N > = \sum_{\text{all paths}} \exp(-S(\{\vec{r}_i(t)\})/\hbar) \qquad (7)$$

Where $S(\{\vec{r}_i(t)\}) = \int_0^{\beta\hbar} dt \left[\sum_j m_j (d\vec{r}_j(t)/dt)^2 \sum_{j\neq k} e_j e_k v_c(r_{jk}(t)) \right]/2$ denotes the classical action of the system

The sum in Equ. (7) runs over all individual paths $\vec{r}_j(t)$ going from the point $\vec{r}_j$ to the point $\vec{r}''_j$ during a virtual "time" $\beta\hbar$. Using the brownian bridge, we can parametrize these paths as $\vec{r}_j(s) = \vec{r}_j \lambda_j \vec{\xi}_j(s)$ with $s = t/\beta\hbar$. $\vec{\xi}_j(s)$ is a dimensionless brownian bridge subject to the constraint $\vec{\xi}_j(0) = \vec{\xi}_j(1) = \vec{0}$ and $\lambda_j = (\beta\hbar^2/m_j)^{1/2}$ is the de Broglie wavelength associated to the particle j. By a specific choice of the new dimensionless variable $\vec{\xi}(s)$, we define a Gaussian measure $\mathscr{D}(\vec{\xi})$ of the brownian bridge process. This measure is normalized to 1 and its covariance is given by $\int \mathscr{D}(\vec{\xi}) \xi_\mu(s) \xi_\nu(t) = \delta_{\mu\nu} \times s(1-t)$ for $s \leq t$ or $\delta_{\mu\nu} \times t(1-s)$ for $t \leq s$

The FK representation of the N-body matrix element reads:

$$< \vec{R}'_N | \exp[-\beta H_{N,\Lambda}] | \vec{R}_N > = \prod_j \frac{1}{(2\pi\lambda_j^2)^{3/2}} \int \prod_j \mathscr{D}(\vec{\xi}_j)$$

$$\times \exp\left[-\frac{\beta}{2} \sum_{j\neq k} e_j e_k \int_0^1 ds\, v_c(|\vec{r}_j(s) - \vec{r}_k(s) \lambda_j \vec{\xi}_j(s) - \lambda_k \vec{\xi}_k(s)|)\right] \qquad (8)$$

where the integral stands over all brownian bridges. This expression takes specific form according to the statistics used. For example, for the MB statistics (Equ. (5)), this multiple integral stands only for closed paths (i.e. $\vec{r}_j(s) = \vec{r}_k(s)$. In the case of the exchange of two particles (Equ. (6)), the latter equation runs first over two opened paths (the exchanged particles) and then it is multiplied by an term taking into account the effect of the $(N-2)$ surrounding particles.

THE PAIR DISTRIBUTION FUNCTION AT THE ORDER $\rho^{5/2}$ IN DENSITY

Each extended filament of the classical system is entirely characterized by its spatial position $\vec{r}$, the brownian path $\vec{\xi}(s)$ describing its shape and its species index α. Two filaments in states $\mathscr{E}_j = (\alpha_j, \vec{r}_j, \vec{\xi}_j)$ and $\mathscr{E}_k = (\alpha_k, \vec{r}_k, \vec{\xi}_k)$ interact via the two-body potential $e_{\alpha_j} e_{\alpha_k} v(\mathscr{E}_j, \mathscr{E}_k)$ with $v(\mathscr{E}_j, \mathscr{E}_k) = \int_0^1 ds \, v_c(|\vec{r}_j(s) - \vec{r}_k(s)\lambda_j \vec{\xi}_j(s) - \lambda_k \vec{\xi}_k(s)|)$.

From Equ. (5), the corresponding Maxwell-Boltzmann (MB) grand-canonical partition function Ξ_Λ^{MB} reads:

$$\Xi_\Lambda^{MB} = \sum_{N=0}^{\infty} \frac{1}{N!} \int \prod_{k=1}^{N} d\mathscr{E}_k z(\mathscr{E}_k) \prod_{k<l} [1(\mathscr{E}_k, \mathscr{E}_l)], \tag{9}$$

where the measure $d\mathscr{E} = d\vec{r}\mathscr{D}(\vec{\xi})$ represents the volume element of phase space accessible to a filament and $f(\mathscr{E}_j, \mathscr{E}_k) = [\exp[-\beta e_{\alpha_j} e_{\alpha_k} v(\mathscr{E}_j, \mathscr{E}_k)] - 1]$ represents the generalized Mayer bond corresponding to the potential $v(\mathscr{E}_j, \mathscr{E}_k)$. Here, every Mayer bond diverges because of the long-range nature of the filament-filament potential $v(\mathscr{E}_j, \mathscr{E}_k)$ at large distances (terms of the form $\int d\vec{r} \exp(\pm 1/r)$). It can be shown that these long-range divergencies can be obviated using an extension [15] of the Abe-Meeron cluster resummation method [20, 21].

MB Statistics

In the MB statistics case, the truncated pair distribution function $\rho_T(\alpha_a \vec{r}_a, \alpha_b \vec{r}_b)$ is obtained after integration of Equ. (4) over the shapes $\vec{\xi}$ of the filaments. It comes:

$$\rho_T(\alpha_a \vec{r}_a, \alpha_b \vec{r}_b) = \int \mathscr{D}(\vec{\xi}_a) \mathscr{D}(\vec{\xi}_b) z_\alpha(\vec{\xi}_a) z_\beta(\vec{\xi}_b) \frac{\partial^2 \ln \Xi_\Lambda^{MB}}{\partial z_\alpha(\mathscr{E}_a) \partial z_\beta(\mathscr{E}_b)} \tag{10}$$

where Ξ_Λ^{MB} is the partition function given in Equ. (9) as a virial expansion of generalized Mayer bonds. In order to keep out the long-ranged divergencies, the resummation procedure consists in introducing the truncated bond $f_T(\mathscr{E}, \mathscr{E}')$ defined through the following decomposition

$$f(\mathscr{E}, \mathscr{E}') = f_c(|\vec{r} - \vec{r}'|) \frac{1}{2} f_c^2(|\vec{r} - \vec{r}'|)$$
$$\int_0^1 ds [\lambda_\alpha \vec{\xi}(s).\vec{\nabla}\lambda_{\alpha'} \vec{\xi}'(s).\vec{\nabla}'] f_c(|\vec{r} - \vec{r}'|) T(\mathscr{E}, \mathscr{E}') \tag{11}$$

where f_c is the regular Coulomb shape-independent bond $f_c(|\vec{r}-\vec{r}'|) = -\beta e_\alpha e_{\alpha'} v_c(|\vec{r}-\vec{r}'|)$. By construction, f_T decays as $1/|\vec{r}-\vec{r}'|^3$ when $|\vec{r}-\vec{r}'| \to \infty$. Formally, the replacement of $f(\mathscr{E},\mathscr{E}')$ by any of the four bonds appearing in the decomposition of Equ. (11) generates new classes of graphs, the so-called Γ graphs. In a Γ graph, any single bond connecting two filaments is either the f_T bond or one of the three other non-integrable bonds present in the r.h.s. of the latter equation. Divergencies induced by the non-integrable bonds can be eliminated through the summation of specific convoluted direct chains built within the Coulomb bonds [15]. The remaining classes of graphs, those constituted with f_T bonds alone, are integrable by construction. Additional classes are built up from four different resumed bonds whose combination in the same graph obeys to specific topological rules. These bonds can be expressed explicitly in terms of the MB particle densities. Two of them are short-ranged and are proportional to the screened charge-charge potential (i.e. the pure Debye bond) and the screened dipole-charge potential [15]). The other resumed bonds are the so-called f_{dip} and f_R bonds and decay at least like $1/|\vec{r}-\vec{r}'|^3$ when $|\vec{r}-\vec{r}'| \to \infty$.

Up to the order $\rho^{5/2}$, the MB truncated pair distribution function is built up exclusively with graphs composed with the association of one or more pure Debye bonds $f_D = -\beta e_\alpha e_\beta \exp(-\kappa r)/r$, f_{dip} and f_R bonds [15, 22]. For illustration, the first graphs beyond the Salpeter approximation are purely classical and involve the association of single Debye bond f_D convoluted with a double Debye ladder bond $f_D^2/2$. It comes, according to the nomenclature adopted in Ref. [15]:

$$g_{(N=0,M_R=1)}(r) =$$
$$- 2\beta^3 e_\alpha e_\beta (e_\beta e_\alpha) \frac{\pi}{\kappa} \frac{e^{-\kappa r}}{r} \left[e^{2\kappa r} Ei(-3\kappa r) - Ei(-\kappa r) - \ln 3 \right] \sum_\gamma \rho_\gamma e_\gamma^3$$
$$2\beta^4 e_\alpha e_\beta \frac{\pi^2}{\kappa^3} \frac{e^{-\kappa r}}{r} \sum_{\gamma,\delta} \rho_\gamma \rho_\delta e_\gamma^3 e_\delta^3$$
$$\times \left[e^{2\kappa r}[1-\kappa r] Ei(-3\kappa r) - [\kappa r][Ei(-\kappa r)\ln 3] \frac{4}{3} [\frac{e^{-\kappa r}}{r} - 1] \right] \quad (12)$$

where the function $Ei(-\alpha x) = -\int_x^\infty dx e^{-\alpha x}/x$ and $\kappa = (4\pi\beta \sum_\alpha e_\alpha^2 \rho_\alpha)^{1/2}$ is the inverse Debye screening length.

Similarly, the convolution of two directional f_{dip} bonds leads to the a graph contributing like $\rho^{5/2}$. This term describes the first diffraction term and reads:

$$g^{(2)}_{(N=1,M'_D=2)}(|\vec{r}'-\vec{r}|) = \frac{\pi}{6}\beta^2 e_\alpha e_\beta (\kappa r - 2) \frac{e^{-\kappa r}}{r} \sum_\alpha \lambda_\alpha^2 e_\alpha^2 \rho_\alpha \quad (13)$$

where λ_α is the thermal De Broglie wavelength. This term is the first long-ranged quantum correction and is proportional to $\hbar^2$, which is consistent with the Wigner-Kirkwood expansion of the weakly quantum gas limit. Extensive expression of the various corrections of the MB pair distribution function are given in [22].

Fermi-Bose Statistics

According to the scheme introduced in Equ. (4), we need to include the effects of Fermi-Dirac and/or Bose-Einstein statistics. The FK representation allows this along an elegant method consisting in implementing statistics as a perturbation onto the Maxwell-Boltzmann pair distribution function. As exchange contributions are short-ranged, the sole contribution to be retained up to order $\rho^{5/2}$ is the one describing the exchange of two opened filaments immersed in a Boltzmann bath of $(N-2)$ closed filaments. It comes:

$$g_{exch}(|\vec{r}'-\vec{r}|) = \sum_\alpha (-1)^{2\sigma_\alpha} \frac{(2\sigma_\alpha)}{(2\pi\lambda_\alpha^2)^3} \int \mathcal{D}(\vec{\xi}_1) \int \mathcal{D}(\vec{\xi}_2)$$
$$\times \left[1\beta e_\alpha^2 \kappa(\rho) \right] \exp\left[-\frac{|\vec{r}'-\vec{r}|^2}{\lambda_\alpha^2} - \beta e_\alpha^2 v(\mathcal{F}_{12}^\alpha, \mathcal{F}_{21}^\alpha) \right] \quad (14)$$

In this expression, $v(\mathcal{F}_{12}^\alpha, \mathcal{F}_{21}^\alpha)$ represent the exchange potential between two opened filaments $\mathcal{F}_{12}^\alpha$ and $\mathcal{F}_{21}^\alpha$. The overall term in the exponential evaluates the statistical weight of the exchange of two particles in the vacuum. The prefactor term $(1\beta \kappa e_\alpha^2)$ is the correction to the exchange when screening of surrounding charged particles is considered. Writing backward the path integral formulation in term of a matrix element, Equ. (14) gives:

$$g_{exch}(|\vec{r}'-\vec{r}|) = (1\beta \kappa e_\alpha^2)(-1)^{2\sigma_\alpha} \frac{2\sigma_\alpha}{(2\pi\lambda_{\alpha\alpha}^2)^{3/2}} <\vec{r}'\,\vec{r}|\exp(-\beta h_{\alpha\alpha}|\vec{r}\,\vec{r}'> \quad (15)$$

Collecting the different contributions arising from the Maxwell-Boltzmann statistics and from the Fermi-Bose statistics, the expression of the overall pair distribution function at short distance ($\varepsilon << 1/\kappa$) up to the order $\rho^{5/2}$ in density reads:

$$g_{\alpha,\beta}(\varepsilon) = 1 - \beta e_\alpha e_\beta \frac{\exp(-\kappa\varepsilon)}{\varepsilon}$$
$$\beta^4 e_\alpha^2 e_\beta^2 \sum_\gamma \rho_\gamma e_\gamma^3 (e_\alpha e_\beta) \frac{e^{-2\kappa\varepsilon}}{\varepsilon^2} \left[e^{2\kappa\varepsilon} Ei(-3\kappa\varepsilon) - Ei(-\kappa\varepsilon) - \ln 3 \right]$$
$$\beta^4 e_\alpha^2 e_\beta^2 \sum_\gamma \rho_\gamma e_\gamma^4 I_4'(\varepsilon,\kappa)$$
$$- \beta^5 \frac{e_\alpha e_\beta}{\kappa^2} \sum_{\gamma,\delta} \rho_\gamma \rho_\delta e_\gamma^3 e_\delta^3 [e_\gamma(e_\alpha e_\beta) I_5(r,\kappa) e_\alpha e_\beta I_5'(\varepsilon,\kappa)]$$
$$\beta^6 \frac{e_\alpha e_\beta}{\kappa^4} \sum_{\gamma,\delta,\varepsilon} \rho_\gamma \rho_\delta \rho_\varepsilon e_\gamma^3 e_\delta^3 e_\varepsilon^3 [(e_\alpha e_\beta) I_6(\varepsilon,\kappa) \gamma I_6'(\varepsilon,\kappa)]$$
$$- \beta^7 \frac{e_\alpha e_\beta}{\kappa^6} \sum_{\gamma,\delta,\varepsilon,\eta} \rho_\gamma \rho_\delta \rho_\varepsilon \rho_\eta e_\gamma^3 e_\delta^3 e_\varepsilon^3 e_\eta^3 I_7(\varepsilon,\kappa)$$

$$\frac{\pi \kappa}{3}\beta^2 e_\alpha e_\beta \sum_\alpha \lambda_\alpha^2 e_\alpha^2 \rho_\alpha$$

$$(2\pi\lambda_{\alpha\beta}^2)^{3/2} < \vec{\varepsilon}|e^{-\beta h_{\alpha\beta}}|\vec{\varepsilon}> -1\frac{\beta e_\alpha e_\beta}{\varepsilon}\frac{\beta^2 e_\alpha^2 e_\beta^2}{2\varepsilon^2}[e^{-2\kappa\varepsilon}-1]$$

$$\beta e_a e_b(\frac{1-e^{-\kappa\varepsilon}}{\varepsilon})$$

$$\times\left[(2\pi\lambda_{\alpha\beta}^2)^{3/2} < \vec{\varepsilon}|e^{-\beta h_{\alpha\beta}}|\vec{\varepsilon}> -1\frac{\beta e_\alpha e_\beta}{\varepsilon}\frac{\beta^2 e_\alpha^2 e_\beta^2}{6\varepsilon^2}[e^{-\kappa\varepsilon}-2\kappa\varepsilon-2]\right]$$

$$-4\pi\beta\sum_\gamma \rho_\gamma e_\gamma\int dr e^{-\kappa r} r\left[e_\alpha(2\pi\lambda_{\gamma\alpha}^2)^{3/2}<\vec{r}|e^{-\beta h_{\gamma\alpha}}|\vec{r}>\right.$$

$$\left. e_\beta(2\pi\lambda_{\gamma\beta}^2)^{3/2}<\vec{r}|e^{-\beta h_{\gamma\beta}}|\vec{r}>\right]$$

$$\frac{8\pi^2}{\kappa}\beta^2 e_\alpha e_\beta \sum_{\gamma,\delta}\rho_\gamma\rho_\delta e_\gamma e_\delta\int drr^2 e^{-\kappa r}(2\pi\lambda_{\gamma\delta}^2)^{3/2}<\vec{r}|e^{-\beta h_{\gamma\delta}}|\vec{r}>$$

$$-\beta\kappa e_\alpha e_\beta 2\pi\beta^3 e_\alpha e_\beta (e_\alpha e_\beta)\ln 3\sum_\gamma \rho_\gamma e_\gamma^3 - \frac{16\pi^2}{3\kappa^2}\beta^4 e_\alpha e_\beta \sum_{\gamma,\delta}\rho_\gamma\rho_\delta e_\gamma^3 e_\delta^3$$

$$(1\beta\kappa e_\alpha^2)\frac{(-1)^{2\sigma_\alpha}}{2\sigma_\alpha}(2\pi\lambda_{\alpha\alpha}^2)^{3/2}< -\vec{\varepsilon}|e^{-\beta h_{\alpha\alpha}}|\vec{\varepsilon}>$$

$$-\frac{2\pi\beta^2 e_\alpha e_\beta e^{-\kappa\varepsilon}}{\kappa\varepsilon}\sum_\alpha \frac{(-1)^{2\sigma_\alpha}}{2\sigma_\alpha}e_\alpha^2\rho_\alpha^2(2\pi\lambda_{\alpha\alpha}^2)^{3/2}\int d\vec{r}< -\vec{r}|e^{-\beta h_{\alpha\alpha}}|\vec{r}>$$

(16)

The four first terms of the latter equation correspond to the classical contributions of order $\rho^{5/2}$ in density beyond the Debye-Huckel term $g_{DH}(\varepsilon)=-\beta e_\alpha e_\beta e^{-\kappa\varepsilon}/\varepsilon$. The structure of these contributions to the pair distribution function doesn't give simple analytical calculation. As a consequence, we have carried out numerical calculations [22] of the different coefficients $I_n(\kappa,r)$ and $I'_n(\kappa,r)$ ($n=[4,7]$). The term proportional to $\hbar^2$ appearing in the fifth line consists in the long-ranged quantum term given in Equ. (13). This term is of order $\rho^{5/2}$.

In the last six lines of Equ.(16), the short-ranged quantum contributions and corrections arising from the Fermi-Bose statistics are given. The ρ^2 short-ranged term expresses essentially the quantum pair distribution function of the two fusing particles regardless of the screening of the surrounding charged particles. The following term corrects this situation by taking into account the screening mechanism (term proportional to $\beta e_a e_b (2\pi\lambda_{\alpha\beta}^2)^{3/2}<\vec{\varepsilon}|\exp(-\beta h_{\alpha\beta})|\vec{\varepsilon}>(1-e^{-\kappa\varepsilon}/\varepsilon)$ and followings). For the exchange terms, we have essentially two distinct contributions. The first one is describe the pair distribution function coming from the exchange of two similar fusing particles at short distance when the screening of the surrounding charged particles is felt. The second contribution appears as a statistical correction of the Debye term.

We have checked that we retrieve the expression of the free energy up to order $\rho^{5/2}$ given in Ref. [16] when we perform an integration over the space variable of the pair distribution function multiplied by the Coulomb potential.

EXPRESSION OF THE $P-P$ CHANNEL SCREENING FACTOR TO THE SOLAR FUSION REACTIONS UP TO ORDER $\rho^{5/2}$

From the expression of the pair distribution function given in Equ. (16), we can derive the expression of the enhancement screening factor $f(\kappa,\varepsilon)$ for thermonuclear rate occurring for two fusing protons. The definition given in Equ. (2) leads to:

$$f(\kappa,\varepsilon) = \frac{g_{pp}(\varepsilon)}{g_{0,pp}(\varepsilon)}$$
$$= \exp(\beta e^2 \kappa)$$
$$\times \exp\left(\frac{8\pi^{5/2}\beta^2 e^2}{\kappa} \sum_\alpha \frac{(-1)^{2\sigma_\alpha}}{2\sigma_\alpha} e_\alpha^2 \rho_\alpha^2 \lambda_\alpha^3 \int d\vec{r} < -\vec{\varepsilon}|\exp(-\beta h_{\alpha\alpha})|\vec{\varepsilon}>\right)$$
$$\times \exp\left(4\pi^{3/2}\lambda_p^3 < -\vec{\varepsilon}|\exp(-\beta h_{pp})|\vec{\varepsilon}>\right)_{5/2}(\kappa,\varepsilon)$$
(17)

In the numerator, we have carried out the calculation of the pair distribution function $g_{pp}(\varepsilon)$ at a distance ε when the two fusing protons are surrounded by charged particles enclosed in the volume Λ. Indeed, we have made a simplification and have considered that the surrounding particles located at a distance smaller than the Debye radius $1/\kappa$ have reached their thermodynamical equilibrium. In Equ.(17), the denominator term $g_{0,pp}(\varepsilon)$ stands for the pair distribution function of the two fusing protons in the limit of a infinite dilute gas. From Equ. (16), it comes out that: $g_{0,pp}(\varepsilon \to 0) = (2\pi\lambda_{pp}^2)^{3/2} < \vec{\varepsilon}|\exp(-\beta h_{0,pp})|\vec{\varepsilon}>$.

The exponential form taken by the enhancement factor in the latter equation has been obtained by considering only partial resummations of terms in Equ. (16) and using Expr. (2.7) in Ref. [15].

The remaining contributions arising from the virial expansion of Equ. (16) (e.g. diffraction and higher order classical terms) have been collected in the function $f_{5/2}(\kappa,\varepsilon)$.

Quantitative estimation of the screening factor $f(\kappa,\varepsilon)$ is discussed in [22] and show that, in the inner layers of the sun, the first leading contributions beyond the Debye screening factor $f_{DH} = \exp(\beta e^2 \kappa)$ arises from statistics effects.

REFERENCES

1. Schatzman E., *J. Phys. Rad.* **9**, 46, (1948)
2. Salpeter E. E. , *Australian J.Phys.*, **7**, 373 (1954);
3. H. Dzitko, S. Turk-Chieze ,*ApJ*, **447**, 428 (1995);
4. Bahcall J.N , Krastev P.I. , SmirnovA. Yu. ,*Phys. Rev. D* **58**,096016 (1998)

5. DeWitt H.E., Graboske H.C., Cooper M.S., *ApJ*, **181**, 439, (1983)
6. Bruggen M., Gough D.O., *ApJ* **488**,867 (1997)
7. Bruggen M., Gough D.O., *Journal of Mathematical Physics* **41**,260 (2000)
8. Bahcall J.N, Brown L.S., Gruzinov A., Sawyer R.F., *A&A* **383**,291 (2002)
9. Opher M., Opher R., *ApJ* **331**,473 (2000)
10. Gruzinov A.V., *ApJ* **496**,503 (1998)
11. Shaviv G., Shaviv N., *Ap&SS Lib.* **214**,43 (1997)
12. Mitler E.H., *ApJ* **212**,513 (1997)
13. Carraro C., Schafer A., Koonin S.E., *ApJ* **331**,567 (1988)
14. Shaviv G., Shaviv N., *ApJ*, **529**, 1054 (2000)
15. Alastuey A., Cornu F., Perez A., *Phys. Rev. E* **51**, 1725 (1995).
16. Alastuey A., Perez A., *Phys. Rev. E* **53**, 5714 (1996)
17. Alastuey A., Jancovici B., *ApJ* **226**, 1034, (1978)
18. Alastuey A., Cornu F., Perez A., *Phys. Rev. E* **49**, 1077 (1994).
19. Simon B., in "Functional Integration and Quantum Physics" (*Academic, NY* (1979)
20. Meeron E.,*J. Chem. Phys.* **28**, 630 (1958);
21. Abe R., *Progr. Theor. Phys.* **22**, 213 (1959)
22. Perez A., Chabrier G., *in preparation for ApJ*

Isolating the effects of chemical composition in the equation of state

Chia-Hsien Lin* and Werner Däppen[†]

*Armagh Observatory, College Hill, Armagh, BT61 9DG, N. Ireland, U.K.
[†]Department of Physics and Astronomy, University of Southern California, Los Angeles, California, 900089-1342, U.S.A.

Abstract. The effects of different chemical compositions and of different equation-of-state formalisms on a solar structure are often mixed. The reason being that the chemical composition in modern, sophisticated equations of state cannot be easily adjusted. Therefore, most of related studies simply consider the combined effect of the two, rather than individual effect. The actual error in the formalism of the equation of state is thus inevitably obscured.
  The aim of this paper is to examine and isolate the effects of varying chemical compositions, in hope to help to extract the effects purely from the formalism of equation of state.
  We present a strategy and the results of examining the effects of the ionization of individual element.

INTRODUCTION

The development of helioseismology and the quality of the helioseismic data have enabled the detailed study of the microphysical structure of the solar interior. In our other paper in this proceeding, we demonstrated our results of examining the equation of state (EOS) and chemical abundance in the Sun through an inversion procedure. We used the adiabatic gradient, $\gamma_1 \equiv (\partial \ln P / \partial \ln \rho)_s$, as the probe to detect the discrepancy in the EOS and chemical abundance between the Sun and a solar model. The advantage of using γ_1 is that it is a direct output from an EOS and is independent of macroscopic structural properties. Specifically, we exploited the fact that γ_1 deviates from the isentropic value, 5/3, at the element ionization zones, which are determined by the EOS formalism and the element abundance. The results in that paper led to the interest to examine the effects of different chemical abundances on γ_1 at a fixed EOS formulation.

The chemical composition is often considered as an integrated part of an EOS although, strictly speaking, an EOS is simply the relation among pressure, density and temperature. The latest spectroscopic data [1] revealed that the currently accepted value of the composition might need to be modified. In other words, the error in EOS between the Sun and a solar model could be due to both an incorrect formalism of EOS and incorrect chemical abundances. This leads to a question: how much does the incorrect chemical composition contribute to the discrepancy in EOS between the Sun and a solar model?

To answer this question, a straightforward experiment is to assess the effect of tuning the value of chemical composition at a fixed EOS formalism. However, many modern

EOSs are so computationally heavy that the chemical composition cannot be easily adjusted. Because of this reason, we chose a simple yet sufficiently accurate EOS, CEFF [2], for this study.

Among a number of EOS derivatives, we chose the adiabatic gradient, $\gamma_1 \equiv (\partial \ln P/\partial \ln \rho)_s$ (s being entropy), as the indicator of the effects. Our reason is that there exist inversion procedures to reveal the discrepancy of this quantity between different solar structures. Hence, the result of this study can be of practical value.

In practice, an inversion routine compares γ_1 at the same *depth* of two structures. However, to correctly represent the discrepancy in the EOS and chemical abundances, γ_1 must be compared at the same *thermodynamic point* of the two structures. To convert $\delta\gamma_1/\gamma_1$ at the same depth to $\delta\gamma_1/\gamma_1$ at the same thermodynamic point, we followed the equation proposed by [3]:

$$\frac{\delta\gamma_1}{\gamma_1} = \left(\frac{\partial \ln \gamma_1}{\partial \ln P}\right)_{Y,\rho} \frac{\delta P}{P} + \left(\frac{\partial \ln \gamma_1}{\partial \ln \rho}\right)_{Y,P} \frac{\delta\rho}{\rho} + \left(\frac{\partial \ln \gamma_1}{\partial Y}\right)_{P,\rho} \delta Y + \frac{\delta\gamma_{1,\mathrm{int}}}{\gamma_{1,\mathrm{int}}} \quad (1)$$

The last term in the right-hand-side of the equation, $\delta\gamma_{1,\mathrm{int}}/\gamma_{1,\mathrm{int}}$, (or, "intrinsic γ_1 difference"), is the relative difference at the same *thermodynamic point*.

The result in our other paper in these proceedings demonstrated that tuning the abundances in models does show distinguishable effects in inverted $\delta\gamma_{1,\mathrm{int}}/\gamma_{1,\mathrm{int}}$. The results triggered the interest to further examine the effect from individual element.

Since CEFF does not consider excited states of chemical elements, the value of γ_1 is mainly affected through the ionization of element. Our strategy to isolate the effect of one element is to prohibit the ionizations of all other elements by setting the ionization potentials of them to infinity.

THE EXPERIMENTS

The models

The models in this study are *seismic envelope models*. A seismic model is simply a snapshot of the present Sun, and does not contain the evolutionary history of the Sun, A seismic envelope model is a seismic model that satisfies all the boundary conditions at the surface but deviates from the reality near the core. All of our models are implemented with CEFF equation of state and OPAL opacity table [4]. The chemical composition is assumed to be constant through the entire region of computation. The assumption is acceptable in the convective zone where all the elements are well mixed by the convection, but would become unrealistic in the deeper region. Our envelope program [described in 5] computes the models from the surface inward to the region where the assumption of a constant composition can no longer be valid.

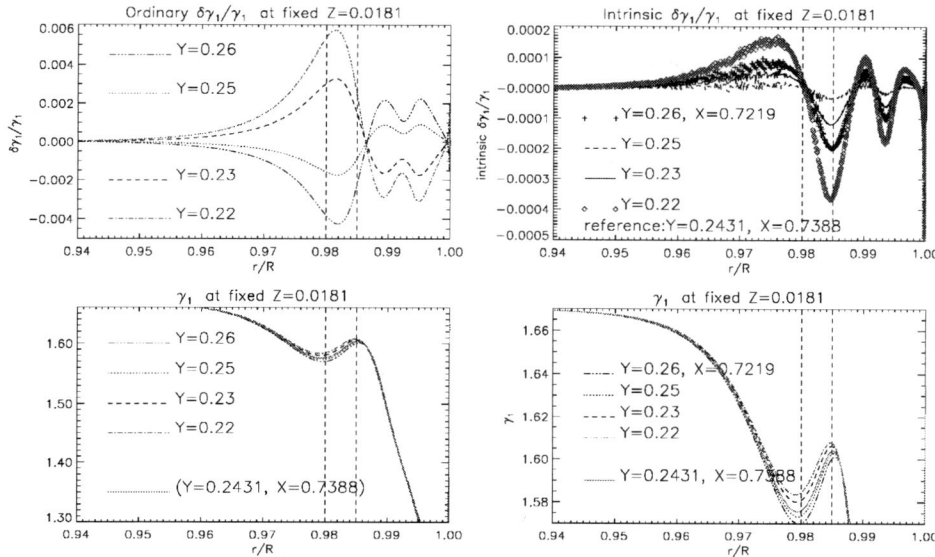

FIGURE 1. A comparison of the profile of $\delta\gamma_1/\gamma_1$ and of $\delta\gamma_{1,int}/\gamma_{1,int}$. The dashed vertical lines are plotted to help to identify the same locations in different figures.

The experiments

The experiments can be categorized into four groups:

1. All heavy elements are forced to remain neutral, but hydrogen and helium are allowed to ionize;
2. Setting one heavy element to be fully ionized through the entire structure;
3. One heavy element, in addition to H and He, is allowed to ionize at its ionization zones;
4. Two heavy elements, in addition to H and He, are allowed to ionize at their ionization zones;

In each category, we tuned X (hydrogen abundance), Y (helium abundance) or Z (heavy element abundance) in various ways. The purpose is to look for the effects of the neutral particles, the extra electrons, the ionizations and the interference/entanglement among the above and among multiple ionizations of elements.

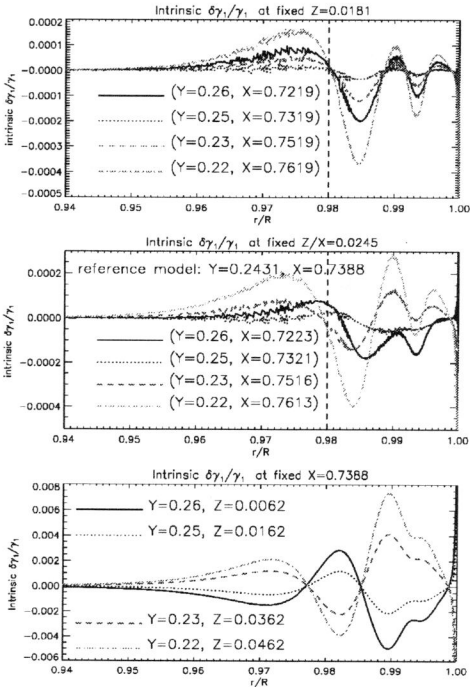

FIGURE 2. A comparison of the effects of tuning the chemical composition at fixed Z, fixed Z/X and fixed X. Heavy elements are *not* ionized.

RESULTS AND DISCUSSION

No ionization of heavy elements

If the thermal structure (i.e, $P(r), \rho(r), T(r)$, where r is the distance from the solar center) of a model does not change much by the adjustment of chemical abundances, the profiles of $\delta\gamma_1/\gamma_1$ and $\delta\gamma_{1,\text{int}}/\gamma_{1,\text{int}}$ would resemble each other. Most of our results from the experiments in this category showed otherwise (see the example in Fig. 1). In addition, the dips and bumps in $\delta\gamma_{1,\text{int}}/\gamma_{1,\text{int}}$ do not always coincide with the ionization zones. This could be due to the following reason: Varying the chemical composition alters the thermal structure (for example, the temperature at the same depth is changed), and, thus, shifts the physical locations of ionization zones.

Fig. 2 illustrates how $\delta\gamma_{1,\text{int}}/\gamma_{1,\text{int}}$ varies when the chemical composition is tuned at fixed Z, X and Z/X. There are two interesting features in this figure: One is the occurrence of the nodes at $\delta\gamma_{1,\text{int}}/\gamma_{1,\text{int}} \approx 0$ whenever one abundance (X or Z) is fixed (cf. the top and bottom panels). The other is that in the first two panels (i.e., fixed Z and fixed Z/X) the variation of $\delta\gamma_{1,\text{int}}/\gamma_{1,\text{int}}$ is not linearly related to the variation of the abundances. In other words, $\delta\gamma_{1,\text{int}}/\gamma_{1,\text{int}}$ does not monotonically vary with the

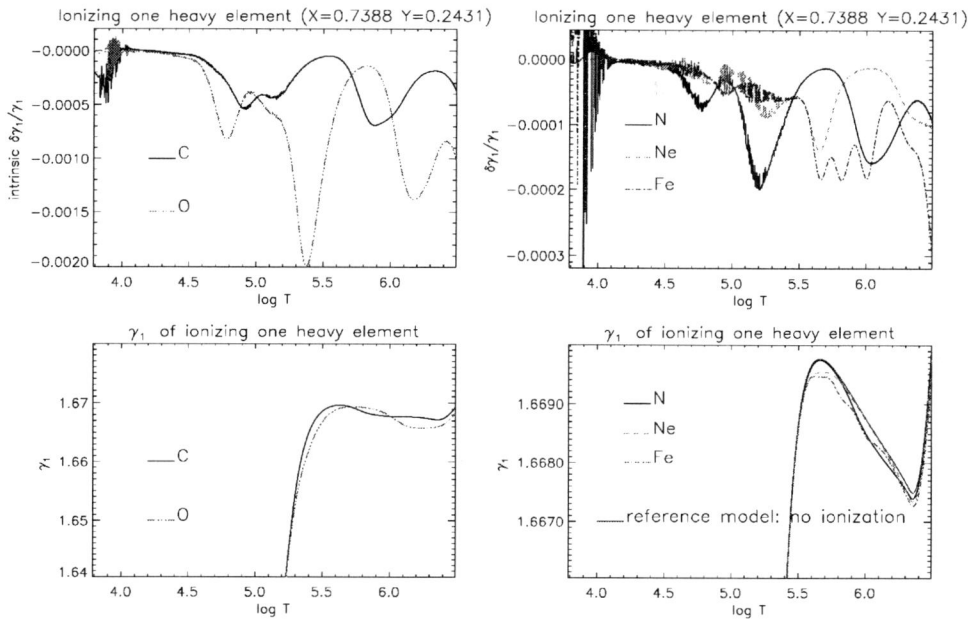

FIGURE 3. The effect of ionizing one heavy element. The locations of dips/bumps in $\delta\gamma_{1,\text{int}}/\gamma_{1,\text{int}}$ coincide the locations of the respective ionization zones. Note that the unit of horizontal axis is $\log T$ rather than r, as in many other plots.

variation of the magnitude of the abundance difference between two models. However, the variations of $\delta\gamma_{1,\text{int}}/\gamma_{1,\text{int}}$ and abundances are linearly correlated in the bottom panel, where X is fixed. The bottom panel also shows the largest magnitude of $\delta\gamma_{1,\text{int}}/\gamma_{1,\text{int}}$.

The influence of one heavy element

Ionizing one heavy element. To allow the ionization of only one heavy element, we allow the ionization potentials of the selected element to be their theoretical values, but set the ionization potentials of all other elements to infinity. Therefore, the selected element would ionize at appropriate ionization zones while all other elements would remain as neutral particles everywhere in the structure. The result of this experiment is illustrated in Fig. 3. The figure shows that the dips in $\delta\gamma_{1,\text{int}}/\gamma_{1,\text{int}}$ do coincide with the ionization zones of the element, in contrary to the results shown in the previous experiment. For instance, the right column in Fig. 1 shows that the ionization zone 0.98R (cf. lower panel) does not coincide with any dips/bumps (cf. upper panel).

Adding one fully ionized heavy element. In this experiment, we set the ionization potential of one heavy element to zero. In other words, the element is fully ionized

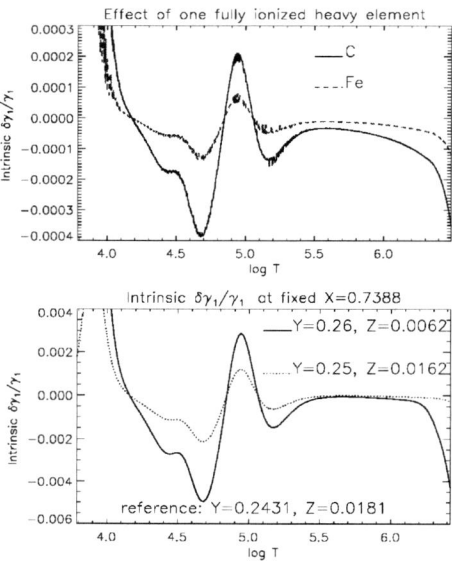

FIGURE 4. A comparison of the effect of adding one fully ionized heavy element and the effect of tuning Y at fixed X. Note that the unit of horizontal axis is $\log T$, as in Fig. 3.

everywhere in the structure. Therefor, all of the electrons associated with this element are added to the composition.

The upper panel in Fig. 4 compares $\delta\gamma_{1,\text{int}}/\gamma_{1,\text{int}}$ of fully ionized C and FE. The main features of both curves appear in H and HE ionization zones. The curves are essentially featureless in the ionization zones of C and FE (compare Fig. 4 with C and FE curves in Fig. 3). This is because adding a fully ionized element is equivalent to increasing the number density of electrons, which, consequently, affects the ionization of H and HE, the only ionizable elements in the mixture.

We note that the effect of fully ionized C is greater than that of fully ionized Fe. The reason is that C is much more abundant than Fe although Fe has more electrons than C does.

In addition, our results show that the effect of these additional electrons is similar to the effect of increasing Y at fixed X (compare the upper and lower panels in Fig. 4). This implies that increasing Y at fixed X might affect the thermodynamics of the structure by increasing the number density of electrons.

The influence of two heavy elements

Here, we allow two selected heavy elements to ionize. Some results from this experiment are shown in Fig. 5. In the upper row of the figure, the solid lines are $\delta\gamma_{1,\text{int}}/\gamma_{1,\text{int}}$ curves of ionizing two elements, and the dash and dash-dot lines are $\delta\gamma_{1,\text{int}}/\gamma_{1,\text{int}}$ of ion-

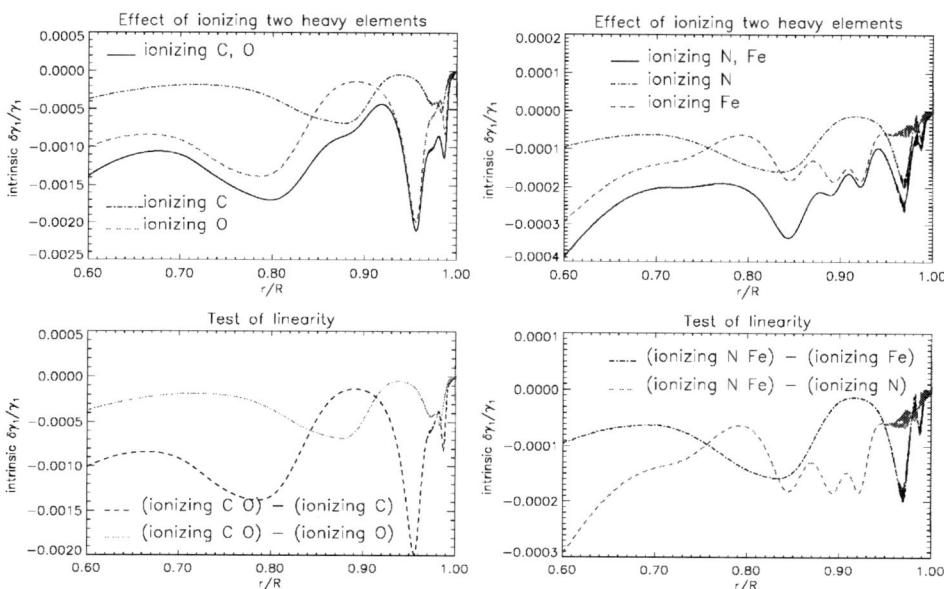

FIGURE 5. The effect of ionizing two heavy element. The locations of dips/bumps in $\delta\gamma_{1,\text{int}}/\gamma_{1,\text{int}}$ coincide the locations of the respective ionization zones.

izing individual element. From the figure, we observe that the effect of multiple ionization satisfies the linear superposition relation, that is, the effect of ionizing two elements is equal to the summed effect of ionizing individual element. Our further tests showed that the superposition relation remains valid even when Z is unrealistically increased to be comparable to X or Y. The validity of the linear superposition indicates that varying Z possibly imposes equal influence on all the heavy elements.

To see how the profile of $\delta\gamma_{1,\text{int}}/\gamma_{1,\text{int}}$ evolves as the abundances, especially Z, vary, we carried out two experiments with either Y or X fixed. Although the abundance of heavy elements is only 2% in a realistic solar structure, we exaggerated Z to a value comparable to either X or Y, in order to magnify the effects from heavy elements and to see the occurrence of non-linearity.

Fig. 6 shows the results from the exaggerated experiments. The left column in the figure is the results of setting $Y = 0.001$ and varying Z from 0.4 to 0.6. We chose $Z = 0.5$ as the reference model, which means the abundance difference between two structures varies between 10% and 20%. The figure shows that the high abundance and large variation indeed result in irregular, non-linear behaviors of $\delta\gamma_{1,\text{int}}/\gamma_{1,\text{int}}$ profile. The right column in the same figure shows the results of setting $X = 0.4$ and varying Z between 0.6 and 0.56. $Z = 0.575$ was chosen as the reference; hence, the abundance difference is less than 5%. In contrast to the irregularity seen in the left column, the right column shows a linear, regular variation of $\delta\gamma_{1,\text{int}}/\gamma_{1,\text{int}}$ profile as the abundances are varied.

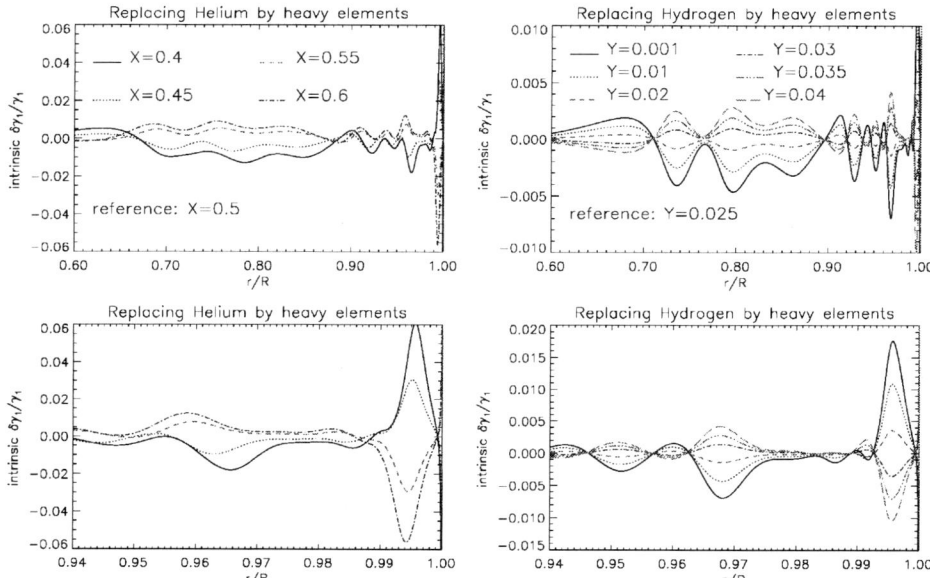

FIGURE 6. The effect of heavy-element ionization with large Z. The ionizing elements are Ne and Fe. The left column shows the result of replacing helium by heavy elements. The composition is: $Y = 0.001$ and X varies as indicated in the figure. The right column illustrates the results of replacing hydrogen by heavy elements. The composition is: $X = 0.4$, and Y varies as indicated in the figure. The plots in the lower row are a zoomed-in version of the plots in the upper row.

The difference between the two columns suggests that the non-linearity occurs when the abundance *difference* between two structures is large.

EXAMINING THE NON-LINEARITY OF $\dfrac{\delta\gamma_{1,\text{INT}}}{\gamma_{1,\text{INT}}}$

Our experiments revealed that the profile of $\delta\gamma_1/\gamma_1$, (*i.e.*, comparing γ_1 at a same *depth*), varies linearly as the composition is varied. An example is illustrated in the left column of Fig. 1. However, the profile $\delta\gamma_{1,\text{int}}/\gamma_{1,\text{int}}$ shows non-linear behavior when the abundance difference between two models is high. The purpose of this section is to identify where the discrepancy between $\delta\gamma_1/\gamma_1$ and $\delta\gamma_{1,\text{int}}/\gamma_{1,\text{int}}$ results from.

From Eq. 1,

$$\frac{\delta\gamma_1}{\gamma_1} = \left(\frac{\partial \ln \gamma_1}{\partial \ln P}\right)_{Y,\rho} \frac{\delta P}{P} + \left(\frac{\partial \ln \gamma_1}{\partial \ln \rho}\right)_{Y,P} \frac{\delta \rho}{\rho} + \left(\frac{\partial \ln \gamma_1}{\partial Y}\right)_{P,\rho} \delta Y + \frac{\delta\gamma_{1,\text{int}}}{\gamma_{1,\text{int}}},$$

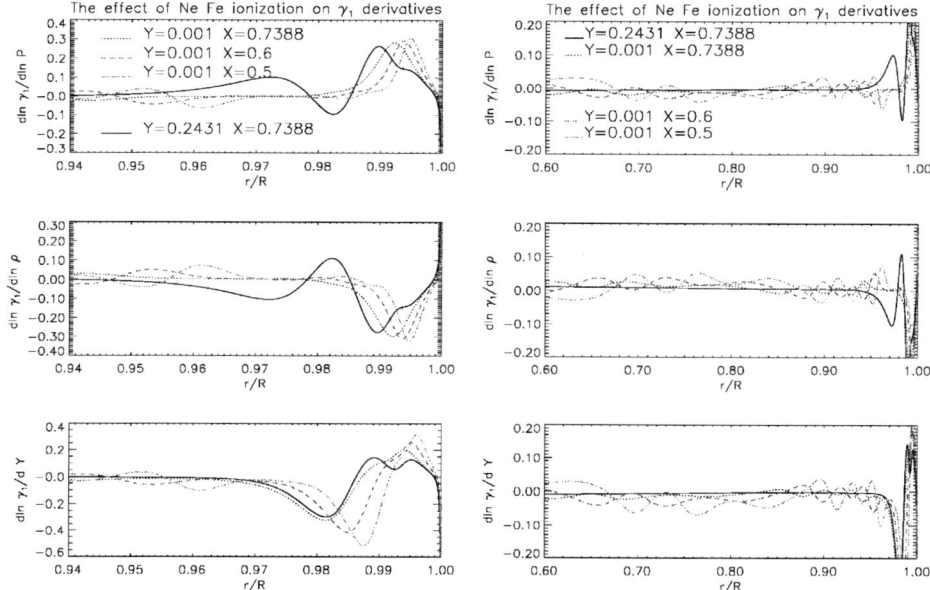

FIGURE 7. The profiles of the three derivative terms that correspond to $Y = 0.001$. The left column is simply a zoom-in version of the right column. The top row is $(\partial \ln \gamma_1 / \partial \ln P)_{Y,\rho}$, the middle row illustrates $(\partial \ln \gamma_1 / \partial \ln \rho)_{Y,P}$ and the bottom row shows $(\partial \ln \gamma_1 / \partial Y)_{P,\rho}$

we see that there are three partial derivative terms that could contribute to the discrepancy, $\left(\dfrac{\partial \ln \gamma_1}{\partial \ln P}\right)_{Y,\rho} \dfrac{\delta P}{P}$, $\left(\dfrac{\partial \ln \gamma_1}{\partial \ln \rho}\right)_{Y,P} \dfrac{\delta \rho}{\rho}$ and $\left(\dfrac{\partial \ln \gamma_1}{\partial Y}\right)_{P,\rho} \delta Y$.

Therefore, we computed each derivative to see how the profile of the term is altered by the adjustment of composition (see Fig. 7 and 8).

Comparing Fig. 7 and 8, we observed that the modification of the abundance affects the derivatives mainly by *shifting* the locations of the features (*i.e.*, dips and bumps), which are manifestations of element ionizations. The relative amplitudes of the features are related to the abundance of the element. The results suggest that varying the abundance changes the thermal structure (e.g., the temperature through the structure), which results in shifting of the ionization zones.

The results in the two figures indicate that reducing Z, in our two experiments, shifts the curves deeper, which could translate to lowering the temperature at a given depth.

CONCLUSION

The adiabatic gradient, γ_1, has been a useful tool to examine the EOS and chemical composition. In this paper, we carried out a series of experiments to study the response of

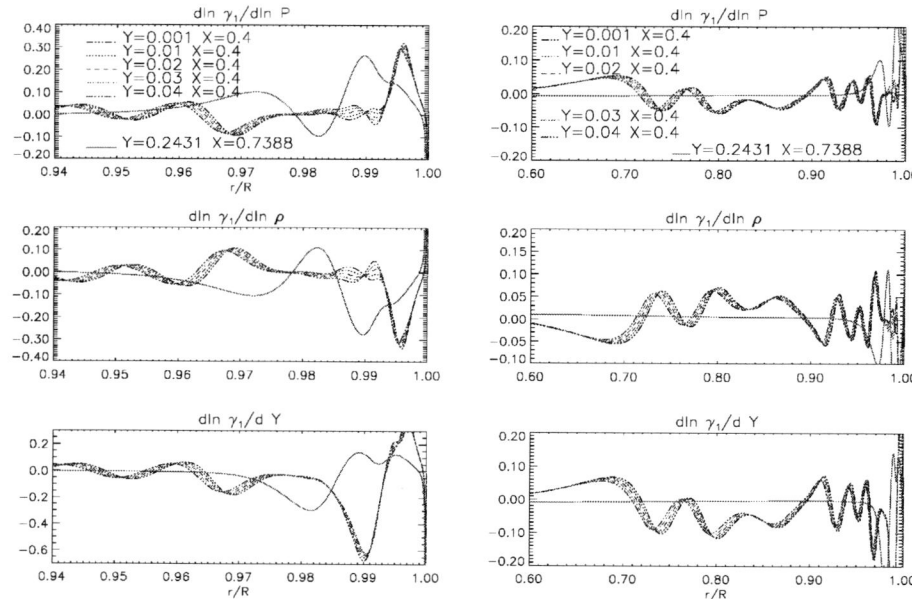

FIGURE 8. Same as in Fig. 7. The profiles of the three derivative terms at $X = 0.4$.

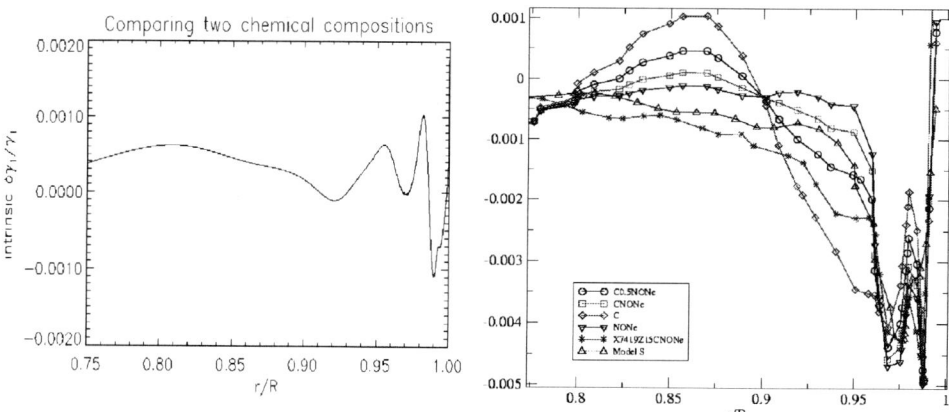

FIGURE 9. The left panel: $\delta\gamma_{1,\mathrm{int}}/\gamma_{1,\mathrm{int}}$ between a model with the old chemical composition [6] and a model with the latest chemical composition [1]. The right panel: Inverted $\delta\gamma_{1,\mathrm{int}}/\gamma_{1,\mathrm{int}}$ between the Sun and the models. The squares are the result of using a reference model implemented with the old chemical composition [6].

γ_1 when the chemical composition is varied at a fixed EOS formalism. We found that $\delta\gamma_{1,\text{int}}/\gamma_{1,\text{int}}$ is not always linearly correlated to the magnitude of the abundance difference between two models. In contrast to this non-linearity exhibited in the experiments on varying the abundance difference between two models, the effects from the ionizations of multiple elements satisfy the linear superposition relation. That is, the total effect of multiple ionizations is equal to the sum of the effect from individual ionization. In addition, we also found that increasing Y at fixed X creates similar effects as adding a fully ionized element does.

An examination on the derivative terms in Eq. 1 suggested that the linear behavior is *not* affected by the abundance of the element(s) in examination, but by the abundance *difference* between two models. When the abundance difference between two models is large (*e.g.*, $> 10\%$ of the abundance of the reference model), the curves of the derivatives would be significantly shifted, which leads to the irregular behavior in $\delta\gamma_{1,\text{int}}/\gamma_{1,\text{int}}$. However, the variation of $\delta\gamma_{1,\text{int}}/\gamma_{1,\text{int}}$ remains linear if the abundance difference is less than 5%.

The experiments in this paper showed that adjusting the abundance of a *single* element within realistic range results in a relatively small $\delta\gamma_{1,\text{int}}/\gamma_{1,\text{int}}$ of the order of magnitude 10^{-4}, compared to 10^{-3} found between the Sun and a solar model. However, the combined effect of adjusting the abundance of multiple (or all) elements can lead to a significant effect. In Fig. 9, we computed $\delta\gamma_{1,\text{int}}/\gamma_{1,\text{int}}$ between a model using previously accepted chemical composition [6] and a model implemented with the latest chemical composition [1]. The figure shows that the adjustment of the chemical composition alone produced $\delta\gamma_{1,\text{int}}/\gamma_{1,\text{int}}$ comparable to that between the Sun and a model. Specifically, the positive values in the region above 0.9R can greatly compensate the negative values in the sun–model inversion. This indicates the importance of separating the effect from chemical composition and the effect from EOS formalism. This study is the first step to achieving this ultimate goal.

ACKNOWLEDGMENTS

This work was supported by a PRTLI research grant for Grid-enabled Computational Physics of Natural Phenomena (Cosmogrid) and by the grant AST-0307578 of the National Science Foundation.

REFERENCES

1. Asplund, M., Grevesse, N., Sauval, A. J., Allende Prieto, C., and Kiselman, D., *The Astronomy and Astrophysics Journal*, **417**, 751–768 (2004).
2. Eggleton, P. P., Faulkner, J., and Flannery, B. P., *The Astronomy and Astrophysics Journal*, **23**, 325–+ (1973).
3. Basu, S., and Christensen-Dalsgaard, J., *The Astronomy and Astrophysics Journal*, **322**, L5–L8 (1997).
4. Iglesias, C. A., and Rogers, F. J., *The Astrophysical Journal*, **443**, 460–463 (1995).
5. Christensen-Dalsgaard, J., and Däppen, W., *The Astronomy and Astrophysics Review*, **4**, 267–361 (1992).
6. Grevesse, N., and Sauval, A. J., *Space Science Reviews*, **85**, 161–174 (1998).

The chemical composition and equation of state of the Sun inferred from seismic models through an inversion procedure.

Chia-Hsien Lin* and Werner Däppen[†]

*Armagh Observatory, College Hill, Armagh BT61 9DG, N. Ireland, U.K.
[†]Department of Physics and Astronomy, University of Southern California, Los Angeles, California 90089-1342, U.S.A.

Abstract. The present helioseismic data, with their remarkable precision, provide the possibility to investigate the microscopic physics of the solar interior. In this study, we are specifically interested in the chemical composition and equation of state (EOS). Since these properties are not directly measurable, we chose the adiabatic gradient, $\gamma_1 \equiv \partial \ln P / \partial \ln \rho |_s$, as the probe to examine these properties. Specifically, we infer the discrepancies in the chemical composition and EOS by the discrepancy in γ_1, i.e., $\delta \gamma_1 / \gamma_1$, between two solar structures.

Even though the heavy elements only constitute less than 2% of the solar material, our inversion results showed that the variation in the relative abundances among heavy elements would result in discernible features in $\delta \gamma_1 / \gamma_1$.

INTRODUCTION

In the study of the solar structure, the EOS not only plays a role in the hydrostatic balance, where it links the density and pressure, but also provides essential information (*e.g.*, the ionization degree of the chemical elements) for the computation of the opacity.

Through most part of the solar interior, however, the effects of EOS are inseparable from other microphysical properties, such as, opacity and nuclear reaction rate. Fortunately, the solar convective zone provides a possibility for the separation. One reason is that the nuclear reaction is negligible in this region. The second reason is that the opacity does not affect the structure in the convective zone because energy is mainly transported by convection rather than radiation. Therefore, the structure of the convection zone is mostly determined by EOS and the chemical composition.

The chemical composition on the solar surface has been determined by spectroscopy with high precision [see *e.g.* 1]). However, the helium abundance and the chemical composition below the surface can only be determined by helioseismology. The abundance in the solar interior can provide crucial information for the mechanisms altering the chemical composition over the life of the Sun.

The goal of our study is to reveal the errors in the currently accepted values of the interior chemical composition and the equation of state through an inversion procedure, based on the discrepancy in the adiabatic gradient, γ_1, at same thermodynamic points between the Sun and a solar model. Once the errors in the equation of state are reduced, the helium abundance, which is an essential parameter in solar evolution, can be better determined.

THE INVERSION METHOD

Our inversion method is based on a differential inversion scheme. A detailed description can be found in, [e.g., 2, 3]. The basics of the method is presented in the following.

The linear relation between the relative frequency difference, $\delta\omega/\omega$, and the relative structural differences is:

$$\frac{\delta\omega_i}{\omega_i} = \int_0^R K_{u,Y}^i(r)\frac{\delta u}{u}(r)dr + \int_0^R K_{Y,u}^i(r)\delta Y(r)dr$$
$$+ \int_0^R K_{c^2,\rho}^i(r)\frac{\delta\gamma_{1,\text{int}}}{\gamma_{1,\text{int}}}(r)dr + \frac{F_{\text{surf}}(\omega_i)}{Q_i}. \quad (1)$$

K : kernels; $u \equiv P/\rho$; Y : Helium abundance; c : adiabatic sound speed $\delta\gamma_{1,\text{int}}/\gamma_{1,\text{int}}$: The relative γ_1 difference at same thermodynamic points (intrinsic γ_1 difference).

$$\frac{F_{\text{surf}}(\omega_i)}{Q_i} = \frac{\Sigma_\lambda^\Lambda P_\lambda(\omega_i)}{Q_i} \quad (2)$$

Q_i is the mode inertia normalized by the inertia of the radial mode with same frequency.

The aim is to find a set of coefficients $\{c_i(r_0)\}$ such that an average of the property of interest (i.e., $\delta\gamma_{1,\text{int}}/\gamma_{1,\text{int}}$) around a target point r_0 can be determined. That is,

$$\left\langle \frac{\delta\gamma_{1,\text{int}}}{\gamma_{1,\text{int}}}(r_0) \right\rangle = \sum_i c_i(r_0)\frac{\delta\omega_i}{\omega_i}$$
$$\approx \int_0^R \mathcal{T}(r_0,r)\frac{\delta\gamma_{1,\text{int}}}{\gamma_{1,\text{int}}}(r)dr \quad (3)$$

where, $\left\langle \frac{\delta\gamma_{1,\text{int}}}{\gamma_{1,\text{int}}}(r_0) \right\rangle$ is the average of $\delta\gamma_{1,\text{int}}/\gamma_{1,\text{int}}$ centered at r_0, $\mathcal{T}(r_0,r)$ is a normalized target function which peaks at r_0 and is small elsewhere.

REFERENCE MODELS

Our reference model is a seismic model which satisfies all the boundary conditions at the surface but deviates from the reality near the core; hence, the model is called *envelope model*. The equation of state implemented in our models is CEFF [4], and the opacity table implemented is OPAL [5]. Each model was assigned a specific chemical composition, which is set to be constant through the entire region of computation. Therefore, our envelope program [described in 6] computes models from the surface inward to the region where the assumption of constant chemical composition can no longer be valid.

The chemical composition of each model is reflected in the name of the model (cf. Table 1).

TABLE 1. The specifics of reference models

CNONe

EOS: CEFF	
Y: 0.2431 Z/X:0.0245	
Heavy element	Relative abundance
C	0.1907
N	0.0558
O	0.5430
Ne	0.2105

C

EOS: CEFF	
Y: 0.2431 Z/X:0.0245	
Heavy element	Relative abundance
C	1.0
N	0
O	0
Ne	0

NONe

EOS: CEFF	
Y: 0.2431 Z/X:0.0245	
Heavy element	Relative abundance
C	0.0
N	0.0690
O	0.6709
Ne	0.2601

X7419Z15CNONe

EOS: CEFF	
Y: 0.2431 Z/X:0.0202	
Heavy element	Relative abundance
C	0.1907
N	0.0558
O	0.5430
Ne	0.2105

C0.5NONe

EOS: CEFF	
Y: 0.2431 Z/X:0.0245	
Heavy element	Relative abundance
C	0.5
N	0.0340
O	0.3355
Ne	0.1301

Model S

EOS: OPAL	
Y: 0.2431 Z/X:0.0245	
Heavy element	Relative abundance
C	0.1907
N	0.0558
O	0.5430
Ne	0.2105

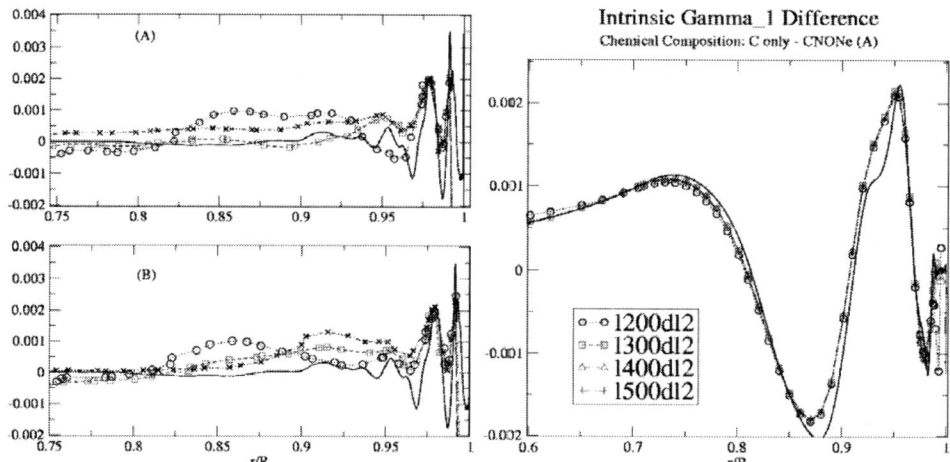

FIGURE 1. Inverted results of **ModelS - CNONe** (left panel) and of **C - CNONe** (right panel). Symbols represent the results from using different inversion parameters.

RESULTS

Model study

The purpose is to assess the quality of the inversion code. Two selected results are shown in Fig.1.

Real-Sun study

Our inversion results for each reference model are plotted in Fig.2. To demonstrate the interesting physics indicated by these inversions, the inverted curves of all the models are plotted together in Fig.3.

We have made the following observations:

Between 0.75R and 0.95R.

- There are two locations, 0.8R and 0.9R, where $\delta\gamma_{1,int}/\gamma_{1,int}$ is unchanged by the variation of the carbon abundance (cf. circles, square, diamond and triangle down).
- As the carbon abundance decreases, the profile becomes flatter (compare, for instance, **Sun - CNONe** and **Sun - NONe**).
  Indicating that the carbon abundance might be lower than the previously determined value.
- **Model S** (triangle up in the graph) appears to be worse than **CNONe**(square) and **NONe**(triangle down) (*i.e.*, larger deviation from zero).

Suggesting that a seismic model can be more accurate than an evolution model.
- The curve of (**Sun-X7419Z15CNONe**) is more negative than the curve of (**Sun-CNONe**). That is, a lower heavy-element abundance, Z, results in a larger adiabatic gradient. This can be explained as follows: Lower heavy-element abundance, Z, leads to less ionization of heavy elements, which, therefore, causes less reduction in the adiabatic gradient.

Between 0.95R and 0.99R.

- All CEFF models show a "W" feature in the region between 0.96R and 0.99R. This feature possibly results from the difference in the equations of state.
- The W feature can be slightly flattened by either reducing Z or reducing the abundance of carbon (compare, *e.g.*, curves of **CNONe, NONe** and **X7419Z15CNONe**).
- Our study has shown that varying the abundance of a given heavy element leads to a lowering or elevating of the magnitude of $\delta\gamma_{1,int}/\gamma_{1,int}$.
- Such an adjustment of the heavy-element composition could become a measurement of the heavy-element abundance in the Sun, provided that at the same time, the uncertainty in equation of state can be reduced, either with the help of theoretical studies or laboratory experiments.

DISCUSSION AND CONCLUSION

Conclusions from the model study:.

- The inversion procedure is sufficiently sensitive and accurate to reveal the small effects ($\sim 10^{-4}$) resulting from the modification of heavy-element abundance.
- The procedure is robust against the current observational error: Our inverted results were unaffected by gaussian-distributed errors ($\sim 10^{-1} - 10^{-2} \mu Hz$)
- If inverted results are sensitive to tuning inversion parameters, it is likely that the function to be inverted contains features narrower than or comparable to the width of averaging kernels. Nevertheless, the accuracy of such unstable inversion can be greatly improved by implementing an appropriate surface term.

Conclusions from the real-sun study:.

- Our inverted $\delta\gamma_1/\gamma_1$ between the Sun and model S is consistent with the results by Basu, Däppen and Nayfonov (1999).
- $\delta\gamma_1/\gamma_1$ between the Sun and all of our CEFF models show two prominent dips with the magnitude $\approx 0.004 - 0.005$.
- Reducing carbon abundance in CEFF models reduces $\delta\gamma_1/\gamma_1$ below 0.95R, which could indicate that carbon in the deeper region is less abundant than at the surface. This finding is in the same direction as suggested by very recent spectroscopic studies [7].

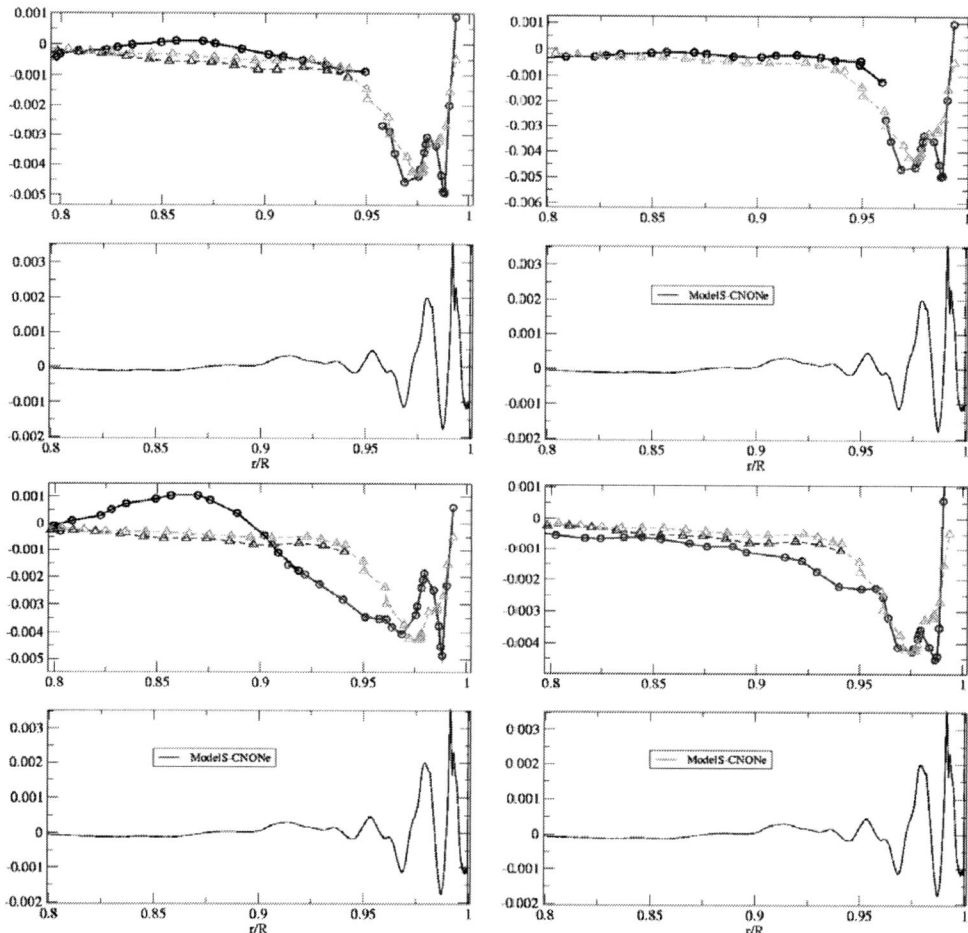

FIGURE 2. The plots illustrate the resemblance of the inverted $\delta\gamma_{1,int}/\gamma_{1,int}$ profile of **Sun-various CEFF models** and the profile of **ModelS-CNONe**. (a): **Sun - CNONe**; (b): **Sun - NONe**; (c):**Sun - C**; (d): **Sun - X7419Z15CNONe** In each plot, the upper panel: the circles are the inverted $(\delta\gamma_{1,int}/\gamma_{1,int})$, and triangles are the inverted result of **Sun-ModelS**. the lower panel: The expected $(\delta\gamma_{1,int}/\gamma_{1,int})$ curve.

The ultimate goal of our study:

Surface composition of heavy elements (determined by spectroscopy) $\Rightarrow$ **the determination of the correct EOS** $\Rightarrow$ **the determination of Y and chemical composition in the deeper region**

$\Rightarrow$ **Revealing the process of solar evolution**

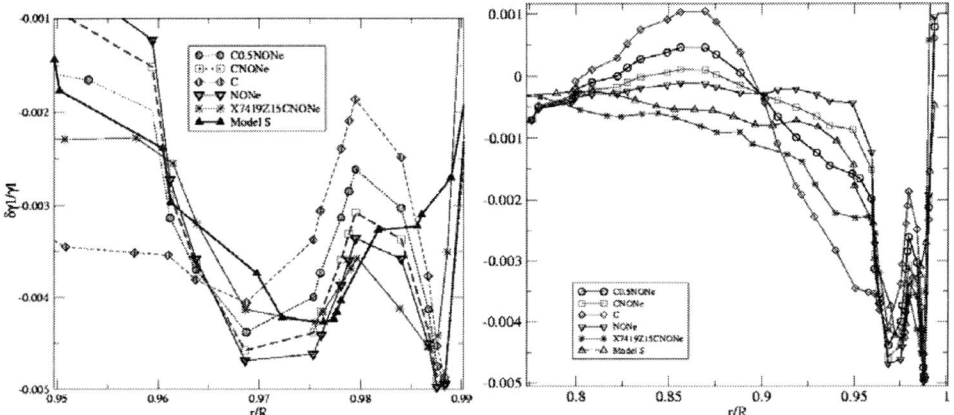

FIGURE 3. Inverted $(\delta\gamma_{1,\text{int}}/\gamma_{1,\text{int}})$ between the Sun and various reference models.

ACKNOWLEDGMENTS

This work was supported by a PRTLI research grant for Grid-enabled Computational Physics of Natural Phenomena (Cosmogrid) and by the grant AST-0307578 of the National Science Foundation.

REFERENCES

1. Grevesse, N., and Sauval, A. J., *Space Science Reviews*, **85**, 161–174 (1998).
2. Rabello-Soares, M. C., Christensen-Dalsgaard, J., and Thompson, M. J., "Seismic Constraints on Sound Speed in the Solar Core," in *Structure and Dynamics of the Interior of the Sun and Sun-like Stars SOHO 6/GONG 98 Workshop Abstract, June 1-4, 1998, Boston, Massachusetts, p. 511*, 1998, pp. 511–+.
3. Pijpers, F. P., and Thompson, M. J., *The Astronomy and Astrophysics Journal*, **281**, 231–240 (1994).
4. Eggleton, P. P., Faulkner, J., and Flannery, B. P., *The Astronomy and Astrophysics Journal*, **23**, 325–+ (1973).
5. Iglesias, C. A., and Rogers, F. J., *The Astrophysical Journal*, **443**, 460–463 (1995).
6. Christensen-Dalsgaard, J., and Däppen, W., *The Astronomy and Astrophysics Review*, **4**, 267–361 (1992).
7. Asplund, M., Grevesse, N., Sauval, A. J., Allende Prieto, C., and Kiselman, D., *The Astronomy and Astrophysics Journal*, **417**, 751–768 (2004).

PART IV

FROM WEAKLY TO STRONGLY COUPLED PLASMAS TO POSSIBLE PHASE TRANSITIONS

Phase transitions in dense hydrogen - helium plasmas

Vladimir Filinov*, Pavel Levashov*, Michael Bonitz[†] and Vladimir Fortov*

*Institute for High Energy Densities, Izhorskaya 13/19, 125412 Moscow, Russia
[†]Christian-Albrechts-Universität zu Kiel, Institut für Theoretische Physik und Astrophysik, Leibnizstr. 15, 24098 Kiel, Germany

Abstract. In this paper we study thermodynamic properties of hydrogen-helium mixtures with the help of direct Path-Integral Monte Carlo simulations. Presented are preliminary results of pressure and energy isotherms in the range from 10^4 K to $2 \cdot 10^5$ K in comparison with available theoretical and experimental results. In the density region, where experiments have observed a sharp conductivity rise the simulations yield indications for one or two plasma phase transitions, in accordance with earlier predictions.

INTRODUCTION

Many astrophysical problems require the knowledge of thermodynamic properties of hydrogen and helium [1–5]. To understand different effects in stellar structure and evolution one should provide accurate modelling of the underlying physics including equation of state (EOS) effects. In normal stars where there plasma is fully-ionized and almost ideal the construction of EOS doesn't reveal particular difficulties. However the investigation of the giant planets Jupiter and Saturn, and to a lesser extent brown dwarfs demands thermodynamic information for hydrogen and helium in the approximate range of temperatures 1000 K $< T <$ 100000 K and mass densities $0.01 < \rho < 100$ g/cm^3. In this region the complexity of an EOS calculation increases considerably when nonideal effects are compounded with chemical reactions associated with partial pressure dissociation and ionization equilibria [6–10]. Moreover, in this region the the so-called plasma phase transition (PPT) has been predicted [9, 10]. Significant efforts have been made in the last decades to understand the behaviour of dense fully-ionized and partially-ionized hydrogen and helium (see, for example, [8] and references therein). In these works mostly the chemical picture is applied for the calculation of thermodynamic properties. The chemical picture assumes that bound configurations, such as atoms and molecules, retain a definite identity and interact through pair potentials; in other words, this model is valid only at weak interparticle interactions. However at densities corresponding to pressure ionization the electrons in bound configurations become delocalized and bound species lose their identity [11]. Therefore there is a great interest in direct first-principle numerical simulations of strongly coupled degenerate systems which avoid such approximations.

In this work we use the direct path integral Monte Carlo (DPIMC) method to calculate the thermodynamic properties of hydrogen - helium mixtures. This method is well established theoretically and allows the treatment of quantum and exchange effects without any preliminary physical approximations. Using the results of our simulations we analyze the problem of plasma phase transition in dense hydrogen - helium mixtures.

DIRECT PATH INTEGRAL MONTE CARLO

Path integral Monte Carlo [12, 13] is based upon Feynman's formulation of quantum-statistical mechanics using path integrals [14]. In this work we consider hydrogen - helium mixtures at temperatures from 10^4 K to $2 \cdot 10^5$ K and electron particle densities from 10^{20} to $3 \cdot 10^{24}$ cm^{-3}. Under such conditions electrons are degenerate while protons and α-particles can be treated as classical particles because of their relatively large masses. Thus for the case of electro-neutral hydrogen-helium plasma with volume V the partition function Z is given by

$$Z(N_e, N_p, N_\alpha, V, \beta) = \frac{1}{N_e! N_p! N_\alpha!} \sum_\sigma \int_V dq_p dq_\alpha dr \, \rho(q_p, q_\alpha, r, \sigma; \beta). \tag{1}$$

Here N_e, N_p, and N_α are the number of electrons, protons, and α-particles, $\beta = 1/k_B T$, T is the temperature, $q_p \equiv \{\mathbf{q}_{p1}, \mathbf{q}_{p2}, \cdots, \mathbf{q}_{pN_p}\}$, $q_\alpha \equiv \{\mathbf{q}_{\alpha 1}, \mathbf{q}_{\alpha 2}, \cdots, \mathbf{q}_{\alpha N_\alpha}\}$ are coordinates of protons and α-particles, respectively, $r \equiv \{\mathbf{r}_1, \mathbf{r}_2, \cdots, \mathbf{r}_{N_e}\}$ are the coordinates of electrons, and $\sigma \equiv \{\sigma_1, \sigma_2, \cdots, \sigma_{N_e}\}$ are the spin variables of the electrons. The density matrix in (1) is expressed via a path integral:

$$\int_V dR^{(0)} \sum_\sigma \rho(R^{(0)}, \sigma; \beta) = \sum_P \sum_\sigma (-1)^{\kappa_P} \int_V dR^{(0)} \cdots dR^{(n)} \times$$
$$\rho^{(1)} \rho^{(2)} \cdots \rho^{(n)} S(\sigma, \hat{P}\sigma') \hat{P}) \rho^{(n+1)}, \tag{2}$$

where $\rho^{(i)} = \rho(R^{(i-1)}, R^{(i)}; \Delta\beta) \equiv \left\langle R^{(i-1)} \left| e^{-\Delta\beta \hat{H}} \right| R^{(i)} \right\rangle$, $\Delta\beta \equiv \beta/(n+1)$, $\hat{P}$ is the permutation operator, κ_P is the parity of permutation, S is the spin matrix, $\hat{H}$ is Hamiltonian of the system, $\hat{H} = \hat{K} + \hat{U}_C$, $\hat{K}$ is the kinetic energy, $\hat{U}_C$ is the potential energy, consisting of Coulomb interaction of electrons (e), protons (p), and α-particles (α): $\hat{U}_C = \hat{U}_C^p + \hat{U}_C^e + \hat{U}_C^\alpha + \hat{U}_C^{ep} + \hat{U}_C^{e\alpha} + \hat{U}_C^{p\alpha}$. We denote particle coordinates as follows: $R^{(i)} = (q_p, q_\alpha, r_i)$, $i = 1, \cdots, n+1$, $R^{(0)} \equiv (q_p, q_\alpha, r)$, $R^{(n+1)} \equiv R^{(0)}$, $\sigma' = \sigma$. Thus electrons participating in the simulation are represented as fermionic loops with n vertexes: $[R] \equiv [R^{(0)}; R^{(1)}; \cdots; R^{(n)}; R^{(n+1)}]$. Exchange effects for Fermi statistics are taken into account by the permutation operator $\hat{P}$ and the sum over the permutations with parity κ_P. It is possible to reduce the expression (2) to a form in which the sum over all permutations is replaced by the determinant of the exchange matrix $\psi_{ab}^{n,1}$. This technique allows us to improve the accuracy of simulation for strongly degenerate plasma:

$$\sum_\sigma \rho(q_p, q_\alpha, r, \sigma; \beta) = \frac{1}{\lambda_p^{3N_p} \lambda_\alpha^{3N_\alpha} \lambda_\Delta^{3N_e}} \sum_{s=0}^{N_e} \rho_s([R], \beta),$$

$$\rho_s([R],\beta) = \frac{C_{N_e}^s}{2^{N_e}} \exp\{-\beta U([R],\beta)\} \prod_{l=1}^{n}\prod_{m=1}^{N_e} \phi_{mm}^l \det\left|\psi_{ab}^{n,1}\right|_s. \qquad (3)$$

Here $\lambda_p^2 = 2\pi\hbar^2\beta/m_p$, $\lambda_\alpha^2 = 2\pi\hbar^2\beta/m_\alpha$, $\lambda_\Delta^2 = 2\pi\hbar^2\Delta\beta/m_e$, m_p, m_α, m_e are the masses of proton, α-particle and electron, respectively. In equation (3) $U = U^{pp} + U^{\alpha\alpha} + U^{p\alpha} + \sum_{l=1}^{n}\{U_l^{ee} + U_l^{ep} + U_l^{e\alpha}\}/(n+1)$ and $\phi_{mm}^l \equiv \exp\left[-\pi\left|\xi_m^{(l)}\right|^2\right]$ are functions generated from the kinetic energy density matrix, $\xi^{(1)},\cdots,\xi^{(n)}$ are dimensionless distances between neighbor vertexes of fermionic loops which represent electrons $[R] \equiv [R^{(0)}; R^{(0)} + \lambda_\Delta\xi^{(1)} + \lambda_\Delta(\xi^{(1)} + \xi^{(2)}; \cdots]$. Elements of the exchange matrix $\psi_{ab}^{n,1}$ are defined by the expression:

$$\left\|\psi_{ab}^{n,1}\right\|_s \equiv \left\|\exp\left\{-\frac{\pi}{\lambda_\Delta^2}|(r_a - r_b) + y_a^n|^2\right\}\right\|, \quad y_a^n = \lambda_\Delta \sum_{k=1}^{n} \xi_n^{(k)}.$$

The index s stands for the number of electrons with same spin projection.

As a high-temperature density matrix one can use its asymptote in the limit $T \to \infty$. Every N-particle high-temperature density matrix can be represented as a product of two-particle density matrices. For the two-particle density matrix there is an analytical solution of the Bloch equation by first-order perturbation theory [16]:

$$\rho(\mathbf{r}_a,\mathbf{r}'_a,\mathbf{r}_b,\mathbf{r}'_b,\beta) =$$
$$\frac{m_a m_b}{(2\pi\hbar^2\beta)^3}\exp\left[-\frac{m_a}{2\hbar^2\beta}(\mathbf{r}_a - \mathbf{r}'_a)^2\right]\exp\left[-\frac{m_b}{2\hbar^2\beta}(\mathbf{r}_b - \mathbf{r}'_b)^2\right]\exp\left[-\beta\Phi^{ab}\right],$$

where $\Phi^{ab}(\mathbf{r}_a,\mathbf{r}'_a,\mathbf{r}_b,\mathbf{r}'_b,\beta)$ — nondiagonal effective two-particle pseudopotential:

$$\Phi^{ab}(\mathbf{r}_{ab},\mathbf{r}'_{ab},\beta) = e_a e_b \int_0^1 \frac{d\alpha}{d_{ab}(\alpha)} \mathrm{erf}\left(\frac{d_{ab}(\alpha)}{2\lambda_{ab}\sqrt{\alpha(1-\alpha)}}\right). \qquad (4)$$

Here $d_{ab}(\alpha) = |\alpha\mathbf{r}_{ab} + (1-\alpha)\mathbf{r}'_{ab}|$, $0 \le \alpha \le 1$, $\mathrm{erf}(x) = 2/\sqrt{\pi}\int_0^x \exp(-t^2)\,dt$ denotes the error function, $\lambda_{ab}^2 = \hbar^2\beta/2\mu_{ab}$, e_a, e_b are the charges of particles, m_a, m_b are the masses of particles, $\mu_{ab}^{-1} = m_a^{-1} + m_b^{-1}$ are reduced mass. In the limit of high temperature two-particle nondiagonal effective potential can be approximated by a half-sum of diagonal pseudopotentials (4):

$$\Phi^{ab}(|\mathbf{r}_{ab}|,\Delta\beta) = \frac{e_a e_b}{\lambda_{ab} x_{ab}}\left\{1 - \exp(-x_{ab}^2) + \sqrt{\pi}x_{ab}[1 - \mathrm{erf}(x_{ab})]\right\}, \qquad (5)$$

where $x_{ab} = |\mathbf{r}_{ab}|/\lambda_{ab}$. It is worth to underline that $\Phi^{ab}(|\mathbf{r}_{ab}|,\Delta\beta)$ tends to a finite value at $x_{ab} \to 0$ and to the Coulomb potential $e_a e_b/x_{ab}$ at $x_{ab} \to \infty$. It is proved that the pseudopotential (5) coincides with an exact quantum potential at temperatures $T > 2\cdot 10^5$ K [17]. U^{ln} and U_l^{ln} in Eq. (3) is the sum of the effective quantum pair interaction between two charged plasma particles described by the Φ^{ab}.

All thermodynamic properties can be expressed through the partition function derivatives. For example, pressure and total energy are given by the formulas:

$$E = -\beta \partial \ln Z / \partial \beta,$$
$$\beta P = \partial \ln Q / \partial V = [\eta/3V \partial \ln Q / \partial \eta]_{\eta=1} . \qquad (6)$$

Multiple integrals in the formulas (6) are calculated with the help of the standard Metropolis technique in a cubic cell with periodic boundary conditions [12]. The accuracy ε of the calculations depends on the number of factors n in the equation (2), temperature T and electron degeneracy parameter $\chi = n_e \lambda_e^3$ and is given by expression $-\varepsilon \sim (\beta \text{Ry})^2 \chi / (n+1)$, where n_e is the particle density of electrons, $\lambda_e^2 = 2\pi \hbar^2 \beta / m_e$ [18]. According to this estimation to simulate a Coulomb system at the temperature 10^4 K it is sufficient to choose $n = 20$. High temperature density matrix in Eq. (2) relates in this case to temperature higher than Ry.

SIMULATION RESULTS

We tested our computational scheme by many ways. First of all we calculated thermodynamic properties of ideal hydrogen plasma [19] and found very good agreement up to degeneracy parameter $\chi = 10$. To extend the region of degeneracy parameter we improved the treatment of exchange effects. Unlike the previous version of the method [12] in this work we take into account exchange effects not only inside the main Monte Carlo cell but also with the neighboring periodic images. It is necessary to include such procedure into the algorithm if the electron thermal wavelength is comparable or larger than the size of the Monte Carlo cell. Thus the exchange interaction was calculated in the nearest 3^3, 5^3 etc. Monte Carlo cells in accordance with the value of electron thermal wavelength. The accuracy of exchange effects treatment was controlled by comparing the results of calculations with analytical dependences for pressure and energy of ideal degenerate plasma.

We also studied interacting hydrogen plasma in a wide range of temperatures and particle densities [20, 21]. The DPIMC method allowed us to investigate the effects of temperature and pressure dissociation and ionization *ab initio*. From the analysis of pair distribution functions we observed the formation and break-up of molecules and atoms under different conditions. At very high density we also observed the effect of proton ordering indicating the formation of Coulomb crystal. We found rather good agreement with the calculations performed by other methods at small and medium densities. However at high values of plasma density in the region of pressure ionization no reliable analytical methods exist.

The simulation results for hydrogen plasma in the region of temperatures from $T = 10^4$ K to 10^6 K and electron particle densities from $n_e = 10^{22}$ cm^{-3} to 10^{24} cm^{-3} allowed us to calculate the deuterium shock Hugoniot [22]. It is interesting to note that the resulting curve is located between the experimental data of Knudson *et al.* [23] and Collins *et al.* [24].

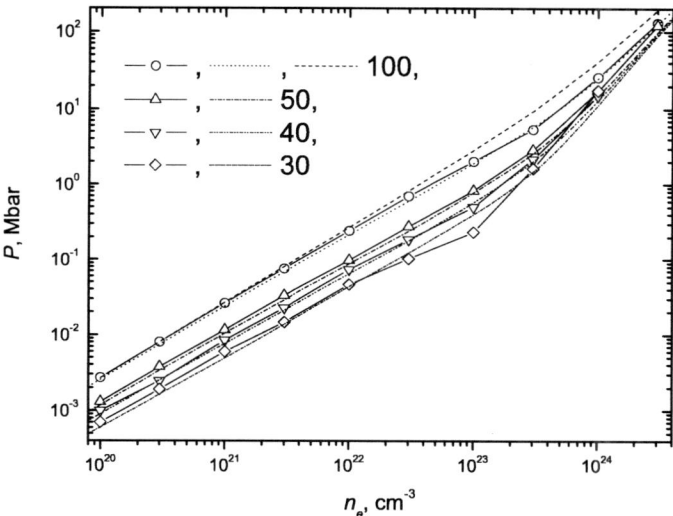

FIGURE 1. Pressure in a hydrogen-helium mixture with the mass concentration of helium $Y = 0.234$. Shown are isotherms calculated with the help of DPIMC method (lines with symbols) and related isotherms from [2, 7] (lines without symbols). Shown on the picture numbers are temperatures in units of thousand Kelvin. For $T = 100000^0$ K the dashed line presents the EOS for ideal plasma.

In this work we apply our computational scheme to the thermodynamic properties simulations of hydrogen-helium mixture with a composition corresponding to that of the outer layers of the Jovian atmosphere. During the mission of the Galileo spacecraft the helium abundance in the atmosphere of Jupiter was determined as $Y = m_{He}/(m_{He} + m_H) = 0.234$ and was close to the present-day protosolar value $Y = 0.275$. As the model of the Jupiter is significantly determined by its composition and EOS it is interesting to simulate the thermodynamic properties of the mixture with such composition in the region of pressure dissociation and ionization where traditional chemical models of plasma fail.

We carried out the calculations of thermodynamic properties of hydrogen - helium mixture in the region of temperatures from $T = 10^4$ K to $2 \cdot 10^5$ K and electron densities from $n_e = 10^{20}$ cm^{-3} to $3 \cdot 10^{25}$ cm^{-3}. First of all we compare our results with available data on hydrogen-helium mixture of [2, 7]. This model is based on the chemical picture with classical statistics for molecules and ions and Fermi-Dirac statistics for the electrons. It takes into account a lot of physical effects including dissociation and ionization, interactions between charged particles and neutral atoms and molecules, neutral-neutral interactions, high-pressure screening effects, excited electronic states of molecules as well as a number of "second-order" phenomena. Owing to the complexity of of the

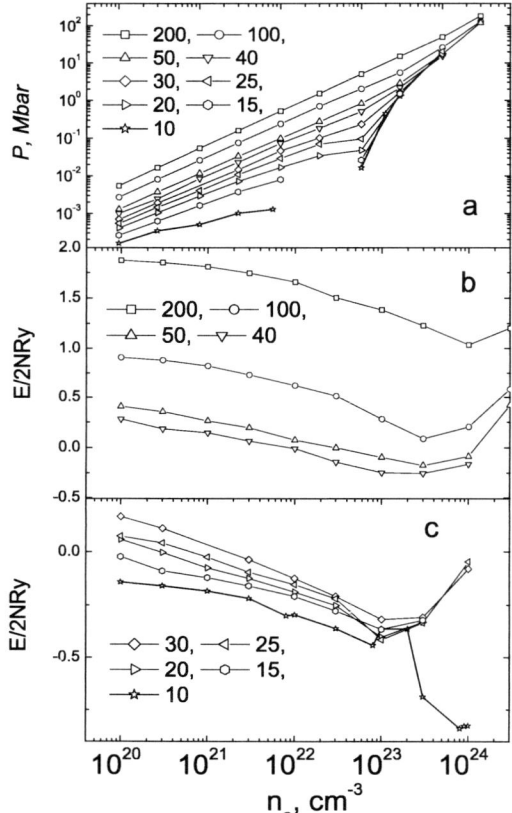

FIGURE 2. Pressure a) and energy b,c) in a hydrogen-helium mixture with the mass concentration of helium $Y = 0.234$. Shown are isotherms calculated with the help of DPIMC method. Temperature is given in units of thousand Kelvin.

model [2, 7] equations of state for hydrogen and helium are presented in tabular form [2]. Thermodynamic properties of hydrogen-helium mixtures can then be calculated by interpolation in composition between the two pure EOS. Using the so-called "linear mixing" it is possible approximately to calculate the density $\rho(P,T)$ of the hydrogen-helium mixture with the mass fraction of helium Y at pressure P and temperature T:

$$\frac{1}{\rho(P,T)} = \frac{1-Y}{\rho^H(P,T)} + \frac{Y}{\rho^{He}(P,T)}.$$

The results of comparison are shown in Fig. 1. The agreement between our calculations and the model [2] along the isotherms $T = 3 \cdot 10^4, 4 \cdot 10^4, 5 \cdot 10^4$, and 10^5 K is quite good and becomes better with the increase of temperature. The smaller values of pressure on the DPIMC isotherm $3 \cdot 10^4$ K near the particle density value 10^{23} cm^{-3} can be explained by a strong influence of bound states in this region; these effects are taken into account only approximately in the model [2]. The formation of atoms and molecules is also the reason of the pressure reduction along the 10^5 K isotherm with respect to the isotherm of non-interacting hydrogen-helium mixture.

In Fig. 2 the dependency of pressure vs. electron particle density along isotherms is shown. At high temperatures the isotherms have no peculiarities. However along the isotherm $T = 2 \cdot 10^4$ K there is a region for $n_e > 3 \cdot 10^{23}$ cm^{-3}, where our simulations do not reach the equilibrium state, the pressure strongly fluctuates and even becomes negative. Also along the isotherms $T = 10^4$ K and $T = 1.5 \cdot 10^4$ K there are two such regions 10^{22} cm$^{-3} < n_e < 10^{23}$ cm^{-3} and $n_e > 3 \cdot 10^{23}$ cm^{-3}. Earlier we found a similar effect for pure hydrogen at $T = 10^4$ K in the region of pressure ionization and showed that in the transition region a number of large clusters (droplets) were formed [25, 26]. Such behavior is typical for Monte Carlo simulations of metastable systems. In this region of pressure ionization the PPT was predicted by many authors [6, 7, 9–11, 27–29] and moreover a sharp electrical conductivity rise was measured in [30]. These instabilities in our calculations indicate the existence of PPT in dense hydrogen. Later [13] we found the PPT and the formation of droplets in electron-hole plasma of germanium semiconductor at low temperature and found good agreement with experimental data [31]. The appearance of clusters (droplets) in plasma leads to a drop of total energy of the system and Fig. 2 illustrates this fact. On isotherms one can see minima corresponding to the region where recombination of hydrogen molecules occurs. Under these conditions we observed the formation of molecules from the pair distribution functions analysis. The number of hydrogen molecules increases with the temperature drop. At low temperatures $T = 10^4$ and $2 \cdot 10^4$ K we observe the formation of large clusters of atoms and molecules.

The problem of PPT in a hydrogen-helium mixture is significantly determined by the composition of the mixture [2, 7, 29]. From shock-wave experiments one can estimate the range of temperature and density where a sharp electrical conductivity rise takes place. In quasi-isentropic compression the transition from a low-conductivity state to a high-conductivity one for hydrogen occurs at $T \sim 3 - 15$ kK and $\rho = 0.4 - 0.7$ g/cm^3 [30, 32] whereas for helium at $T = 15 - 40$ kK and $\rho = 0.7 - 1.25$ g/cm^3 [33]. However it is not enough to determine the region of existence of the PPT. According to theoretical equations of state for hydrogen and helium based upon the quantum statistical approach the critical point of the PPT in pure hydrogen is $T_{crH}^{(1)} = 14.9$ kK, $P_{crH}^{(1)} = 0.723$ Mbar in Ref. [29] and $T_{crH}^{(2)} = 15.3$ kK and $T_{crH}^{(2)} = 0.61$ Mbar in Refs. [2, 7]. In pure helium the critical point was found to be $T_{crHe}^{(1)} = 17$ kK, $P_{crHe}^{(1)} = 7.22$ Mbar [29]. At helium mass concentration $Y < 0.93$ and temperature less than both critical temperatures the properties of hydrogen-helium mixture are determined mostly by hydrogen and only one PPT exists. At high values of $Y > 0.93$ both the hydrogen and helium PPT can occur for the same temperature. In our DPIMC simulations we have observed one PPT at $T = 2 \cdot 10^4$ K and two PPTs at $T = 10^4$ K even at $Y = 0.234$. The results of our simulation are shown in Fig. 3 together with experimental data and theoretical predictions.

Along the isotherm $T = 2 \cdot 10^4$ K we found the region with bad convergence in the

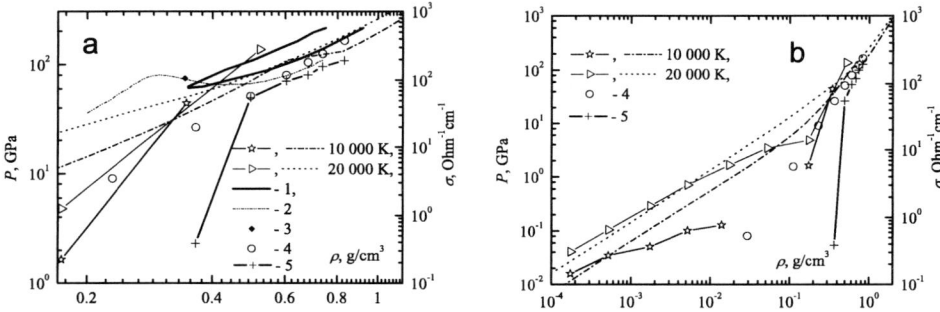

FIGURE 3. The DPIMC (lines with asterisks and triangles) and theoretical (dashed and dash-dotted lines) [2, 7] isotherms for pressure in a hydrogen-helium mixture with the mass concentration of helium $Y = 0.234$ vs. density. 1 — phase boundaries of PPT in hydrogen [2, 7], 2 — undercritical metastable isotherm $T = 1.2 \cdot 10^4$ K and $Y = 0.308$ [29], 3 — critical point of PPT in hydrogen-helium mixture with $Y = 0.308$ [29]. Also shown are experimental results of Ref. [34]: 4 — quasi-isentrope of hydrogen-helium mixture, $T \sim 5000$ K, 5 — electrical conductivity of hydrogen along the quasi-isentrope (right axis). Left picture is the enlargement of the high density part of the right figure.

range of densities between 0.5 and 5 g/cm^3. Along the isotherm $T = 1.5 \cdot 10^4$ K and $T = 10^4$ K such region is even wider and begins from 0.38 g/cm^3. Surprisingly there was another region where pressure became negative: from 0.015 to 0.19 g/cm^3. The nature of this phenomenon is currently unclear for us. From Fig. 3 it can be easily seen that other predictions of PPT in hydrogen or hydrogen-helium mixtures [2, 29] with low mass concentration of helium are located in the beginning of the region where DPIMC simulation fails to converge to the equilibrium state. The sharp rise of electrical conductivity of hydrogen-helium mixture along the quasi-isentrope with the initial state $T = 77.4$ K and $P = 8.1 \cdot 10^{-3}$ GPa is also observed experimentally in the range of densities 0.5–0.83 g/cm^3 [34], see line with crosses in Fig. 3a.

ACKNOWLEDGMENTS

This work is done under financial support of RF President Grant No. MK-1769.2003.08, the RAS program No. 17 "Parallel calculations and multiprocessor computational systems", the grant for talented young researchers of the Science support foundation and the Deutsche Forshungsgemeinschaft under grant BO 1366-2. The research was also partly sponsored by Award No. PZ-013-02 of the U.S. Civilian Research & Development Foundation for the Independent States of the Former Soviet Union (CRDF) and of Ministry of Education of Russian Federation. The calculations were perfomed thanks to a grant for CPU time at the NIC Jülich.

REFERENCES

1. Chabrier, G., Saumon, D., Hubbard, W. B., and Lunine, J. I., *Astrophys. J.*, **391**, 817–826 (1992).
2. Saumon, D., Chabrier, G., and Van Horn, H. M., *Astrophys. J. Suppl. Ser.*, **99**, 713–741 (1995).
3. Gudkova, T. V., and Zharkov, V. N., *Planet. Space Sci.*, **47**, 671–677 (2000).
4. Nellis, W. J., *Planet. Space Sci.*, **48**, 671–677 (2000).
5. Beule, D., Ebeling, W., Förster, A., Juranek, H., Redmer, R. and Röpke, G., *Journal de Physique IV, (Proceedings)*, **10, Pr5**, 295–299 (2000).
6. Saumon, D., and Chabrier, G., *Phys. Rev. A*, **44**, 5122–5141 (1991).
7. Saumon, D., and Chabrier, G., *Phys. Rev. A*, **46**, 2084–2100 (1992).
8. Fortov, V. E., Ternovoi, V. Y., Zhernokletov, M. V., Mochalov, M. A., Mikhailov, A. L., Filimonov, A. S., Pyalling, A. A., Mintsev, V. B., Gryaznov, V. K., and Iosilevskii, *JETP*, **97**, 259–278 (2003).
9. Norman, G. E., and Starostin, A. N., *Sov. Phys. High Temp.*, **6**, 410 (1968).
10. Ebeling, W., Kraeft, W. D., and Kremp, D., *Theory of Bound States and Ionization Equilibrium in Plasmas and Solids*, Akademie-Verlag, Berlin, 1976.
11. Kraeft, W. D., Kremp, D., Ebeling, W., and Röpke, G., *Quantum Statistics of Charged Particle Systems*, Akademie, Berlin, 1986.
12. Zamalin, V. M., Norman, G. E., and Filinov, V. S., *The Monte-Carlo Method in Statistical Thermodynamics*, Nauka, Moscow, 1977.
13. Filinov, V. S., Bonitz, M., Levashov, P. R., Fortov, V. E., Ebeling, W., Schlanges, M., and Koch, S. W., *J. Phys. A.: Math. Gen.*, **36**, 6069–6076 (2003).
14. Feynman, R. P., *Statistical Mechanics*, Advanced Book Program, Reading, Massachusetts, 1972.
15. Kalman, G., editor, *Strongly Coupled Coulomb Systems*, Pergamon, Oxford, 1988.
16. Ebeling, W., Hoffmann, H. J., and Kelbg, G., *Contrib. Plasma Phys.*, **7**, 233 (1967).
17. Filinov, A. V., Bonitz, M., and Ebeling, W., *J. Phys. A: Math. Gen.*, **36**, 5957–5962 (2003).
18. Filinov, V. S., Bonitz, M., and Ebeling, W., and Fortov, V. E. *Plasma Phys. Contr. Fusion*, **43**, 743 (2001).
19. Filinov, V. S., Levashov, P. R., Fortov, V. E., and Bonitz, M., "Thermodynamic properties of correlated strongly degenerate plasmas," in *Progress in Nonequilibrium Green's Functions*, edited by M. Bonitz, World Scientific, Singapore, 2000, pp. 513–520.
20. Filinov, V. S., Bonitz, M., and Fortov, V. E., *JETP Letters*, **72**, 245–248 (2000).
21. Filinov, V. S., Fortov, V. E., Bonitz, M., and Kremp, D., *Phys. Lett. A*, **274**, 228–235 (2000).
22. Bezkrovniy, V., Filinov, V. S., Kremp, D., Bonitz, M., Schlanges, M., Kraeft, W. D., Levashov, P. R., and Fortov, V. E., *Phys. Rev. E* (accepted, 2004).
23. Knudson, M. D., Hanson, D. L., Bailey, J. E., Hall, C. A., and Assay, J. R., *Phys. Rev. Lett.*, **90**, 035505-1 (2003).
24. Da Silva, L. P., Celliers, P., et al., *Phys. Rev. Lett.*, **78**, 483–486 (1997).
25. Filinov, V. S., Fortov, V. E., Bonitz, M., and Levashov, P. R., *JETP Letters*, **74**, 384 (2001).
26. Levashov, P. R., Filinov, V. S., Fortov, V. E., and Bonitz, M., "Thermodynamic Properties of Nonideal Strongly Degenerate Hydrogen Plasma," in *Shock Compression of Condensed Matter — 2001*, edited by M. D. Furnish, N. N. Thardhani, and Y. Horie, AIP, New York, 2002, pp. 119–126.
27. Saumon, D., and Chabrier, G., *Phys. Rev. Lett.*, **62**, 2397 (1989).
28. Yan, X., Tsai, S., and Ichimaru, S., *Phys. Rev. A*, **43**, 3057 (1991).
29. Schlanges, M., Bonitz, M., and Tschttschjan, A., *Contrib. Plasma Phys.*, **35**, 109–125 (1995).
30. Ternovoi, V. Y., Filimonov, A. S., Fortov, V. E., Kvitov, S. V., Nikolaev, D. N., and Pyalling, A. A., *Physica B*, **265**, 6–11 (1999).
31. Thomas, G. A., Rice, T.-M., and Hensel, J. C., *Phys. Rev. Lett.*, **33**, 219–222 (1974).
32. Weir, S. T., Mitchell, A. C., and Nellis, W. J., *Phys. Rev. Lett.*, **76**, 1860–1863 (1996).
33. Ternovoi, V. Y., Filimonov, A. S., Pyalling, A. A., Mintsev, V. B., and Fortov, V. E., "Thermophysical properties of helium under multiple shock compression," in *Shock Compression of Condensed Matter — 2001*, edited by M. D. Furnish, N. N. Thardhani, and Y. Horie, AIP, New York, 2002, pp. 107–110.
34. Ternovoi, V. Y., Kvitov, S. V., Pyalling, A. A., Filimonov, A. S., and Fortov, V. E., *JETP Letters*, **79**, 8–11 (2004).

Quantum statistical approach to dense, weakly coupled plasmas

J. Vorberger*, M. Schlanges* and W.–D. Kraeft*

*Institut für Physik, Ernst-Moritz-Arndt-Universität, 17487 Greifswald, Germany

Abstract. We present a quantum equation of state (EOS) which is applicable to weakly coupled systems of any density and temperature. As an example, we consider the EOS of hydrogen. We include strong ion-ion correlations through HNC–calculations. We study the influence of correlation effects on the mean kinetic energy.

INTRODUCTION

Dense plasmas can be found in nature as well as under laboratory conditions. We want to mention particularly dense hydrogen which is of special interest in astrophysics and for laser or heavy ion induced fusion experiments.

Due to the long range character of the Coulomb potential and Fermi statistics, dense plasmas show many interesting physical effects and are a very active field of research. There are very interesting experimental investigations which require a better theoretical description of matter under such conditions. In Fig. 1, some physical objects and experiments together with characterizing parameters are shown.

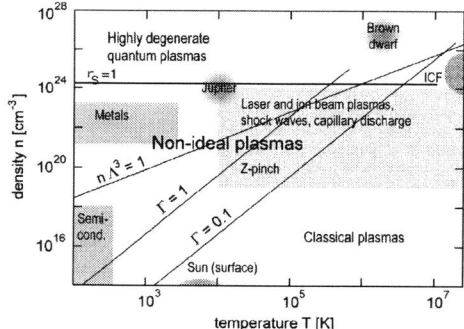

FIGURE 1. Density–temperature plane. $\Gamma = e^2/k_B T d$ – classical coupling parameter, $n\Lambda^3$ – degeneracy parameter, $r_s = d/a_B$ – Brueckner parameter ($d = (3/4\pi n)^{1/3}$ – mean particle distance, $\Lambda = \sqrt{2\pi\hbar/mk_B T}$).

There exists a variety of simulation techniques and analytical models for theoretical investigations of the equation of state (EOS) of dense plasmas. Analytical quantum statistical theories, e.g. the method of Green's functions (GF), provide reliable results in the limiting cases of low and high degeneracy, since no restrictions have to be made concerning the degeneracy of the plasma. Numerical simulation methods are expected to correctly include strong correlations. Both kinds of methods can complement each other and an essential task is to compare analytical and numerical calculations in parameter regions where agreement has to be expected.

In this paper, we calculate thermodynamic properties as well as single particle properties for the one component electron gas and for dense fully ionized hydrogen. We apply the method of Green's functions to these systems. The approximation used is valid for weakly coupled plasmas at any degeneracy. Additionally, we take strong ion-ion correlations into account. This is done using classical HNC techniques. We compare our results with results from PIMC and WPMD simulations.

BASIC THEORY & APPROXIMATIONS

We consider a multi component many particle system consisting of electrons and different nuclei. In the framework of quantum statistical theory, the pressure p is given by the charging formula [1]

$$(p-p_0)\Omega = -\int_0^1 \frac{d\lambda}{\lambda} \langle \lambda V \rangle \quad (1)$$

with $\langle \lambda V \rangle$ being the mean value of the potential energy. p_0 is the pressure of an ideal gas, Ω the volume of the system, λ the charging parameter. In Coulomb systems, screening of the bare two particle Coulomb potential has to be taken into account. An appropriate expression for $\langle \lambda V \rangle$ is then given by [1]

$$\langle \lambda V \rangle = \frac{1}{2} \sum_{a\sigma_a} \sum_{b\sigma_b} \int d1 dr_2 \{\lambda V_{ab}(12) G_a(11^{++}) G_b(22^+)$$
$$+ V_{ab}^s(12,\lambda) \Pi_{ab}(121^{++}2^+)\} \quad (2)$$

where V_{ab}^s is the dynamically screened potential, Π_{ab} is the polarization function, and G_a is the one particle Green's function. The sums run over the species and the spins. **1** means $(\mathbf{r}_1, t_1, \sigma_1)$, $\mathbf{1}^+$ denotes the same variables, but at an infinitesimal later time than **1**: $t_1^+ = t_1 + 0$.

Using Feynman graphs, the following expansion for the polarization function can be given in the case of weak coupling [3]

$$\Pi(12) \approx \quad \rangle\!\langle + \rangle\!\!\!\!\!\rangle\!\langle + \rangle\!\!\!\!\!\rangle\!\langle + \rangle\!\!\!\!\!\rangle\!\langle \quad . \quad (3)$$

Here, a line corresponds to a free Green's function. A wavy line denotes a dynamically screened potential. Insertion of Eq. (3) into Eq. (2) results in an expression for the mean potential energy

$$\langle \lambda V \rangle \approx \quad + \quad + \quad + \quad + 2 \quad . \quad (4)$$

This is the approximation of the EOS we use. It includes all terms with two interaction lines or less. The Hartree term (H) (leftmost) considers mean field effects. The next, the Hartree–Fock term (HF), accounts for exchange effects and so do the last two terms of Eq. (4), the normal and anomal e^4 exchange contributions. The only term including a screened potential is the Montroll–Ward term (MW) which is the third diagram in Eq. (4). It accounts for scattering and collective effects of the plasma.

EQUATION OF STATE

After evaluation of all terms of Eq. (4) without restrictions, we get, using Eq. (1), the pressure as a function of the fugacity. A corresponding expression for the density as function of the fugacity is also obtained. We invert this equation to get the pressure–density dependence.

We use an incomplete inversion (golden rule) [2, 3] and may, therefore, omit the anomal e^4 exchange term. In the $T=0$ limit this term is exactly compensated by further terms of the expansion, and in the non-degenerate limit, it is of higher order in the density. Finally, we arrive at the following expression for the EOS

$$p(\{\beta\mu_c^0\}) = \sum_a \frac{2\sigma_a+1}{\beta\Lambda_a^3} I_{3/2}(\beta\mu_a^0) + \sum_a \frac{(2\sigma_a+1)e_a^2}{\Lambda_a^4} I_{HF}(\beta\mu_a^0)$$
$$+ p^{MW}(\{\beta\mu_c^0\}) + \sum_a p_a^{e^4n}(\beta\mu_a^0) - \sum_a \frac{(2\sigma_a+1)e_a^2}{\Lambda_a^4} I_{-1/2}(\beta\mu_a^0) I_{1/2}(\beta\mu_a^0)$$
$$- \sum_a k_B T \frac{I_{1/2}(\beta\mu_a^0)}{I_{-1/2}(\beta\mu_a^0)} \frac{\partial}{\partial \beta\mu_a} \left(\beta p^{MW}(\beta\mu_a) + \beta p_a^{e^4n}(\beta\mu_a) \right) \Big|_{\beta\mu_a^0}. \quad (5)$$

The density is calculated from $n_a(\beta\mu_a^0) = 2/\Lambda_a^3 \cdot I_{1/2}(\beta\mu_a^0)$. $\beta = 1/k_B T$, μ_a^0 is the free chemical potential of species a, I_ν denotes the Fermi integral of order ν, $I_{HF} = \int_{-\infty}^{\beta\mu_a} d\alpha\, I_{-1/2}^2(\alpha)$. This equation represents what we will call e^4-approximation.

We applied our EOS to dense hydrogen plasmas (see Fig. 2). We choose parameter regions such that bound states do not play an essential role. Comparing our results with those from different other theoretical approaches, we see a quite good agreement in the low density region. In the region of higher coupling or higher degeneracy, respectively, some deviations can be found and have to be discussed further. We want to mention in this context that, although the electron–electron and the electron–ion couplings are weak, the ion–ion coupling still can be strong.

STRONG ION-ION-CORRELATIONS

The area of strong correlations in the ionic subsystem is, because of the ion mass, much larger than the region of strong electron–electron coupling. At high densities the ions form a strongly coupled, classical Coulomb liquid or a lattice screened by the degenerate electron background. To account for strong ion-ion Coulomb correlations, we make use

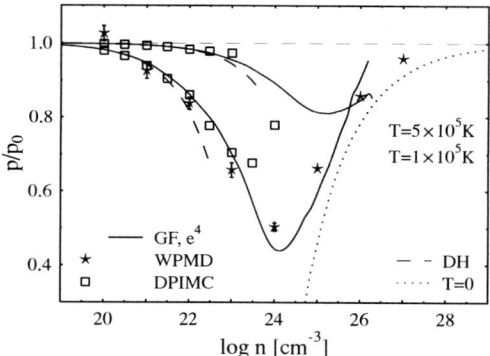

FIGURE 2. Ratio of pressure to ideal pressure for hydrogen as function of the density for various temperatures. Our results (solid) compared to DPIMC results [6] (squares connected by lines), WPMD data [4] (stars, $T = 10^5 K$), and limiting results (DH – Debye Hueckel, $T = 0$ – Gell-Mann & Brueckner).

of the high ion mass and describe the ionic subsystem with the help of methods of classical statistics.

Using the Ornstein-Zernicke (OZ) integral equation and the hypernetted chain closure relation one can sum up infinite Mayer clusters and consider strong correlations. The OZ relation (in HNC approximation) [8]

$$h(\mathbf{r}_{12}) = c(\mathbf{r}_{12}) + N(\mathbf{r}_{12}), \qquad N(\mathbf{r}_{12}) = n \int d\mathbf{r}_3 \, c(\mathbf{r}_{13}) h(\mathbf{r}_{32}) \qquad (6)$$

with h being the total correlation function and c the direct correlation function serves as defining equation for the binary distribution function g

$$g(\mathbf{r}_{12}) = h(\mathbf{r}_{12}) + 1 = e^{-\beta \phi(\mathbf{r}_{12}) + N(\mathbf{r}_{12})}, \qquad (7)$$

which can be used to compute the pressure of the ionic subsystem

$$p_{pp} = nk_B T - \frac{2\pi n^2}{3} \int_0^\infty dr \, r^3 \frac{\partial \phi(r)}{\partial r} (g(r) - 1) . \qquad (8)$$

The only input quantity is the ion-ion potential ϕ. This is the bare Coulomb potential, which was used here. Combining Eq. (8) with our original EOS, one has to avoid double counting of contributions to the EOS. Both, MW–term and the ion-ion HNC term contain the low density Debye-Hueckel limiting law, which, therefore, was subtracted from one of the contributions.

Results for the EOS, now including strong ion-ion correlations, can be seen in Fig. 3. Due to the inclusion of this correlations the depth of the minimum is reduced and shifted to smaller densities. Now our curve agrees better with the simulation data.

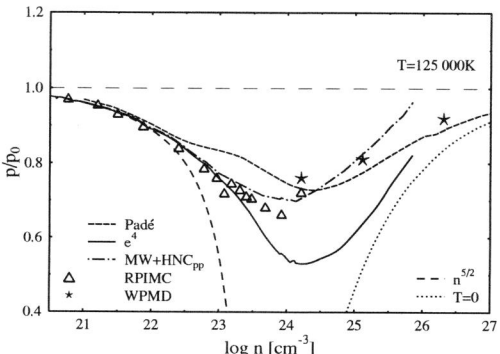

FIGURE 3. Ratio of pressure to ideal pressure for hydrogen as function of the density at temperature $T = 1.25 \times 10^5 K$. RPIMC data from [5], WPMD data from [4], and low density expansion up to $n^{5/2}$ due to [7], our results: (Eq. (5)) is denoted as e^4, "MW+HNC$_{pp}$" additionally includes strong ion-ion correlations.

CORRELATION EFFECTS ON THE MEAN KINETIC ENERGY

Now we want to analyze in detail how the single particle properties in a dense plasma are influenced by quantum and correlations effects. The internal energy U consists of the sum of the mean kinetic $\langle K \rangle$ and mean potential energy $\langle V \rangle$

$$U = \langle H \rangle = \langle K \rangle + \langle V \rangle = \langle K \rangle^{id} + U^{corr} \tag{9}$$

where U^{corr} contains all correlation effects. These can be split in contributions due to the mean potential energy and in correlation corrections to the mean kinetic energy $\langle K \rangle^{corr}$

$$U^{corr} = \langle K \rangle^{corr} + \langle V \rangle, \quad \langle K \rangle = \langle K \rangle^{id} + \langle K \rangle^{corr}. \tag{10}$$

In classical systems, $\langle K \rangle = \langle K \rangle^{id}$ is always valid. In contrast, in interacting quantum systems, $\langle K \rangle$ includes correlations as the momentum distribution deviates from that of an ideal gas [9]

$$\langle K \rangle = \int \frac{d\mathbf{p}}{(2\pi)^3} \frac{p^2}{2m} f(p). \tag{11}$$

The behavior of the mean kinetic energy is defined by the Wigner function $f(p)$

$$f(p) = \int \frac{d\omega}{2\pi} A(p,\omega) f(\omega) \quad f(\omega) \ldots \text{Fermi-function}, \tag{12}$$

where correlation effects enter by the spectral function $A(p,\omega)$

$$A(p,\omega) = \frac{2\,\text{Im}\Sigma(p,\omega)}{\left[\omega - \frac{p^2}{2m} - \text{Re}\Sigma(p,\omega)\right]^2 + \left[\,\text{Im}\Sigma(p,\omega)\right]^2}. \tag{13}$$

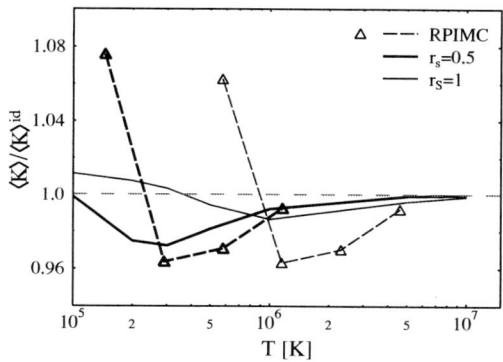

FIGURE 4. Ratio of the mean kinetic energy to its ideal part versus temperature for an electron gas at two densities (red: $r_s = 0.5$, black: $r_s = 1$). Comparison of our results with RPIMC data [10].

The self energy Σ has to be determined in some approximation leading to the inclusion of correlations in the Wigner function and thus in the mean kinetic energy. This method is numerical more difficult than to use our EOS scheme to calculate the internal energy and the mean potential energy instead. The latter scheme was usesd here. The difference between the two quantities is the mean kinetic energy.

As can be seen in Fig. 4, this quantity can be bigger or smaller than the ideal kinetic energy. This is a pure quantum mechanical effect, which cannot be found in classical systems, because in this case the distribution function $f(p)$ is not changed even for systems with correlations. We find some agreement with RPIMC simulation of Militzer et al. [10]. At $r_s = 0.5$ nearly the same lowering is found from both calculations. At $r_s = 1$ our approach gives only about one third of the lowering of Militzers result.

ACKNOWLEDGMENTS

We would like to acknowledge fruitful discussions with V. Filinov, B.Militzer, H.E. De-Witt, D.O. Gericke, and D. Kremp. We thank V. Filinov, M. Knaup, and B. Militzer for providing us with data. This work was supported by the Deutsche Forschungsgemeinschaft through SFB 198 "Kinetik partiell ionisierter Plasmen".

REFERENCES

1. W.D. Kraeft, D. Kremp, W. Ebeling and G. Röpke, *Quantum Statistics of Charged Particle Systems*, Akademie–Verlag Berlin, 1986
2. W. Stolzmann, W.D. Kraeft, Annalen der Physik **36**, 388 (1979)
3. J. Vorberger, M. Schlanges, W.D. Kraeft, Phys. Rev. E **69**, 046407 (2004)
4. M. Knaup, P.-G. Reinhard, C. Toepffer, Contr. Plasma Phys. **41**, 159 (2001), M. Knaup, private communication

5. B. Militzer, PhD, University of Illinois, Urbana, 2000,
   B. Militzer, D.M. Ceperley, Phys. Rev. E **63**, 066404 (2001)
6. V.S. Filinov, M. Bonitz, W. Ebeling, V.E. Fortov, Plasma Phys. Control. Fusion **43**, 743 (2001),
   S.A. Trigger, W. Ebeling, V.S. Filinov, V.E. Fortov, M. Bonitz, arXiv 0110013 (2001),
   V.S. Filinov, private communication
7. A. Alastuey, A. Perez, Europhys. Lett. **20**, 19 (1992)
   J. Riemann, M. Schlanges, H.E. DeWitt, W.D. Kraeft, Physica A **219**, 423 (1995)
8. L.E. Reichl, *A modern course in statistical physics* 2nd ed., Wiley–Interscience New York, 1998
9. W.D. Kraeft, M. Schlanges, J. Vorberger, H.E. DeWitt, Phys. Rev. E **66**, 046405 (2002)
10. C. Pierleoni et al., Phys. Rev. Lett. **73**, 2145 (1994)
    B. Militzer, E.L. Pollock, Phys. Rev. Lett. **89**, 280401 (2002)

Spinodal Decomposition of Metastable Melting in the Zero-Temperature Limit

Igor L. Iosilevski and Alexander Yu. Chigvintsev

Moscow Institute of Physics and Technology (State University) Dolgoprudny 141700, Russia

Abstract. "Conventional" scenario of metastable melting in ordinary substances in the limit of zero temperature assumes that the melting curve reaches the matter zero isotherm ("cold curve"). The same is true for standard variant of one-component plasma model on rigid compensating background in both limits: classical and "cold" quantum melting. The modified OCP on uniform, but compressible background shows the completely different scenario of the metastable melting closure. The remarkable feature of this scenario is that the liquid freezing curve terminates at liquid spinodal curve of 1st-order liquid gas phase transition, which takes place in this type of OCP models ("spinodal decomposition").

INTRODUCTION

Melting and freezing of Coulomb particles is common phenomena in many astrophysical objects [1,2,3]. One-component plasma model (OCP) in its standard variant with rigid compensating background (notified below as OCP(#)) is widely used for theoretical description of melting process of Coulomb (Wigner) crystal [4,5]. The modified OCP on uniform, but compressible background (notified below as OCP(~)) enriched noticeably combination of phase transitions in the model due to addition of 1st-order phase transitions of gas-liquid and gas-solid type with upper critical point [6]. Study of phase transitions in OCP(~) makes it possible to learn general picture of thermodynamically stable and metastable melting and freezing in wide range of thermodynamic parameters including in particular deeply metastable coexistence of extended crystal with overcooled liquid [7]. Recent progress in the technique of dynamic experiments of reaching deep negative pressure in metastable extended condensed matter in crystal and fluid states [8] raises a question about behavior of metastable melting in the limit of zero temperature. Important data to analyze this problem can be obtained from a numerical simulation of phase transitions in sufficiently realistic many-particles systems. Besides the analysis of the problem could be supplemented the study of idealized modeling systems where the features of phase transitions can be calculated directly due to simplified modeling nature of these systems. Present paper is devoted to study of various scenarios of hypothetical closure of zero-temperature limit melting. The analysis is based on the features of idealized models. First of all it is the line of so-called "non-associative" coulomb models [6,7,9] the number of modified versions of well-known prototype – one-component plasma model with rigid uniform compressible background.

"Conventional" Scenarios of Metastable Melting in Ordinary Substances in the Limit $T \Rightarrow 0$

It is assumed according to this scenario (for example [10]) that metastable melting curve reaches the matter 'cold curve' (isotherm $T = 0$). It is assumed also that this hypothetical metastable branch of melting curve is in close agreement with so-called Simon's law (1). Phase diagram with such behavior is shown on the Figure 1.

$$P_{melting} = AT^C + P^*; \qquad (A, C \text{ and } P^* = \text{const.}) \qquad (1)$$

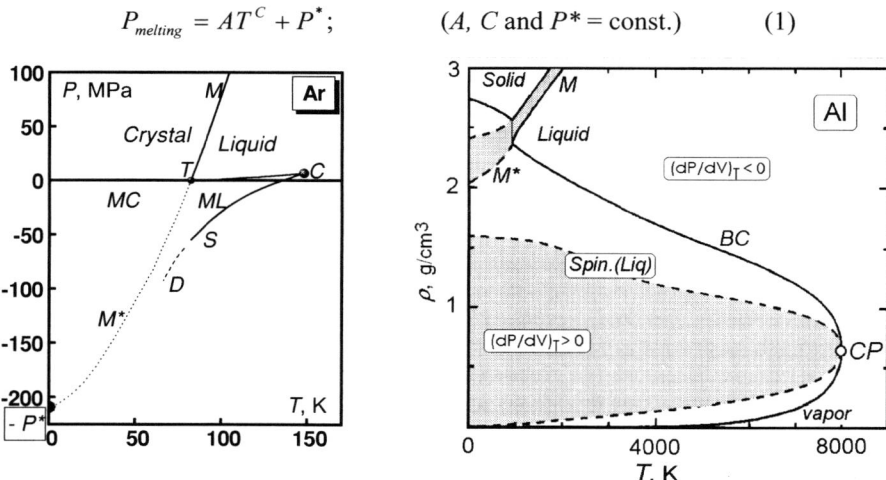

FIGURE 1. Phase diagram of argon with hypothetical metastable melting (Figure from [10]): *TM* – melting curve; *TC* – boiling curve; *T* – triple point; *C* – critical point; *MC* – metastable crystal; *ML* – metastable liquid; *CSD* – liquid spinodal; *TM*P** – metastable branch of melting curve (Simon's Law); *P** – hypothetical zero-temperature limit of metastable melting.

FIGURE 2. Density-temperature phase diagram of Al via semi-empirical EOS [11] constructed under recommendations [10] (Figure from [11]): *BC* – boiling curve; *CP* – critical point; *M* – melting; *M** – hypothetical metastable melting; *Spin(Liq)* – liquid spinodal $\{(\partial P/\partial V)_T = 0\}$.

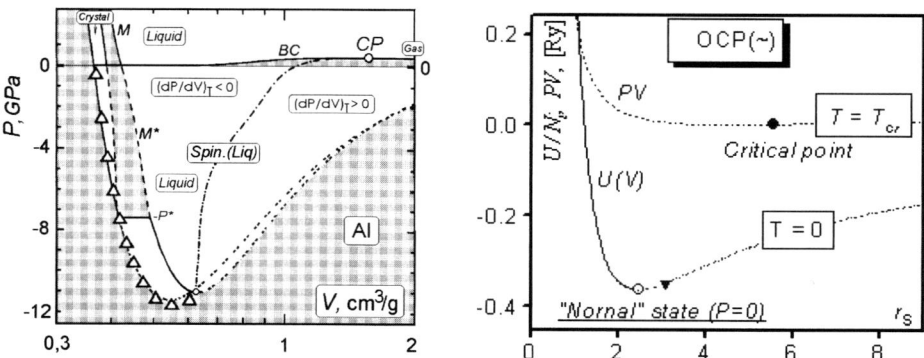

FIGURE 3. The same as Fig. 2 in *P,V*-plane (pressure-specific volume): *CP* – gas-liquid critical point; *Triangles* – ab initio calculation of crystal phase at $T = 0$ [12] (Figure from [11]).

FIGURE 4. "Cold curve" ($U(V)$ at $T = 0$) and critical isotherm (PV at $T = T_{cr}$) vs. Bruckner parameter of background electrons $\{r_S \sim V^{1/3}\}$ in modified OCP(~) model (Figure from [7]). *Solid circle* – critical point; *Open circle* – 'normal' state ($T = 0, P = 0$); *Solid triangle* – solid spinodal $\{(\partial P/\partial V)_T = 0\}$

The same scenario is incorporated in construction of some semi-empirical equation of state (EOS) of metals [11]. Hypothetical multi-phase density-pressure diagram of Al exposed at Figure 2 as an example of such EOS. Fundamental feature of this scenario of zero-temperature metastable melting is a location of crystal-fluid phase transition with a finite density gap just on the thermodynamically stable repulsive part of "cold curve" i.e. isotherm $T = 0$. Such hypothetical 'cold curve' for Al [11] is shown on Figure 3. It should be stressed that isotherm $T = 0$ coincides with the isoentrope $S = 0$, therefore the exposed 'cold' melting at $T = 0$ [11] occurs without entropy change.

Melting in Standard One Component Plasma Model OCP(#)

Standard variant of one component plasma model on rigid background (following notation – OCP(#)) is a system of ions on rigid compensative background. It contains only one phase transition – Wigner crystallization, which takes place without any density gap i.e. melting zone is just a one-dimensional curve. Melting line of crystal and freezing line of fluid coincides with each other. There are two well-known regime of melting in the model OCP(#):

1) Classical melting of non-degenerated ions in low density limit ($\theta \equiv kT/\varepsilon_F \gg 1$) In this case thermodynamic properties of OCP(#) depend only on dimensionless 'non-ideality' parameter $\Gamma \equiv (4\pi n/3)^{1/3}(Ze)^2/kT$. Melting corresponds to: $\Gamma = \Gamma_{melt} \approx 175$

2) "Cold" quantum melting of highly degenerated ions in high density limit [13] ($\theta \equiv kT/\varepsilon_F \gg 1$). In this case thermodynamic properties of OCP(#) depend on Bruckner 'non-ideality' parameter $r_S \equiv (3/4\pi na)^{1/3}$ Melting corresponds to: $r_S \approx \text{const} \approx 100$ [14,15].

General diagram of melting in OCP(#) is shown at Figure 5. It should be stressed that melting in OCP(#) takes place at any temperature and formally follows Simon's law (1) with ($A = \text{const.}, C = 4$ and $P^* \ 0$).

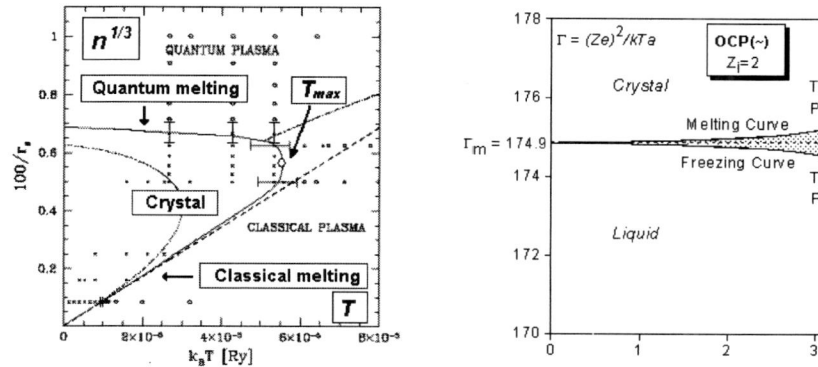

FIGURE 5. Boundary of Wigner crystal at OCP(#) with 'classical' and 'quantum' melting and hypothetical pseudocritical point of maximal melting temperature T_{max}. *Solid* – melting after [14]; *dash-dotted line* – melting line after [15]; *dashed line* – classical melting limit $\Gamma_{melt} \approx 175$ (Figure from [14]).
FIGURE 6. Melting zone of Wigner crystal in modified OCP(~) model [6] with splitted off melting and freezing curves and finite density gap, and low-density melting 'spinodal catastrophe' (From [7])

Scenarios of Metastable Melting in Modified OCP Model with Uniformly Compressible Background – OCP(~).

More realistic than OCP(#) is a model of one-component plasma with *uniform*, but *compressible* background. This property of the background can be defined self-consistently [16] and can be achieved, for example, with the use of *uniform* ideal Fermi-gas of electrons as a background for OCP positive ions. Due to *uniform compressibility* of background not one, but three 1st-order phase transitions appears in this model namely melting, evaporation and sublimation [6]. Melting in OCP(~) became a two-dimensional zone bounded by melting and freezing lines with finite density gap (Figure 6). Modified OCP model (OCP(~)) introduces principally new behavior of melting 'strip' in the limit $T \to 0$ i.e. in all the cases (see below) the melting zone *does not reach* the $T = 0$ isotherm (!). As a result the 'cold curve' in OCP(~) is smooth and continuous without any phase transition (Figure 4). As for the closure of metastable melting, there may be three variants of its termination in OCP(~) depending on charge number of ions, Z. All the variants occur at finite temperature. In first two variants (A) and (B), metastable melting terminates at the liquid or crystal spinodals. In third variant (C) melting transmitted continuously into sublimation, so that there is no metastable melting at all.

The formal definition of Helmholtz free energy in the OCP(~) model is [6]:

$$\left(\frac{F}{NkT}\right)_{OCP(\sim)} \equiv f = \frac{F^{id(e)\,gas}}{N_e kT} + \frac{\Delta F^{OCP(\#)}}{NkT} = f^{ig(e)\,gas} + \Delta f^{OCP(\#)} \tag{2}$$

The advantage of the OCP(~) model is due to prohibition (by definition) of any individual ion-electron correlation features of all phase transitions including characteristic of metastable branches can be directly calculated if properties of two individual subsystems namely ions with rigid background i.e. OCP(#) of ions and that of ideal Fermi-gas of electrons are known. Currently both the constituents are well studied via Monte Carlo and Molecular Dynamics simulations [5,14,15] etc. and effective analytical fits are exist, so that all the properties in 'non-associative' OCP(~) models could be calculated explicitly [6, 7, 9, 17].

Standard Scenario of Metastable Melting Termination in OCP(~)

Along with conventional scenario of metastable melting exposed at Figures 1-3 there is a fundamentally different ones which occur in the OCP(~) model. The first one (**A**) corresponds to the low values of ionic charge number $Z \ll Z^* \approx 35$ (see [18]) According to this variant in the limit of $T \to 0$ the boundary of freezing of metastable liquid touches a liquid spinodal at finite temperature. While the melting zone is approaching a boundary of liquid state absolute instability, i.e. liquid spinodal, its width is dramatically increasing. Figure 7 and 8 show brief and close view of this event i.e. crossing the freezing curve of metastable fluid and fluid spinodal. At all temperatures below the temperature of this intersection fluid is absolutely thermodynamically instable and doesn't exist even in metastable state. Thus the possibility of melting as physical phenomenon vanishes. The only possibility in this case is spontaneous decomposition of metastable system to thermodynamically stable two-phase mixture of solid and vapour at *finite temperature and pressure*!

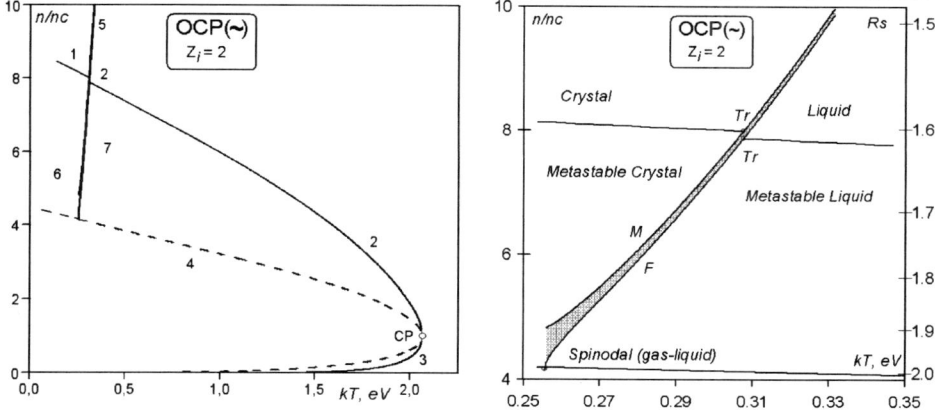

FIGURE 7. Spinodal decomposition of metastable melting in the limit of zero temperature. *Notation*: *1* – crystal; *2* – liquid; *3* – gas; *4* – liquid spinodal $\{(\partial P/\partial V)_T=0\}$; *5* – melting zone; *6*- metastable crystal; *7* – metastable liquid; *CP* – critical point (Figure from [7])
FIGURE 8. The same as Figure 6 in details. *Notation*: *M* – melting curve; *F* – freezing curve; *Tr* –triple point; *Spinodal(g-l)* – liquid spinodal $\{(\partial P/\partial V)_T=0\}$ (Figure from [7]).

Anomalous Scenarios of Metastable Melting Termination in OCP(~)

One of the methodical advantages of the OCP(~) model is additional model parameter – charge of ions Z. While the ion charge is increasing the topology of phase diagram is dramatically changing. At the small value of Z the model shows traditional view of phase diagram (see Figure 1 in [18]). But as the ionic charge number Z increases the melting zone moves toward the critical point of the fluid-crystal phase transition, passes it and finally shifts on the binodal of low-density phase.

In these cases there are additional two completely new scenarios (B) and (C) of metastable melting:

Scenario B. At large values of charge number Z ($Z >> Z^{**} >> 45$) the melting zone crosses a low-density slope of fluid-crystal co-existence curve (see Figure 1 in [18]). In this cases, as well as in the case (A) described above, the metastable melting zone also doesn't reach the zero isotherm (T=0) due to intersection of *melting line* of metastable crystal with *vapor spinodal* (line $\Gamma \approx 6$ at Figure 1 [18]).

Scenario C. This case corresponds to intermediate values of ionic charge number $35 \approx Z^* > Z > Z^{**} \approx 45$ (see [18]). In this case the melting line $\Gamma \approx 175$ of model-prototype OCP(#) falls just into the region of critical point of new 1st order phase transition of gas-liquid type in OCP(~). The model shows an exotic 'unified' boundary of unique crystal-fluid phase coexistence, which is continuous superposition of normal melting and sublimation equilibrium (crystal-fluid and crystal-gas) with common boundary of crystal-fluid two-phase area when melting smoothly turns to sublimation (see Figure 4 in [18]). In context of presently discussing problem of possible variants of metastable melting termination it means that the metastable melting as a phenomenon is absent in the system.

CONCLUSIONS

It should be emphasized that all the details of discussed above hypothetical scenarios of termination of metastable melting in the limit $T \to 0$ could be directly examined in numerical Monte Carlo and Molecular Dynamics simulations. The main problem, which should be overcome on this way, is the necessity of creating and preserving during a finite period of time a deep metastable state for both the competing phases, solid and liquid, with a deep negative pressure in the investigated simulation cell. Experience currently accumulated in such numerical simulation of simultaneous two-phase coexistence (see for example [19]) allows for hope that quick progress in this problem will be achieved soon.

ACKNOWLEDGEMENTS

The authors are grateful to Hugh DeWitt, G. Kanel, D. Yakovlev and A. Potekhin for helpful discussions. The work was supported by Grant CRDF № MO-011-0, Grant ISTC 2107, and by RAS Scientific Programs "Physics and Chemistry of Extreme States of Matter" and "Thermophysics ad mechanics of matter under high energy impact".

REFERENCES

1. Shapiro S. and Teukolsky S., *Black Holes, White Dwarfs, and Neutron Stars*, NY: Willey, 1983.
2. Chabrier G., Ashcroft N., DeWitt H. *Nature* **360**, 48 (1992).
3. Van Horn H. in *Strongly Coupled Plasma Physics*, edited by S. Ichimaru, Elsevier, 1990, p. 3.
4. Baus M., Hansen J.P, *Appl. Phys. Reports* **65**, 1 (1980).
5. Stringfellow G., DeWitt H., Slattery W., *Appl. Phys. Rev A* **41**, 1105 (1990).
6. Iosilevski I., *High Temperatures* **23**, 807 (1985).
7. Iosilevski I. and Chigvintsev A., in *Physics of Strongly Coupled Plasmas*, Edited by W. Kraeft and M. Schlanges., New Jersey-London: World Scientific, 1996, p.145.
8. Kanel G., Razorenov S., Baumung K. and Singer J. *J. Appl. Phys.* **90**(1), 136-143 (2001).
9. Iosilevski I., Chigvintsev A., "Phase Transition in Simplest Plasma Models" in *Physics of Nonideal Plasmas,* edited by W. Ebeling, A. Förster, R. Radtke, Teubner Verlagsgesellschaft, 1992, p. 87.
10. Skripov V. Fizullinn M., *High Temperature* **65** (5), 814 (1999); M.: Fizmatlit, 2003, 160pp.
11. Khishenko K., Fortov V., in *Physics of Matter Under Extreme Conditions*, Edited by V. Fortov, Chernogolovka: IPCP RAS, 2002, p.68.
12. Sinko G., Smirnov N. in *Physics of Matter Under Extreme Conditions,* Edited by V. Fortov, Chernogolovka: IPCP RAS, 2002, p.19.
13. Kirzhnits D., *Soviet JETP,* **38**, 503 (1960).
14. Jones M. Ceperley D., *Phys. Rev. Lett.* **76**, 4572 (1996).
15. Chabrier G., Douchin F., Potekhin A., *J. Phys. Condense Matter* **14**, 9133 (2002).
16. Iosilevski I., in *Strongly Coupled Plasma Physics*, edited by H.M. Van Horn and S. Ichimaru, Rochester: University of Rochester Press, 1993, p.343.
17. Iosilevski I., Chigvintsev A., *Journal de Physique* **IV** 10, 451 (2000).
18. Iosilevski I., Chigvintsev A., "Anomalous Phase Diagram in Simplest Plasma Model" in *this Issue*.
19. Morris J., Wang C., Ho K.M., Chan C., *Phys. Rev.* **B 49** (5), 3109 (1994).

ANOMALOUS PHASE DIAGRAMS IN THE SIMPLEST PLASMA MODELS

Igor L. Iosilevski and Alexander Yu. Chigvintsev

Moscow Institute of Physics and Technology, (State University) Dolgoprudny 141700, Russia

Abstract. Remarkable feature of new first-order phase transitions of gas-liquid gas-crystal types in combination with traditional solid-liquid transition are under consideration in a modified one-component plasma model (OCP) with uniform, but compressible background. Structure and parameters of this phase transition strongly depend on the value of charge number Z. Under high values of Z the model shows remarkable and completely unusual topology of phase diagram.

INTRODUCTION

Problem of Phase Transition (PT) is of traditional great interest in astrophysics [1,2,3] as well as in general theory of so-called Strongly Coupled Coulomb Systems (SCCS) during very long time. Problem of Wigner crystallization in mixture of C(6+) and O(8+) nuclei during cooling of white dwarfs [2] there exists very interesting problem of stratified layers of high-Z crystals in outer core of neutron stars [4]. Besides crystallization the so-called hypothetical 'Plasma Phase Transition' (PPT) [5] and He/H2 phase decomposition [6] are examples of fluid-fluid phase transitions, which are of great interest for the theory of cooling of interiors of Giant Planets (GP) and Brawn Dwarfs (BD) [7] (see also paper [8]).

Besides the study of hypothetical PT in real plasmas a complementary approach is developing where the main subject of interest is definitely existing PT-s in simplified plasma models [9,10]. The well-known prototype model is OCP with a rigid background (notation – OCP(#)). This variant of OCP is studied carefully nowadays. The system cannot collapse or explode spontaneously. The only phase transition – crystallization – occurs in OCP(#) without any density change. More realistic model is One Component Plasma on uniform, but compressible compensating background (following notation – OCP(~)). One of the simplest example of OCP(~) is the model of classical point charges on uniform compressible background if ideal fermi-gas of electrons.

Transition to the OCP on uniform and compressible background leads to appearance of a new first-order phase transitions of gas-liquid type [10,11]. New phase diagram combines previous crystallization, now with a finite density change, with a qualitatively different coexistence curve of the new phase transition [12]. Obviously the structure and parameters of this phase diagram strongly depend on the value of charge number Z. This fact is illustrated at general phase diagram Figure 1.

TOPOLOGY OF PHASE DIAGRAMS IN OCP(~) MODEL

Four qualitatively different situations should be distinguished for the OCP(~) depending on the value of charge number Z:

1) Low value of charge number $\qquad Z < Z_1^* \approx 35$
2) High value of charge number $\qquad Z > Z_2^* \approx 45$
3) Intermediate value of charge number $\qquad Z_1^* < Z < Z_2^*$
4) Boundary value of intermediate charge number interval $\qquad Z=Z_1^*$ and $Z= Z_2^*$

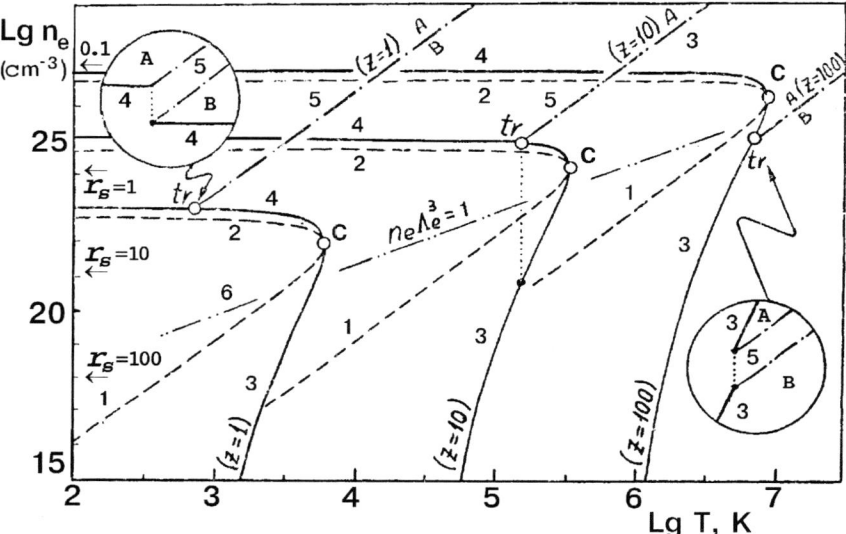

FIGURE 1. Phase diagram of OCP classical point ions on the compensating background of ideal fermi-gas of electrons in T–n_e plane (temperature - background electron density) for $Z = 1$, 10 and 100. *Notations*: Spinodals (*1,2*) и binodals (*3,4*) of two-phase coexistence curve of condensed (*2,4*) and gaseous (*1,3*) phases; *5* – melting $\{\Gamma \equiv (Ze)^2/kTa \approx 175\}$, (*A* – crystal, *B* – fluid); Critical (*C*), triple (*tr*) points; *6* – boundary of electron degeneracy. Position of constant electron Brueckner parameter $r_S = 0.1$, 1, 10, and 100 are shown ($r_S = 100$ corresponds to the cold melting of electron Wigner crystal [13]).

Low Values of Charge Number ($Z \sim 1$)

Phase diagram of the model was carefully studied in [10,11,12]. The *ordinary* structure of global phase diagram was obtained in this case: i.e. the relative position of critical and triple points, of melting zone and gas-liquid and gas-crystal coexistence, all are totally equivalent to those for ordinary substances (see Figure 2 and Figure 3).

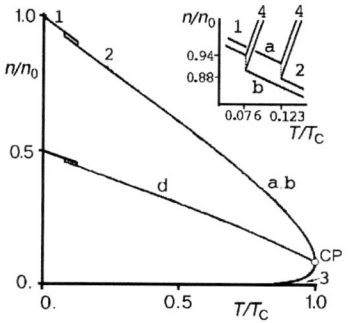
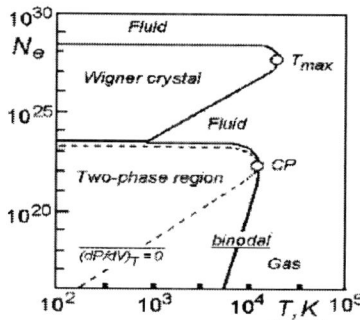

FIGURE 2. Density-Temperature phase diagram in two OCP(~) models at low $Z \sim 1$ in the relative coordinates n/n_0 and T/T_C where n_0 is the density of crystal under 'normal condition' ($P = 0$ and $T = 0$) and T_C is critical temperature of fluid-gas phase transition. *Notations*: a and b mass-asymmetrical model of electron-ion plasma: a – Single-OCP(~) and b – Double-OCP(~) respectively; 1 – crystal; 2 – fluid, 3 – gas; 4 – melting zone; d – 'diameter' of coexistence curves model a and b; *Insertion* – structure of evaporation, sublimation and melting bounds near triple point in a and b models (Figure from [11]).

FIGURE 3. Global phase diagram in mass-asymmetrical ion-electron Double-OCP(~) models of low value of charge number Z. Hypothetical boundary of ionic Wigner crystal and boundaries of new gas-liquid and gas-solid phase transitions are noted. *Notations*: CP – critical point; T_{max} – pseudocritical point of maximum melting temperature; *dashed lines* – spinodals $\{(\partial P/\partial V)_T = 0\}$ (Figure from [9,10]).

OCP(~) Phase Diagram at High Values of Charge Number

Highly anomalous structure of global phase diagram was discovered at previous study of the OCP(~) at very high values of charge number $Z \gg Z_2^* \approx 45$ [10,11]. In this case the melting «stripe» ($\Gamma \equiv (Ze)^2/kTa \approx 175$) crosses low-density 'slope' of two-phase coexistence domain of the new phase transition. Following features are typical for this type of phase diagram (see the variant $Z = 100$ at Figure 1 for details):
- Triple point is placed at low-density 'slope' of two-phase boundary.
- Critical point is placed at crystalline part of two-phase coexistence domain.
- Crystal-crystal coexistence of two crystalline phases, dense and expanded ones, with the same lattice occurs in OCP(~) at such high values of charge number Z.

Intermediate Values of Charge Number ($Z_1^* < Z < Z_2^*$)

The most remarkable anomalous phase diagram corresponds to the case when the melting line of prototype OCP(#) model ($\Gamma \sim \Gamma_{melt} \approx 175$) crosses coexistence curve of the new gas-liquid phase transition just closely to its critical point. As a result of this coincidence:
- The only phase transition exists in the model. It corresponds to the unified global crystal–fluid coexistence, i.e. continuous superposition of melting and sublimation (see Figure 4).
- There is no true critical point.
- There is no triple point.
- Coexistence curve in P-T (pressure-temperature) plane is a continuous, infinite curve (Figure 5).

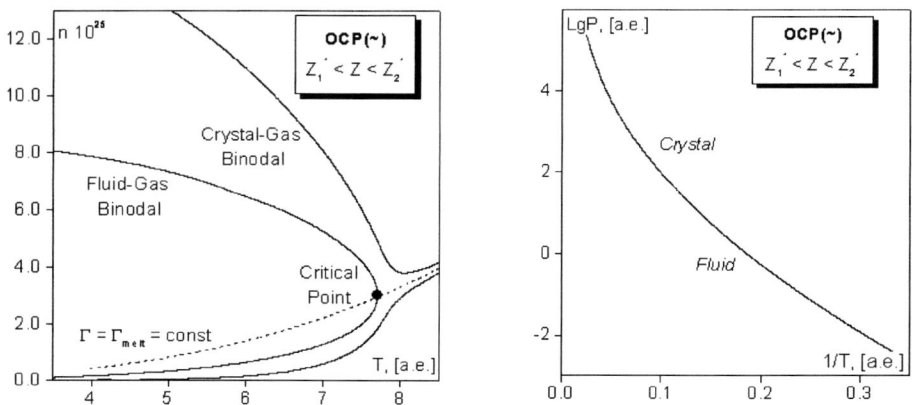

FIGURE 4. Anomalous type of density-temperature phase diagram in OCP(~) model at $Z_1^* < Z = 40 < Z_2^*$. (Figure from [12])

FIGURE 5. Anomalous type of pressure-temperature phase diagram in OCP(~) model at $Z = 40$.

Boundary Values of Charge Number Interval ($Z = Z_1^*$ & Z_2^*)

Remarkable feature of phase diagram of OCP(~) model in the case $Z = Z_1^*$ or $Z = Z_2^*$ is an existence of pseudo-critical point where the well-known standard conditions are fulfilled: $(\partial P/\partial V)_T = 0$ and $(\partial^2 P/\partial^2 V)_T = 0$: at $Z = Z_1^* \approx 34.6$ – on fluid part of crystal-fluid binodal (Figure 6) and at $Z = Z_2^* \approx 45.4$ – on crystalline part of crystal-fluid binodal (Figure 7). When we use the same as in [10,11,12] analytical fits for equation of state of both subsystems, OCP(#) and background, we obtain following parameters of the both pseudo-critical points:

TABLE 1. Parameters of pseudo-critical point in OCP of classical point charges on the uniform and compressible background of ideal fermi-gas of electrons ($Z = Z_1^*$ or Z_2^*).
$\Gamma \equiv Z^2 e^2/a_i kT$; $r_S \equiv a_e/a_B$; $\theta \equiv kT/\varepsilon_F \equiv 4/(9\pi)^{1/3}(n_e \Lambda_e^3)^{2/3}$; $\Lambda_e^2 \equiv 2\pi\hbar^2/m_e kT$; $a_j^3 \equiv 4\pi n_j/3$

	Z	T_C, a.u.	$(n_e)_C$, cc^{-1}	P_C, a.u.	Γ_C	$(r_S)_C$	$(n_e \Lambda_e^3)_C$	$(\theta)_C$
$Z = Z_1^*$	34.6	6.38.	$2.24 \cdot 10^{25}$	11.4	140	0.416	3.30	2.91
$Z = Z_2^*$	45.4	9.29	$3.96 \cdot 10^{25}$	28.4	181	0.344	3.26	2.89

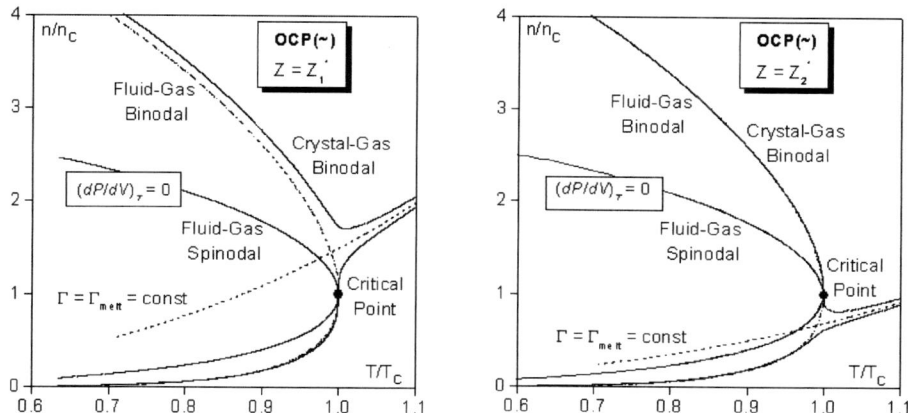

FIGURE 6. Anomalous type of density-temperature phase diagram in O CR(~) model at $Z = Z_1^* \approx 34.6$
FIGURE 7. The same as Figure 6 for the case $Z = Z_2^* \approx 45.4$ (Figures from [12]).

Critical exponents

Remarkable feature of two discussed pseudo-critical points at $Z = Z_1^*$ or $Z = Z_2^*$ is the non-standard values of all critical exponents in comparison with the ordinary (van der Waals like) critical exponents, which correspond to the case of OCP(~) with the charge number Z beyond the discussed interval $Z_1^* \div Z_2^*$. For example, at the latter case $Z < Z_1^*$ or $Z > Z_2^*$, the standard density-temperature relation is valid for the density at coexistence curve $\rho(T)$ $[\rho(T) - \rho_C] \sim |T - T_C|^{1/2}$. At the same time in the case $Z = Z_1^*$ or $Z = Z_2^*$ at pseudo-critical points the following relation may be proved $[\rho(T) - \rho_C] \sim |T - T_C|^{1/3}$ (see [12]).

GENERAL CHARACTER OF ANOMALOUS PHASE DIAGRAMS

Anomalous phase diagram with an unique phase equilibrium crystal–fluid is not a exclusive feature of OCP(~) model. The same behavior is quite common for some systems with traditional inter-particles interaction, which combines intensive shot-range repulsion and finite in depth and spread attraction. For instance, transition from "normal" phase diagram to anomalous one has been observed in the one component system of hard spheres with additional short-range Yukawa-like attraction:

$$V(r) = \begin{cases} \infty & r < \sigma \\ -\varepsilon \dfrac{\exp[k\sigma(1 - r/\sigma)]}{r/\sigma} & r \geq \sigma \end{cases}$$

Analytical and numerical modeling of this system along with experimental results (large colloidal particles in polymer solution) show the effect qualitatively similar to one in OCP(~) model when ions charge Z_i tend to the interval $[Z_1^* \div Z_2^*]$. That is gradual closing triple and critical points right up to their merging and formation a single crystal-fluid phase boundary (Figure 8(a,b)).

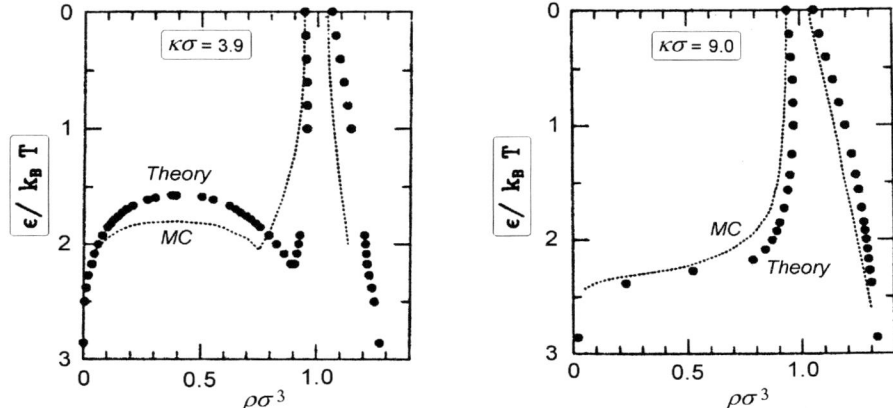

FIGURE 8(*a,b*). Phase diagram of hard sphere system with additional Yukawa attraction. *a* – standard type of gas-liquid-solid coexistence; *b* – anomalous phase diagram with unique crystal-fluid transition. Notations: *Solid circles* – calculation via thermodynamic perturbation theory (TPT); *dotted line* – numeric simulations results (Figure from [14]).

CONCLUSIONS

Anomalous phase diagrams are quite common phenomena in idealized Coulomb systems with high values of charge number Z. Similar topology of phase diagram could be expected at some conditions at astrophysical objects. Using rather simple but modified plasma models it is possible to calculate explicitly all parameters of such anomalous phase transitions and clarify the peculiar topology of its phase diagrams.

The authors are grateful to H. DeWitt, D. Yakovlev and A. Potekhin for fruitful discussions. The work was supported by Grant CRDF № MO-011-0, Grant ISTC-2107, and by RAS Scientific Programs "Physics and Chemistry of Extreme States of Matter".

REFERENCES

1. Shapiro S. and Teukolsky S., *Black Holes, White Dwarfs, and Neutron Stars*, NY: Willey, 1983.
2. Chabrier G., Ashcroft N., DeWitt H. *Nature* **360**, 48 (1992).
3. Van Horn H. in *Strongly Coupled Plasma Physics*, edited by S.Ichimaru, Elsevier, 1990, p 3.
4. Haensel P., Potekhin A., Yakovlev D., *Neutron Stars, Vol.1. Equation of State and Structure*, Dorderecht: Kluwer, 2005 (in press).
5. Norman G., Starostin A., *High Temperature* **6** (3), 410 (1968).
6. Fortney J., Hubbard W., *Icarus*, 2004.
7. Chabrier G., Saumon D., Hubbard W., Lunine J., *Appl. Astr. J.* **381** 817 (1992).
8. Iosilevski I., "Unexpected Features of Phase Transitions in Astrophysical Objects" in *This Issue*.
9. Iosilevski I, *Equation of State under the Extreme Conditions*, Ed. G.Gadiyak, Novosibirsk, 1981, p.20.
10. Iosilevski I., *High Temperatures* **23**, 807 (1985).
11. Iosilevski I. and Chigvintsev A., in *Physics of Strongly Coupled Plasmas*, edited by W. Kraeft and M. Schlanges., New Jersey-London: World Scientific, 1996, p.145.
12. Iosilevski I., Chigvintsev A. in *Physics of Strongly Coupled Coulomb Systems*, edited by G. Kalman, K. Blagoev, J. Rommel, New York: Plenum Press, 1998, pp.135.
13. Ceperley D., Alder B., *Phys. Rev. Letters* **45**, 566 (1980).
14. Mederos L., Navascues G., *J. Chem. Phys* **101**, 9841 (1994).

PART V

FROM PLASMA TO SOLID STATE:
Astrophysical Implications

Selected results and open problems in a semiclassical theory of dense matter

Vladan Čelebonović

Inst. of Physics, Pregrevica 118,11080 Zemun-Beograd, Serbia and Montenegro
vladan@phy.bg.ac.yu

Abstract. Studies of the behavior of materials under high external pressure have started in Serbia shortly after the middle of the last century. The aim of this lecture is to review the theoretical foundations of this work, present a selection of the results, and indicate some open problems within the theory.

INTRODUCTION

Systematic studies of the behavior of materials under high external pressure have started near the end of the *XIX* century. The main "driving force" behind the rise of interest in this field was a professor of physics at Harvard, P.W. Bridgman (1882-1961). In 1946, he was awarded the Nobel prize in physics for *"for the invention of an apparatus to produce extremely high pressures, and for the discoveries he made therewith in the field of high pressure physics"*. Presses used by Bridgman were large, expensive to build and use, and once enclosed in them, specimen were no longer visible. Experimental work in the field was later facilitated with the discovery of the diamond anvil cell (DAC). The DAC fits into a hand, it is much cheaper, simple to operate, the upper limit of pressure is much higher, and the specimen is visible throughout the experiment.

Modern theoretical interest in the field started in the thirties of the last century, when Fermi hinted that increased pressure leads to changes in the atomic structure. The basic aim of these studies is to solve the eigen-problem of a general, non-relativistic many-body Hamiltonian, defined as follows:

$$H = -\frac{\hbar^2}{2m}\sum_{i=1}^{N}\nabla_i^2 + \sum_{i=1}^{N}V(|\vec{x}_i|) + \sum_{i,j=1}^{N}v(|\vec{x}_i - \vec{x}_j|) \quad (1)$$

According to rigorous quantum mechanics, the energy of the system having a Hamiltonian H is given by

$$E = \frac{<\Psi|H|\Psi>}{<\Psi|\Psi>} \quad (2)$$

All the symbols in these equations have their standard meaning. Sums in eq. (1) go over the number of particles which, in real physical situations, is of the order of Avogadro's number. This implies that these sums can not be calculated exactly. The calculation of such sums is a serious problem in many-body physics. Once the eigenproblem of H

defined in eq. (1) is somehow solved, the thermodynamic potentials,phase transition points (or regions) and the phase diagram in general follow by application of standard "prescriptions" of statistical mechanics.

This lecture is devoted to a review of a simple theory of dense matter,whose "founding fathers" are two professors of the University of Belgrade: Pavle Savić and Radivoje Kašanin. As a consequence of the first letters of their family names, it is often called "the SK theory". It is also often called "the semiclassical theory"' because it is founded on laws of classical physics and just a few ideas of atomic physics. The next section contains a review of the basic physical ideas of their theory,while selected results of various applications are outlined in the third part. The fourth part contains a discussion of some of the open possibilities for future work on this theory,and the lecture ends with the final comments.

THE BASIC IDEAS

The development of what later became the SK theory started in 1961 when P.Savić published a short paper [1] presenting an unusual idea: the mean planetary densities, as calculated from the masses and radii known at the time,could be linked to the mean solar density by an extremely simple expression:

$$\rho = \rho_0 2^\phi \qquad (3)$$

In this expression ρ_0 denotes the mean solar density,which at the time was estimated at $4/3 \ gcm^{-3}$. By choosing integral values of the exponent ϕ in the interval $\phi \in (-2,2)$ Savić managed to fit the numerical values of the densities of the major planets. No hint of any possible physical explanation of this simple fit was given. In the following 4 years,collaborating with Radivoje Kašanin, [2] he managed to develop a theory of the behavior of materials under high pressure.In later years,it was nicknamed "the SK theory" after the first letters in their family names.

The basic physical idea of their theory is simple.SK assumed that sufficiently high values of external pressure lead to changes of the electronic structure of atoms and/or molecules. The possibility of such an interaction was for the first time hinted by Fermi several decades before SK. However,according to the avaliable literature,he simply noted the possibility of such an influence, but did not find it interesting enough so as to explore it in detail [3]. The possibility of such an influence has also been invoked around that time by P.W.Bridgman [14]. For the special case of a one-dimensional finite potential well this problem has recently been discussed by the present author [4]. An analytical expression for the first pressure derivative of the energy of a massive particle in such a well has been obtained and discussed in that paper.

The *SK* theory is in way similar to the "jellium model" from classical solid state physics.It represents a material as a uniform distribution of particles with the mean interparticle separation a defined by

$$N_A(2a)^3\rho = A \qquad (4)$$

In this expression N_A denotes Avogadro's number ρ is the mass density and A is the mean atomic mass of the material. Having thus introduced a, one can define the "'accumulated"' energy per electron as

$$E = \frac{e^2}{a} \tag{5}$$

Logically, one might expect that a relation such as the last one should contain the ionic charge Z It can be shown, (for example [5]) that a as defined in eq.(4) is a multiple of the radius of a Wigner-Seitz cell, which actually contains Z. The radius of a Wigner-Seitz (WS) cell is defined by

$$\frac{Am_p}{\rho} = \frac{4}{3}\pi r_{WS}^3 = \frac{Z}{n} \tag{6}$$

It follows from eq.(6) that

$$r_{WS} = (\frac{3}{4\pi})^{1/3}(\frac{A}{\rho N_A})^{1/3} = 2a(\frac{3}{4\pi})^{1/3} = (\frac{3}{4\pi})^{1/3}(\frac{Z}{n})^{1/3} \tag{7}$$

The basic premises on which the SK theory is based are the following statements [6]

1. The density of a material is an increasing function of the pressure to which it is exposed.
2. With increasing density, every material undergoes a sequence of first order phase transitions. Phases are numbered by an index i and the phase ending at the critical point is denoted as the zeroth phase.
   In any given phase (i.e., for any value of i) there exist two limiting values of the density ρ such that

$$\rho_i^0 \leq \rho_i \leq \rho_i^* \Leftrightarrow (\frac{1}{\alpha_i})\rho_i^* \leq \rho_i \leq \rho_i^* \tag{8}$$

where $\alpha_i > 1$

3. The maximal densities of two successive phases are related by

$$\rho_{i+1}^* = 2\rho_i^* \tag{9}$$

This expression follows directly from eq.(2) by assuming that $\phi_{i+1} - \phi_i = 1$.

4. It is assumed that

$$\frac{E_i^*}{E_i^0} = \frac{E_{i+1}^0}{E_i^*} \tag{10}$$

Some form of a link between the accumulated energies in two successive phases was needed in order to render the calculations tractable, and this form was accepted because of its simplicity. After some algebra, it follows that $\alpha_i \alpha_{i+1} = 2$ and that

$$\begin{aligned}\alpha_i &= 6/5, i = 1, 3, 5, ..\\ \alpha_i &= 5/3, i = 2, 4, 6, ..\end{aligned} \tag{11}$$

5. The final density of the zeroth phase is

$$\rho_0^* = \frac{A}{3\bar{V}} \tag{12}$$

which is approximately equal to the critical density in the van der Waals theory. $\bar{V}$ denotes the molar volume of the material at $T = 0K$. In the terminology of the van der Waals theory $\bar{V} = b$.

6. Using assumption 3., it can be shown that

$$\frac{A}{\bar{\rho}} = \frac{A}{2}(\frac{1}{\rho_2^0} + \frac{1}{\rho_2^*}) \tag{13}$$

In this expression $\bar{\rho}$ denotes the density at the zero-point, defined as $\bar{\rho} = A/\bar{V}$.

Starting from these premises, the following set of simple analytical relations can be derived, which are used in all subsequent calculations within this theory. Second order phase transitions can be considered as a special case for which $\bar{V}_i^* - \bar{V}_{i+1}^0 \to 0$.

$$\begin{aligned}
\rho_i^* &= 2^i \rho_0^*; \rho_i^0 = \frac{\rho_i^*}{\alpha_i} \\
\bar{V}_0^* &= 3\bar{V}; \bar{V}_i^0 = \alpha_i \bar{V}_i^* 2^{-i}; \bar{V}_i^* = 2^{-i} \bar{V}_0^* \\
r_0^* &= (\frac{15}{4N_A 10^{-23}})^{1/3} \bar{V}^{1/3}; r_i^* = 2^{-i/3} r_0^*; r_i^0 = r_i^* \alpha_i^{1/3}
\end{aligned} \tag{14}$$

The following section of this lecture contains a selection of results obtained in various applications of this theory. The algorithm for the calculation of phase transition pressure is relatively short, so it will be presented in detail. The set of equations needed for modelling the internal structure of celestial objects is much longer, so the reader interested in its full details should consult [2].

SELECTED RESULTS

Laboratory applications

The application of the SK theory which is most easily verifiable in laboratory experiments is the calculation of the value of pressure on which a first order phase transition occurs in a given material. This value of pressure can be calculated by considering the work done by the external pressure in compressing the material,

$$\Delta W = p_i^* (\bar{V}_i^* - \bar{V}_{i+1}^0) = p_i^* \bar{V}_i^* (1 - \frac{1}{\alpha_i}) \tag{15}$$

and equating it to the change of the accumulated energy

$$\Delta W = \Delta E = N_A (E_{i+1}^0 - E_i^*) \tag{16}$$

Starting from eqs.(11)-(13) it can be shown that the maximal pressure in a phase i of a material is given by

$$p_i^* \cong 1.8077 \beta_i \bar{V}^{-4/3} 2^{4i/3} \text{MBar} \tag{17}$$

with

$$\beta_i = 3 \frac{\alpha_i^{1/3} - 1}{1 - 1/\alpha_i} \tag{18}$$

Values of α for various values of the index i are given in eq.(8). Finally, the value of the external pressure needed to "provoke" a first order phase transition from phase i to the phase $i+1$ in a material is given by

$$p_{tr} = p_i^* - p_i^0 = p_i^*(1 - 2^{-4/3} \frac{\beta_{i-1}}{\beta_i}) \tag{19}$$

This expression gives a simple mathematical procedure for the calculation of the sequence of possible values of a phase transition pressure in a given material. Those values of pressure at which a first order phase transition is physically possible are selected by the following criterion:

$$E_0^* + E_I = E_i^* \tag{20}$$

The symbol E_I denotes the ionisation potential, E_0^* and E_i^* can be calculated from eqs. (5) and (12), with $a = 10 \times r$ nm. Applying the procedure described above, an analysis of the applicability of the SK theory to real materials under high pressure was made [7]. A set of 19 materials for which experimental data on phase transitions under high pressure were easily avaliable was analyzed. The aim was to calculate within the SK theory values of pressure at which first order phase transitions could be expected, and then compare the results with the experimental data and analyze possible causes of the discrepancies. It was shown in [7] that the relative discrepancies between the measured values of phase transition pressure and those calculated within the SK theory are material and pressure dependent. Two basic causes of the discrepancies were identified:

• one is due to that fact that the SK theory takes into account only the simplest form of the electrostatic potential, while in reality in atoms and molecules one "deals" with charge distributions;

• the second "source of problems" is represented by the fact that the SK theory neglects the contribution of various non-electrostatic components to the overall intermolecular potential. This is expected from a semiclassical theory, but it clearly increases the discrepancy between the measured and calculated values of the phase transition pressure. More details on these two problems are avaliable in [7].

Another interesting result of the SK theory concerns the establishment of the thermal equation of state of solids under pressure. Namely, the temperature is not explicitly present in the original formulation of this theory. All its calculations are performed in the $P - \rho$ plane. However, the temperature can be introduced in a simple way: by equating the internal energy densities of a solid, as expressed within the SK theory and in standard solid state physics.

The internal energy per unit volume of a solid is, in the SK theory, given by

$$E = 2e^2 Z (\frac{N_A \rho}{A})^{4/3} \tag{21}$$

and in standard solid state physics, it is given by

$$E = \frac{\pi^2(k_BT)^4}{10(\hbar\bar{u})^3} \tag{22}$$

The symbol $\bar{u}$ in the last expression denotes the mean value of the velocity of sound, and all the other symbols have their standard meanings.

Details are avaliable in [8] and the final result is that the equation of state of a solid in the $T - \rho$ plane has the following form:

$$T[K] = 1.4217 \times 10^5 (\frac{\rho}{A})^{7/12} (\frac{m_e}{M})^{3/8} Z^{7/8} \tag{23}$$

The symbol m_e denotes the bare electron mass, M is the ionic mass, A denotes the mass number of the material and Z is the charge of the ions. A result of an important astronomical application of this equation of state is presented in the following subsection of this contribution.

Last, but certainly not the least, we come to the problem of hydrogen under high pressure, and its possible metallization. The SK theory has been applied to this problem [9], and the result obtained at the time was encouraging. Metallisation was predicted to occur at P=300 GPa, which agreed with predictions by various other theoretical methods. Theoretical predictions of the metallisation pressure of hydrogen have been of this order of magnitude for the last 70 years [10].

The experimental situation in work on hydrogen under high pressure is much less well defined. There have been claims at the beginning of the nineties that metallisation occurs at a pressure $P = 150$ GPa. A short time after these results were shown to be incorrect and due to a chemical reaction of ruby with hydrogen [11]. Static experiments performed at values of pressure $P \leq 342$ GPa have not shown signs of metallization of hydrogen [12], so the problem seems (at present) to be completely open. The literature on this topic is huge, but for examples of interesting recent papers see [13] and [14]. One of these papers, [13], invokes the influence of disorder as a possible reason why metallization was not observed at the theoretically predicted value of the pressure.

Applications in astronomy

The theory we are discussing in this lecture can be applied to modelling of the internal structure of celestial objects. To be precise, the interest of Savić and Kašanin in this problem was the initial "grain of salt" which led to its development. As the calculations in this theory do not contain any mention of internal energy generation in celestial objects (i.e., nuclear reactions), it is inapplicable to work on stellar structure.

The complete calculational scheme for astronomical applications of the SK theory is avaliable in [2]. Unfortunately, due to various reasons, it has not been reformulated in a modern way. The only input data needed for making the model of the internal structure of a celestial object within this theory are the mass and the radius of the object.

Starting from these data, it gives the number and thickness of layers which exist in the interior of the object, the distribution of the values P, ρ, T with depth under the surface, the strength of the magnetic field and the interval of the physically allowed values of the speed of rotation. The theory also gives as a result the mean atomic(or molecular) mass of the chemical mixture that the object under study is made of.

The first celestial body to be modelled was logically the Earth and the main characteristics of this model are shown in the following table.

TABLE 1. The interior of the Earth according to SK

depth (km)	0 - 39	39 - 2900	2900 - 4980	4980 - 6371
ρ_{max}[kg/m^3]	3000	6000	12000	19740
P_{max}[GPa]	25	129	289	370
T_{max}[K]	1300	2700	4100	7000

The mean mass number of the material which makes up the Earth is $A = 26.56$. Taking into account the simplicity of the theory, this model is in remarkable agreement with modern knowledge.For a discussion of the temperature in the interior of the Earth,see for example [15].

Note also that a current model of the Earth's crust called CRUST 5.1,[16] gives around 70 km as the maximal value.Here in Holland,the thickness of the Earth's crust according to this model is around 35 km,which is in excellent agreement with the value calculated within SK.

Apart the Earth,the theory was applied to all the other planets except Saturn and Pluto,the Moon,the Galilean satellites of Jupiter,the satellites of Uranus,Neptune's satellite Triton and the asteroids 1 Ceres and 10 Hygiea.

The results are scattered in the literature,but a "safe" general comment is that the agreement with the consequences of observations and with theoretical work of other authors is good [6].

The following table contains the values of the mean atomic masses of various objects in the Solar System,calculated according to the SK theory.

TABLE 2. The composition of the Solar System according to SK

object	A	satellite	A
Sun	1.4	Moon	71
Mercury	113	J1	70
Venus	28.12	J2	71
Earth	26.56	J3	18
Mars	69	J4	19
1 Ceres	96	U1	38
Jupiter	1.55	U2	43
Saturn	/	U3	44
Uranus	6.5	U4	32
Neptune	7.26	U5	32
Pluto	/	Triton	67

It can easily be seen from the preceding table that our planetary system is far from being chemically homogenous;at first sight,the well known division on the terrestrial and Jovian planets is clearly visible. These differences are obviously a "remnant" of various transport and mixing processes which have been active in the formation epochs of the planetary system.Note also that similar differences are visible in the satellite systems which were modeled.

Various conclusions can be drawn from data in Table 2. For example,asteroid 1 Ceres is currently orbiting the Sun between the orbits of Mars and Jupiter.However,by its chemical composition it is similar to the planet Mercury [6]. As chemically similar bodies are expected to have been formed close to each other,this similarity implies that *" once upon a time"* Ceres and Mercury originated in the same region of the protoplanetary system,but that their orbits later diverged.The physical process(es) which have led to this diverging of their orbits can be a subject of further studies.

Concerning asteroids,using the value of A calculated for 1 Ceres,the mass of the asteroid 10 Hygiea was calculated, and the result turned out to be in excellent agreement with the result known in celestial mechanics [17].

Two cases of applications of the SK theory are especially worth mentioning.The composition of the Galilean satellites was determined within this theory in 1987. In 1996.some results of this calculation were confirmed by measurements from the Galileo space probe.Details are given in [18] which contains the reference to the original publication of 1987.

Another interesting application of this theory concerns Neptune's satellite Triton [19].It was shown in that paper that Triton has a composition similar to Mars,which was interpreted as implying that Triton is a captured body,in perfect agreement with earlier work based purely on celestial mechanics.This conclusion is of interest for cosmogony,as it can be regarded as an independent proof that collisions were important in the early phases of existence of our planetary system.For a recent example of observational work on the structure of a protoplanetary cloud see [20].

SOME POSSIBILITIES FOR FUTURE WORK

In this lecture we have so far reviewed the basic ideas and a number of results of applications of the SK theory. However,in spite of successes,there are open possibilities for future work related to the theory we are discussing,and this section is devoted to an outline of these topics.

Paper [1] was devoted to a possible relation tying the mean planetary densities (derived from their masses and radii) with the mean solar density.Eq.(1) was the result,and it was applied in various calculations in its original form ever since.The important detail in that paper was not only the simple form of eq.(1),but also the fact that the exponent ϕ in eq.(1) had *integral* values.However,as nearly 40 years have elapsed since [1],it seemed appropriate to undertake a verification of eq.(1) but with modern data,and on a broader set of objects. Such a check has recently been performed [21] and the results are shown in table 3.

Modern data on densities of 22 planets and their major satellites were used.It was shown that eq.(1) is still valid,but that the exponent ϕ can also take *non-integral* values. The immediate question is of course the interpretation of these non-integral values.Namely,the integral values of ϕ which appeared in [1] were interpreted there and in later publications as a qualitative analogy with atomic quantisation, implying that high

TABLE 3. Modern values of the exponent ϕ

object	ρ_{mean}	ϕ	ρ_{calc}
Sun	1408	0	1408
Mercury	5427	2	5632
Venus	5243	2	5632
Earth	5515	2	5632
Moon	3350	5/4	3349
Mars	3933	3/2	3982
Phobos	1900	2/5	1858
Deimos	1750	1/3	1774
Jupiter	1326	-1/10	1314
JI	3530	7/5	3716
JII	3010	1	2816
JIII	1940	1/2	1991
JIV	1840	2/5	1858
Saturn	687	-1	704
Titan	1881	2/5	1858
Uranus	1270	-1/7	1275
Ariel	1670	1/4	1674
Neptune	1638	1/4	1674
Triton	2050	1/2	1991
Pluto	1750	1/3	1774
Charon	2000	1/2	1991

external pressure leads to changes in the atomic structure.The problem is open,and one of the possibilities would be to try to cautiously pursue the analogies between the atomic structure and the structure of the planetary system.

Another "'open"' astronomical application of the SK theory is the calculation of the angular speed of rotation of a celestial object.Instead of proposing an algorithm for this calculation and giving a unique number at the end,this theory gives a physically allowed interval in which the speed of rotation can be.

It was applied to the Earth and several other planets. The results are very promising,but it would be useful to reduce this allowed interval to a unique number.

A possibility for future work concerns laboratory applications of the SK theory.We have discussed to some extent the applicability of this theory to the calculation of phase transition pressure in various solid materials,and to the establishment of a thermal equation of state of solids. The discrepancies between the predicted values of phase transition pressure and experimental results are material dependent. Work in this direction could in the future advance along several different lines:

At first it would be useful to apply the theory to more materials for which first order phase transition pressures are known.In this way one would obtain a more precise empirical estimate of the systematic trends (if there are any) in the discrepancies between the calculated and measured values of the phase transition pressure.

On the purely theoretical side,in would be necessary to refine the theory with the inclusion of the contribution of more components to the interparticle potential energy.

An even more interesting problem would be to link the SK theory to the well established theoretical framework of statistical physics.Although the SK theory has given good results in a range of applications,its unusual formulation hinders its wider spread in the research community. A preliminary attempt in that direction has recently been performed in [21],where is was shown how the parameters appearing in the SK theory can be connected with those of the Landau theory. It would be interesting trying to reformulate the SK theory along the lines of present day theories of quantum phase transitions (for example,[22]).

A note concerning the equation of state of solids under pressure in $T-\rho$ plane:

This calculation,described in detail in [8] gives physically acceptable numerical values,which is good.However,the calculation invokes the speed of sound in a solid,and (as well known) this invokes the knowledge of $\partial P/\partial \rho$ - that is the equation of state.

In the calculation which has led to the proposed form of thermal equation of state,an approximation called the "'Bohm-Staver"' formula was used [23].It is planned in the future to repeat this calculation,but using some more refined form of the equation of state.

FINAL COMMENTS

This lecture was an attempt to present a "balanced" review of a theory of dense matter proposed a little less than half a century ago by P.Savić and R.Kašanin. The adjective "balanced" simply means that the basic ideas,successful applications,but also problems for future work were discussed.In the past,work within the SK theory has been more oriented towards various applications and to a lesser extent towards its modernization.In future work, efforts will concentrate more on the refinement of this theory and extension of its range of applicability. It is hoped that as result of this workshop some "joint venture" in that direction will be started.

ACKNOWLEDGEMENTS

The author is grateful to Prof. Wim van Saarloos, director of the Lorentz Center, for making the initial proposal for the organization of this workshop, and to all in the Lorentz Center, whose work helped that the workshop runs so smoothly and successfully. The preparation of this contribution was financed by the Ministry of Science and Environment of Serbia within its Project 1231.

REFERENCES

1. Savić, P.: 1961,Bull. de la classe des Sci. Math. et Natur. de l'Acad. Serbe des Sci. et des Arts **XXVI** p.107.
2. Savić, P.and Kašanin, R.: 1962/65, The behaviour of materials under high pressure **I - IV**,Ed. by SANU, Beograd.
3. Schatzman, E.: 1958, White Dwarfs, North Holland Publishing Company, Amsterdam.
4. Čelebonović,V.: 2001, Phys. Low-Dim. Struct., **7/8**,127.
5. Leung,Y.C.: 1984, Physics of Dense Matter, Science Press/World Scientific, Beijing and Singapore.
6. Čelebonović,V.: 1995, Bull. Astron. Belgrade, **151**,37 and astro-ph/9603135.
7. Čelebonović,V. 1992, Earth, Moon and the Planets,**58**,203.
8. Čelebonović,V.: 1991, Earth, Moon and Planets, **54** 145.
9. Čelebonović,V.: 1989, Earth, Moon and Planets, **45** 291.
10. Wigner,E. and Huntington, H.B.: 1935, J. Chem. Phys., **3**,764.
11. Mao,H.K., Hemley,R.J. and Hanfland,M.: 1992, Phys. Rev., **B45**,8108.
12. Narayana,Ch., Luo,H., Orloff,J. and Ruoff,A.L.: 1998, Nature, **393**,46.
13. Nellis, W.J.: 2002, High Pressure Research, **22**,1.
14. Ashcroft,N.W.: 2004, J. Phys.: Condens. Matter, **16**,S945.
15. Duffy,T.S. and Hemley,R.J.: 1995, Rev.of Geophysics-Supplement, **33**.
16. http://quake.wr.usgs.gov/research/structure/
17. Čelebonović,V.: 1988, Earth,Moon and the Planets, **42**,297.
18. Čelebonović,V.: 1998, preprint astro-ph/9807202.
19. Čelebonović,V.: 1986, Earth,Moon and the Planets, **34**,59.
20. Fukugawa,M., Hayashi,M., Tamura,M. et.al.: 2004, Astrophys.J., **605**,L53.
21. Čelebonović,V.: 2004, preprint astro-ph/0405032.
22. Vojta,M.: 2003, Rep.Progr.Phys., **66**,2069.
23. Ashcroft,N.W. and Mermin,N.D.: 1976, Solid State Physics, Holt,Rinehart and Winston,New York.

Basic notions of static high pressure experiments

Vladan Čelebonović

Inst. of Physics, Pregrevica 118, 11080 Zemun-Beograd, Serbia and Montenegro
vladan@phy.bg.ac.yu

Abstract. The first recorded attempt of an experimental study of materials under high pressure dates from the *XVIII* century, but a systematic development of this field has started only towards the end of the *XIX* century. A big leap forward occurred in this kind of experiments with the discovery of the diamond anvil cell around the middle of the last century. The aim of this lecture is to briefly review the historical development of experimental work under static high pressure, discuss the principles of functioning of the diamond anvil cell and experiments using it, and end with a discussion of a few examples of astrophysically and geophysically important materials.

INTRODUCTION

The fact that materials occurring in nature can have widely different densities has (very probably) been discovered in human prehistory, when somebody somewhere for some reasons fell (or was thrown) from a high cliff. No document of any kind exists to prove this statement, but intuitively it can be accepted as realistic. Of course, modern science has advanced considerably since those times,and contemporary experimental methods do not rely upon throwing people from high cliffs.

Apart the introduction, this lecture has three more parts. The next one is devoted to the historical development of high pressure experiments, which in fact to a large extent means the work of P.W. Bridgman. His experiments are compared with work performed with diamond anvil cells. The following section contains a detailed description of the diamond anvil cell (DAC) which is, for several decades already, the main instrument for high pressure work under static pressure. We shall not only discuss the DAC itself,but also the difficult (and interesting) task of the choice of diamonds.The final part of the manuscript deals with several "real life" examples of high pressure studies of astrophysically important materials.

SOME HISTORY

Historically speaking,the first recorded attempt of intentionally compressing a material dates from the XVIII century.Historians of science have noted that a certain British gentleman named Mr.Canton compressed water at room temperature to a pressure of the order of 0.1 GPa [1].To his amazement,Mr.Canton witnessed what we would today call a phase transition,as water transformed into ice.This was just an isolated attempt,and we can only wonder how would Mr.Canton react at the "simple fact" that today we know about the existence of more than 10 kinds of water ice.

Systematic studies of the behaviour of materials under high pressure started towards the end of the *XIX* century,due to efforts of P.W.Bridgman (1882.-1961.)[2]. Bridgman worked as a professor of physics at Harvard University,and was the first to start systematic studies of materials under high pressure.What is perhaps more important is the fact that he was also the first to start systematic planning,design and construction of equipment for experiments under high pressure.His presses were large and complicated to build and maintain operational. Their good side was that they could accept large specimen and that they had very small pressure gradients.However,once enclosed in Bridgman's press a specimen was invisible,and it was impossible to directly observe what was happening to it under high pressure.For his life long work, Bridgman was awarded the Nobel prize for physics in 1946.

A big boost in high pressure experimental research occurred around the middle of the *XX* century with the discovery of the diamond anvil cell (DAC).Lawson and Tang were the first to show that x-ray diffraction studies can be performed by using a diamond with a miniature piston [3].Historically,this was the first attempt,but it was followed by a period of "silence" until 1959.

Two different versions of a DAC,developed by independent research teams,appeared then "on the scene".One was devoted to high-pressure x-ray diffraction studies,while the other was conceived for IR absorption measurements.Details with references are avaliable in [4].For the sake of completeness,at least three important steps should be mentioned here: the introduction of the metallic gasket technique,which facilitated the generation of *hydrostatic* pressure,the introduction of different kinds of hydrostatic liquids in different pressure domains,and the discovery of the ruby scale (and its later modifications) as the method for pressure measurement in DACs.One should not forget here the experiments with heating (and cooling) specimen in DACs.As a result,the DAC has become a sophisticated instrument,giving the possibility to explore the region of $T-P$ phase plane delimited by the values of $4 \leq T[K] \leq 7000$ and $P[GPa] \leq 450$.

THE DIAMOND ANVIL CELL

Starting from the initial efforts nearly 50 years ago,various types of diamond anvil cells have been developed.A cross-section of one of these types (the so called "NBS type") is shown on figure 1.Note the first advantage of a DAC compared to Bridgman's presses:a DAC is small and fits into a hand.It is therefore possible to fill the DAC in a laboratory,and then transport it into another one (even on another continent) in order to perform an experiment.

It is clear from the figure that the DAC is a precise instrument based on simple mechanical principles. The diamonds are a "doubly crucial" part of a DAC. They are hard,and as they are transparent they make possible optical excitation and deexcitation of a specimen.The notion of "hardness" which seems perfectly simple in everyday life,is not precisely defined in science.For a recent study of this problem see,for example,[5].Optical parameters of diamonds depend on the nature and concentration of

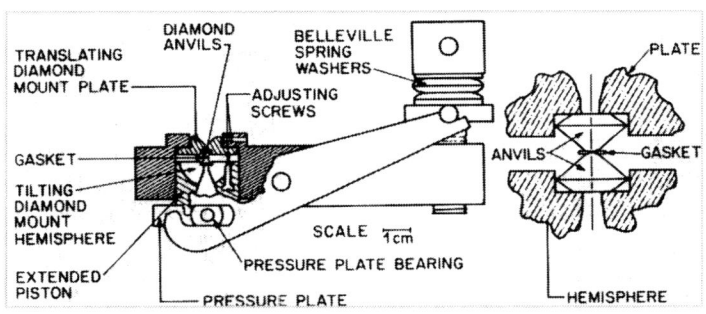

FIGURE 1. The cross section of a "NBS type" DAC

impurities which are present in the stones.The basic impurity is nitrogen,but the optical characteristics of the stones depend on the concentration and location of the impurities.

Diamonds are in the literature usually divided into 2 main types and 2 sub-types,listed in the following table.To the author's amazement,in personal contacts with stone merchants (for example D.Drukker from Amsterdam),it turned out that there exist about 2000 types!The objective criterion for differentiating between stones of different types are their Raman spectra ([6] and later work).The "crude" classification of diamonds is presented in the following table,in which the letter N denotes the chemical symbol for nitrogen.

TABLE 1. The classification of diamonds

type	description
I_a	N impurities in form of platelets
I_b	N in dispersed form, common in synth. stones
II_a	Effectively free of N, very rare in nature
II_b	blue in colour, extremely rare in nature

The DAC is not only small,but it is also lightweight - its mass is of the order of 0.8 kg.These good points (small length and mass) "hide" a potential source of complications.The point is that DAC experiments demand extremely fine and calm preparations.

The two mutually facing surfaces of diamonds have diameters of the order of $0.5 - 0.7$ mm. Details of the "working volume" between the diamonds,in which actual experiments take place,are shown in fig.2.In first experiments specimen were placed between two mutually facing flat surfaces of diamonds.However,the problem with such a setup was that it did not guarantee that the pressure to which a specimen is subdued is hydrostatic.Hydrostatic pressure became achievable on a routine basis with the introduction of the "gasket" [7].

The gasket is a thin metallic plate,inserted between the diamonds. A hole is drilled in the center of a gasket,in which one inserts a hydrostatic liquid,the specimen and the pressure sensor. The initial thickness of a gasket is $0.2 - 0.25$ mm,while the diameter of the hole in it is only about 200 microns.

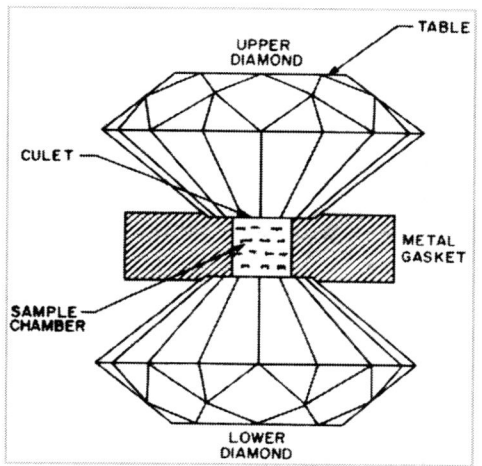

FIGURE 2. The working volume of a DAC

Drilling of such small holes, and inserting into them a hydrostatic liquid, the pressure sensor and of course the specimen, is a complex task. For example, the edges of the hole have to be as smooth as possible, because if they are not the hydrostatic liquid will slowly flow out, the pressure will be non-hydrostatic and a risk of breaking the diamonds will be non-negligible. It usually takes between several hours and a whole day to complete the operation of filling a DAC.

At this point, an appropriate question: How does one measure the pressure in a DAC? A first idea may be to apply the definition of pressure, and use the formula $p = F/S$. However, it appeared early in the development of this field of research that the application of this definition would induce large relative errors in the calculated values of the pressure because of the small value of the surface of the specimen.

Various pressure standards were employed in the "pre-DAC" epoch, with varying degree of precision. A definite (in the sense that is is still being applied) solution to this problem has been proposed in 1973; this was the famous "ruby scale" [8].

The ruby scale is based on an extremely simple idea. It has been shown experimentally that a chip of ruby (the chemical formula of ruby is $Al_2O_3 : Cr^{3+}$) excited by a spectral lamp or laser at a wavelength around 550 nm emits a spectrum consisting of two sharp lines [8]. At the microscopic level, the origin of the stronger ruby line is the electronic transition $^4F_2 \rightarrow\, ^2E$ in Cr^{3+}. The wavelengths of both lines are pressure dependent, but the scale was calibrated by the stronger one (called the $R1$ line). The scale was originally calibrated by measuring the compression of $NaCl$ by x-ray powder diffraction and relating it to the measured shifts of the $R1$ line [9]. The first calibration of the ruby scale was performed sometime before that [10]. In that paper the pressure was calculated as a function of the lattice parameter of $NaCl$ and temperature. $NaCl$ was used because its chemical composition is simple, the inter-atomic forces well known and accordingly its equation of state is also known with satisfactory precision.Interestingly,although the ruby scale is being used for several decades,a quantum mechanical theory of the pressure shifts of the R lines started developing only relatively recently [11] (and later work).The basic physical concept used in [11] was the expansion of the electron wavefunction with increasing pressure.

Over the years the ruby scale was refined and revised.No one has questioned its validity in principle,but work has been going on in several directions.The first quantitative expression [12] of a relationship between the wavelength of the $R1$ line and pressure was:

$$P[\text{GPa}] = 380.8 \left\{ (1 + \frac{\Delta \lambda}{694.2})^5 - 1 \right\} \quad (1)$$

where $\Delta \lambda$ denotes the shift of the wavelength of the $R1$ line at a pressure P compared to its wavelength at normal pressure. In one of the recent proposals for slight revisions of the ruby scale,[13],eq.(1) changes to the following form:

$$P = \frac{A}{B+C} \left[\exp((\frac{B}{C}+1)(1-\frac{\lambda}{\lambda_0})^{-C}) - 1 \right] \quad (2)$$

Numerical values of various coefficients are avaliable in [13].The scale is linear up to approximately $P \leq 30\text{GPa}$ and the pressure coefficient is $0.365 \frac{\text{nm}}{\text{GPa}}$.

The wavelengths of the R lines are also temperature sensitive,but the value of $\partial \lambda / \partial T$ is of the order of $0.0068 \frac{\text{nm}}{\text{GPa}}$ (for example [14]).

Apart widespread use of the ruby scale,parallel work is going on attempting to find alternative pressure sensors.This would demand a review of its own,so only two examples will be mentioned in the following.

It was recently shown [15] that the elasticity of the chemical compound MgO can be used as a pressure standard in high temperature DAC experiments in the region $P \leq 55\text{GPa}$. As discussed in that paper,by combining the elasticity data and the pressure-density equation of state of MgO,it became possible to determine the main elements of the elasticity tensor to $P \leq 55\text{GPa}$ thus nearly doubling the region of pressure in which they are known.This result on its own has interest for astrophysics,because MgO is one of the known constituents of the mantle of our planet.For the purposes of high pressure metrology,[15]is also interesting.Namely,it was shown there that the elasticity of MgO is so well known that it can be used as a primary pressure standard for $P \leq 55\text{GPa}$. A direct consequence is a re-calibration of the ruby scale, which ultimately leads to the increase of its accuracy up to $\pm 1\%$.

Another line of research uses the idea that it is much simpler (and also more precise) to measure short lifetimes than small changes of wavelength.In other words,it would be useful to find an alternative to the ruby scale base on the measurement of the lifetimes of the excited levels in Cr^{3+} as a function of pressure.An example of interesting work in that direction is [16] and other publications by the same author. In [16] the lifetime as a function of pressure was measured for the transition which is at the origin of the $R1$ line in $Al_2O_3 : Mn^{4+}$.

The final question concerning ruby is the region of applicability of this scale. Experiments have shown that with increasing pressure the intensity of the $R1$ line diminishes,and that it is nearly zero for values of pressure of the order of 200GPa, so this is currently considered as the upper limit of the ruby scale.As an interesting reference on this subject, the reader can consult for example [17].Higher values of pressure can be measured in DACs,but there one has to apply x-ray diffraction.

EXAMPLES OF APPLICATIONS

In this section we shall discuss several examples of DAC studies of materials whose behaviour under high pressure is at least of some interest for astrophysics.

The most abundant chemical element in the universe is hydrogen.It enters in the composition of all kinds of celestial objects,ranging from the interstellar medium to the giant stars.In the solar system,it is present in large percentages in the interiors of the giant planets,and that is one of the main reasons for the interest of the astrophysical community in the behaviour of hydrogen under high pressure.Logically,hydrogen being the simplest chemical element,one could expect that its behaviour under high pressure would be relatively easily predictable.

The first theoretical study of the problem of dense hydrogen dates from the thirties of the last century,when it was predicted for the that hydrogen should become metallic at a pressure of the order of 300 GPa [18]. Interestingly,this numerical value did not change much over the years.Different authors by widely different theoretical methods were obtaining values for the metallisation pressure of hydrogen which all converged on a narrow interval centered at 300 GPa.Some details, and a calculation of the metallisation pressure of hydrogen within the SK theory of behaviour of materials under high pressure, are presented in [19].At the time when the first theoretical studies of this problem were published,experimental measurement of such a high value of pressure was in the domain of wild dreams.Just for comparison,note that the highest values of pressure accessible to presses used by P.W. Bridgman was of the order of $10-20$ GPa.But,the community was firm in believing that "once" when this pressure becomes accessible,experiments will confirm the predictions.

It the late nineties of the last century values of pressure of the order of 300 GPa became almost routinely accessible in DAC experiments.It was expected that metallic hydrogen would soon be detected,but this did not happen.Experiments at Cornell University performed at $P \leq 342 GPa$ [20] did not find predicted signs of metallisation of hydrogen.What is even worse,is the fact that no clear interpretation of this negative result exists.Is it due to our misunderstanding of the behaviour of hydrogen under high pressure,or perhaps to the fact that the research community is wrongly predicting the experimentally accessible signs of the metallisation of hydrogen? Some interesting results concerning hydrogen under high pressure were released from LLNL at the time of this meeting,but in spite of that the problem remains open.

Another material,interesting and extremely useful here on Earth,but also important in astrophysics,is water.The importance of water for the life cycle on the Earth is clear.Note that one of the "practical" reasons for the exploration of Mars is related to the disappearance of water from that planet.If this should somehow occur on Earth,it would certainly mark the end of all forms of life. On the "'extraterrestrial"' side,water (and its ice) are present on the surfaces of various satellites and planets.The classical examples are the polar caps of Mars.

The motive for the interest of the planetary science community in water ice is physically simple. Ice is present on the surfaces of various planetary system bodies, and as such contributes to their albedo. This implies that in order to correctly interpret any observed values of the albedo one has to know the crystal structure of the ice on its surface, which (in other words) means the phase diagram of ice. Another problem from planetary science related to ice is the accretion of the icy satellites, which has recently been analyzed in some detail [21]. Turning to diamond anvil cells, water has been compressed to 128 GPa [22]. This experiment was performed with the aim of measuring the molar volume of water as a function of pressure by using the technique of x-ray diffraction. The data were interpreted as showing that ice is less compressible at very high pressure than at some intermediate values, but at the same time more compressible than some theoretical models predict.

Another more recent experimental paper in which water under high pressure was studied is [23]. This experiment was concerned with the measurement of the velocity of sound in liquid water under high pressure at variable temperature. The pressure was $P \leq 3.5 GPa$ and $T \leq 200C$. The equation of state of liquid water under high pressure was derived and compared to the theoretical results of various intermolecular potentials. As a by-product of this study, an interesting result has emerged: it was shown that the mean error of the values of pressure determined by the ruby scale is only $\pm 4 MPa$. The implication of this result is that for most values of pressure encountered in diamond anvil cell experiments, the relative error of the measured values of pressure is very nearly zero.

There are numerous examples of diamond anvil cell experiments with materials of geological importance. For illustration purposes, just two of these will be discussed at some length in the following.

It has been recently experimentally shown [24] that the chemical elements Ni and Co become less siderophile with increasing pressure. The experiment was performed for $P \leq 42 GPa$ and $T \leq 2500K$ in a laser heated diamond anvil cell. A theoretical analysis of the results shows that in the formative stages the magma ocean on the Earth would have extended to the depth of about 1200km in order to get a homogenous equilibrium between the core-forming metals and the silicate mantle of the Earth.

In another recent experiment [25] the elastic anisotropy in the highly textured hcp iron has been determined for the first time in a diamond anvil cell experiment, under room temperature at $P = 112 GPa$. The compressional sound velocity at $50°$ and $90°$ from the c axis was determined, and the anisotropy obtained is of the same order of magnitude as the anisotropy of the Earth's inner core. Implications for the composition of the Earth's core have been derived.

SOME FINAL COMMENTS

This short lecture had a double aim: to discuss the basic notions of methods used in laboratory work on materials under high static pressure and to present some examples of applications of these methods to astrophysically interesting materials.It was prepared with the idea that it would be appropriate to have,in a workshop like this one,a short introductory review of the experimental "side". Of course,the choice of topics is to a certain extent subjective,dictated by the author's research interest.It is hoped that the interested reader will be able to start from in and continue following the ever growing literature in the field,or (even better) start some research work.

ACKNOWLEDGEMENTS

The author is grateful to Prof. Wim van Saarloos, Director of the Lorentz Center, for making the initial proposal for the organization of this workshop, and to all in the Lorentz Center, whose work helped that the workshop runs so smoothly and successfully. The preparation of this contribution was financed by the Ministry of Science and Environment of Serbia within its Project 1231.

REFERENCES

1. S. Block, G. Piermarini and R.G.Munro: 1980, La Recherche, **11**,806.
2. P.W. Bridgman: 1964, Collected Experimental Papers, **I-VIII**, Harvard Univ. Press, Cambridge, Mass.
3. A.W. Lawson and T.Y. Tang: 1950, Rev. Sci. Instrum., **21**,815.
4. A. Jayaraman: 1983, Rev. Mod. Phys., **55**, 65.
5. G. Will and P.G. Perkins: 1999, Materials Letters, **40**,1.
6. D.M. Adams and S.K. Sharma: 1977, J.Phys., **E10**,680.
7. A.A. Van Valkenburg: 1965, Conference Internationale sur les Hautes Pressions, Le Creusot, France.
8. J.D. Barnett, S. Block and G.J. Piermarini: 1973, Rev. Sci. Instr., **44**,1.
9. G. Piermarini, S. Block, J.D. Barnett and R.A. Forman: 1975, J. Appl. Phys., **46**,2774.
10. D.L. Decker: 1965, J. Appl. Phys., **36**,157.
11. Dong-ping Ma, Xi-te Zheng, Yi-sun Xu and Zhang-gang Zhong: 1986, Phys. Lett. **A115**,245.
12. H.K. Mao, P.M. Bell, J.W. Shaner and D.J. Steinberg: 1978, J. Appl. Phys., **49**,3276.
13. W.B. Holzapfel: 2003, J. Appl. Phys., **93**,1813.
14. W.L. Vos and J.A. Schouten: 1991, J. Appl. Phys., **69**,6744.
15. C.S. Zha, H.K. Mao and R.J. Hemley: 2000, Proc. Natl. Acad. Sci. USA, **97**,13494.
16. B.R. Jovanić:1998, J.Phys.: Condens.Matter, **10**,7897.
17. Y.K. Vohra, C.A. Vanderborgh, S. Desgreniers and A.L. Ruoff: 1990, Phys. Rev., **B42**, 9189.
18. E. Wigner and J. Huntington: 1935, J. Chem. Phys., **3**,764.
19. V.Čelebonović: 1989, Earth, Moon and Planets, **45**,291.
20. C. Narayana, H. Luo, J. Orloff and A.L. Ruoff: 1998, Nature, **393**,46.
21. M. Arakawa and D. Tomizuka: 2004, Icarus, **170**,193.
22. R.J. Hemley, A.P. Jephcoat, H.K. Mao et al.: 1987, Nature, **330**,737.
23. S. Wiryana, L.J. Slutsky and J.M. Brown: 1998, Earth and Planetary Sci. Lett., **163**,123.
24. M. Ali Bouhifd and A.P. Jephcoat: 2003, Earth and Planetary Sci. Lett., **209**,245.
25. D.Antonangeli, F.Occeli, H.Roquardt et al.: 2004, Earth and Planetary Sci. Lett., **225**,243.

A model of the internal structure of Titan: first results

G. Pavičić* and V. Čelebonović[†]

*Public Observatory, Gornji Grad 16, 11000 Beograd, Serbia and Montenegro
moskito752000@yahoo.com
[†]Inst.of Physics, Pregrevica 118, 11080 Zemun-Beograd, Serbia and Montenegro
vladan@phy.bg.ac.yu

Abstract. The aim of this short contribution is to report some preliminary results on a theoretical model of the internal structure of Titan, the satellite of Saturn. All the calculations were performed within the *SK* theory of dense matter. The central pressure and temperature, as well as a possible chemical composition of this satellite have been determined, but they will ultimately become verifiable only by data from the *Cassini* mission.

INTRODUCTION

Astronomers, physicists and even biologists have been interested in Titan for many years. The main motivation for their interest is the fact that Titan is the only satellite in the Solar System which has an atmosphere. It is widely thought in the research community that this atmosphere resembles the early atmosphere of the Earth, perhaps in pre-biotic times [1],[2] and earlier references given there. This continuous interest was the main "driving force" for the preparation and launch of the *Cassini* mission. Titan's atmosphere is dense and opaque, so there exists no way to gain any information whatsoever about its surface from Earth based observation. Accordingly, the logical question which emerges concerns the possibility of obtaining some information about the surface and interior of Titan theoretically.

The aim of this note is to briefly propose a theoretical model of the interior of Titan. It was calculated within the so-called *SK* theory of dense matter, proposed by P.Savić and Radivoje Kašanin, [4], and reviewed elsewhere in these proceedings [5]. The following section contains a brief remainder of the main ideas of the *SK* theory and the final one is devoted to results of its applications to Titan.

THE SK THEORY: A QUICK INTRODUCTION

The development of what later became the SK theory started in 1961 when P.Savić published a short paper [3] presenting an unusual idea: the mean planetary densities, as calculated from the masses and radii known at the time, could be linked to the mean solar density by an extremely simple expression:

$$\rho = \rho_0 2^\phi \tag{1}$$

In this expression ρ_0 denotes the mean solar density, which at the time was estimated at $4/3 \, gcm^{-3}$. By choosing integral values of the exponent ϕ in the interval $\phi \in (-2,2)$ Savić managed to fit the numerical values of the densities of the major planets. No hint of any possible physical explanation of this simple fit was given. In the following 4 years, collaborating with Radivoje Kašanin, [4] he managed to develop a theory of the behavior of materials under high pressure. In later years, it was nicknamed "the SK theory" after the first letters in their family names.

The basic physical idea of their theory is simple. SK assumed that sufficiently high values of external pressure lead to changes of the electronic structure of atoms and/or molecules. In the course of time, two kinds of applications of this theory were developed. One concerns modelling of celestial bodies, and the other is oriented towards the analysis of laboratory data on the behaviour of materials under high static pressure.

A final detail: in astronomical applications, this theory needs only a couple of input data: the mass and the radius of the object under study.

APPLICATION TO TITAN

The mass and radius of Titan were taken from the web site of the US National Space Science Data Center, at the address http://nssdc.gsfc.nasa.gov/. After the application of the algorithm of the theory, the following values of various parameters of Titan were obtained:

$$82 \leq A \leq 86$$

$$T^* = 300K$$

$$p^* = 0.01 MBar$$

The central pressure and temperature are denoted by p^*, T^*. The symbol A denotes the mean molecular mass of the mixture of materials that the object under study is made of. It can be calculated within the SK theory, but representing it with any combination of real chemical elements and/or compounds is a difficult task, which has (whenever possible) to be constrained by observed data. As in the case of Titan there exist no such data, the value of A obtained was represented by a combination of materials which seemed reliable, and which is

$$Fe_2SiO_3 + FeS + SO_2 + CH_4 + N_2 + Ir$$

How close this combination is to "'real'" Titan will only become verifiable by forthcoming data from the *Cassini* mission. At the time of writing, the probe is orbiting Saturn, and it has already discovered two new moons. According to predictions, the first close flyby of Titan will occur towards the end of October of 2004.

ACKNOWLEDGEMENTS

The authors are grateful to Prof. Wim van Saarloos, Director of the Lorentz Center, for the initial proposal on the organization of this workshop, and to all in the Lorentz Center whose work helped that the workshop was running so smoothly and successfully. One of the authors (V.Č) was financed by the Ministry of Science and Protection of the Environment of Serbia under its Project 1231.

REFERENCES

1. de Morais, A.: 2004,Lunar and Planetary Science,**XXXV**,*in press*.
2. Roe, H.G.: 2003,Publ. Astr. Soc. Pacific, **115**,1262.
3. Savić, P.: 1961,Bull. de la classe des Sci. Math. et Natur. de l'Acad. Serbe des Sci. et des Arts **XXVI** p.107.
4. Savić, P. and Kašanin, R.: 1962/65,The behaviour of materials under high pressure **I - IV**,Ed. by SANU, Beograd.
5. Celebonović, V.: 2004, these proceedings.

The Narrow Line Region of an AGN Sample

E. Bon[1], D. Ilić[2], L. Č. Popović[1], E. Mediavilla[3], V. Čelebonović[e], G. Pavičić

[1] *Astronomical Observatory, Volgina 7, 11160 Belgrade, Serbia*
[2] *University of Belgrade, Department of Astronomy, Serbia*
[3] *Instituto De Astrofisica De Canarias, La Palma, Tenerife, Spain*
[e] *Institute of Physics, Belgrade*

Abstract. Using high-resolution spectra of 8 Active Galactic Nuclei (AGN) observed with INT we found that shapes of narrow emission lines indicate very complex structure of the Narrow Line Region (NLR). We resolved, at least two kinematically separated emission regions: (i) one with narrower lines (~100 km/s) with shift that follows cosmological red-shift of the galaxies; (ii) and second with broader lines (~400 km/s). The differences between kinematical parameters of these two regions show the physical and kinematical stratification in emitting NLR plasma.

INTRODUCTION

The matter in Active Galactic Nuclei (AGN) emitting region is over wide range of physical and kinematical conditions. Consequently, different phases of matter in the central part are expected. Morever, the emission plasma has very different parameters from the region very close to the Black Hole (e. g. the plasma in the BLR might be in some condition of partial thermodinamical equilibrium, see e.g. Popović 2003) to the Narrow Line Region with low density and high temperature plasma.

Emission line profiles of AGNs have been basic source for determine kinematical properties and structure of the emitting gas. It is well known that AGN lines are coming from two different emission regions; the Narrow Line Region (NLR) and the Broad Line Region (BLR) that are physically and kinematically separated (see e.g. Osterbrock 1989, Krolik 1999). For observational studies of the NLR , the [OIII]4959,5007 emission lines have been typically used, as there are most intense forbidden lines found in AGN spectra (see e.g. Busko and Steiner, 1989). We analyze here the [OIII] emission line profiles from the set of 8 galaxies with the aim to investigate the complex kinematical structure of the NLR.

Narrow emission spectral lines are also characteristic of AGN spectra. They originate in NLR, a low-density ($Ne \approx 10^4$ cm^{-3}) and high temperature ($\sim 10^6$ K) region. It is quite certain that most of the narrow (FWHM about several hundred km/s) lines come from a bi-conical region coaxial with the accretion disk axis. It is very extended - up to about 1 kpc. In several near active galaxies like, for example, NGC 5252 the NLR has been directly observed, usually in the wavelength 5007 Å of a forbidden [OIII] line. The whole region is highly ionized, with abundant free electrons and where high velocity ($\leq$ 500 km /s) corpuscular winds and plasma shocks travel outward - all the way to the interstellar space in the home galaxy.

Here we present our investigation of the NLR of a sample of eight AGNs with strong X-ray emission.

OBSERVATIONS AND DATA REDUCTION

The observations were obtained with 2.5m Isaac Newton Telescope at La Palma Islands in Spain, in the period of 21st to 25th of January 2002. We used the Intermediate Dispersion Spectrograph (IDS) and the 235 camera in combination with the R1200Y grating. The seeing was 1".1 and the slit width 1". The spectral resolution was around 1.0 Å.

Standard reduction procedures including flat-fielding, wavelength calibration, spectral response, and sky subtraction were performed with the help of the IRAF software package. The software package DIPSO was used for reducing the level of the local continuum, by subtracting the N order polynomial, fitted through the dots taken to be on the local continuum in spectral range 4700 - 5100 Å for the Hβ (rest wavelengths).

ANALYSES OF THE DATA

First we fitted each line with a sum of Gaussian components, using X^2 minimization. We have been taking into account some features form atomic physics for the Hβ, that the intensity ratio of the [OIII] narrow lines of 1:3.03. We also fitted the Fe II multiplet template (Korista, 1992). We assumed that the narrow emission lines can be composed by more Gaussian components.

Table 1. The parameters of the Gaussian components of the [OIII] lines. Gaussian widths (W) and $\Delta z = z_{NLR2} - z_{NLR1}$ are given in km/s. The intensity ratio $R_{NLR1} = I_{5007}/I_{4959}$ of the 'blue component' obtained from the Gaussian fit, the measured flux ratio $R_F = F_{5007}/F_{4959}$ and central intensity ratio of the 'blue' and 'central' Gaussian components (I_B/I_C) are also given (columns 2, 3 and 6).

Table 1. The parameters of the Gaussian components of the [OIII] lines.						
Object	W_{NLR1}	R_{NLR1}	R_F	W_{NLR2}	Δz	I_B/I_C
3c120	420	3.0	2.9±0.1	130	-110	0.25
Mrk 1040	330	1.8	2.4±0.1	190	-340	0.30
Mrk 110	300	2.6	3.0±0.1	150	-40	0.17
Mrk 817	510	3.0	3.0±0.1	145	-300	0.43
Mrk 841	215	2.6	2.8±0.1	80	-20	0.45
NGC 3227	400	3.0	3.0±0.1	140	-120	1.87
NGC 4253	340	2.8	3.0±0.1	105	-55	0.55
PG 1211	270	2.8	3.0±0.1	81	-180	1.8

It was found that three broad Gaussians and one narrow component could fit well the profiles of the Hβ lines. The fitting procedure has been described in Popović et al. (2001, 2002, 2003, 2004). It was found that three broad Gaussians and one narrow component could fit well the profiles of the Hβ line. As for the [OIII] lines, they show very extended wings. We noticed that the wings are asymmetrical, being more gently sloped towards the blue. Therefore we fitted each [OIII] line with two Gaussians, one relatively broad and blue-shifted and one narrow component. The results are presented in the Table 1.

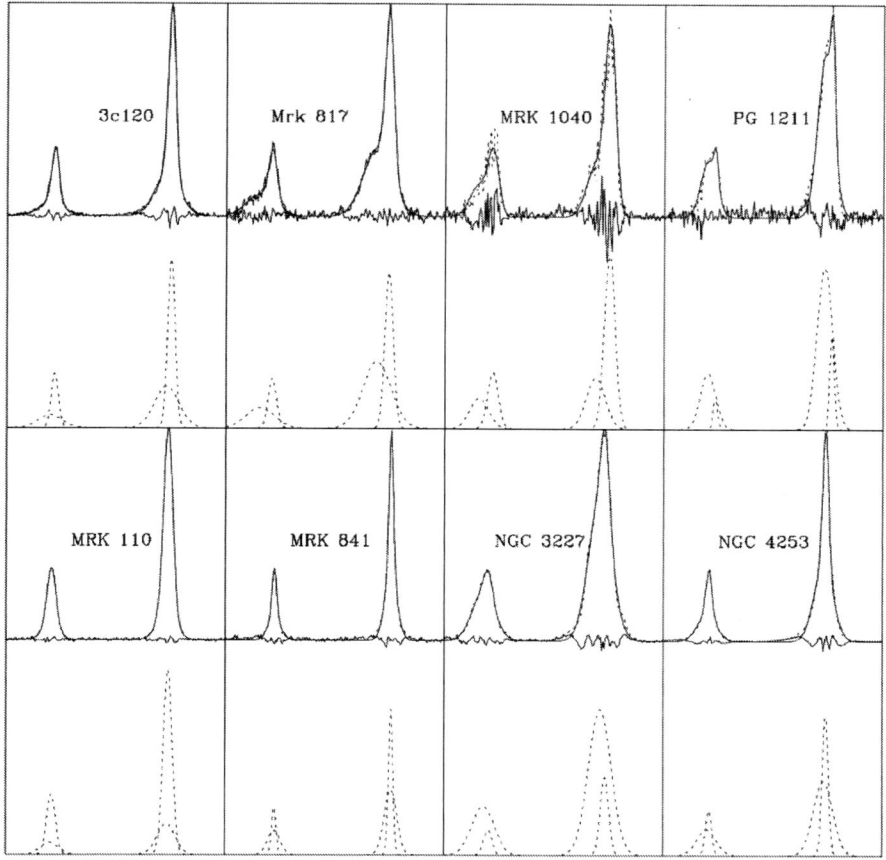

FIGURE 1. : Decomposition of the [OIII] lines of observed AGN.. The Gaussian components are presented at bottom. The [OIII] components are in the red wing of Hβ.

We can clearly see at least two phases of NLR in the [OIII] lines: (i) the NLR1, which has an internal random velocity from ~200 to 500 km/s, and relative approaching velocities from 100-250 km/s with respect to the systemic redshift of the observed galaxy; and (ii) the NLR2 which has an internal random velocity between 100 - 250 km/s, and a redshift equal to the systemic one of the corresponding object.

The intensity ratio of the components from the NLR2 follows the line ratio ($I(5007):I(4959)$= 3.03) (see Table 1). Although a slight difference between the observed and predicted intensity ratio of the [OIII] lines might exist (Storey & Zeippen 2000), this is certainly due to the case of Mrk 1040, to the high residues of the fit (see Figs. 2 and 3). The difference in shifts and widths of the [OIII] λλ4959,5007 between these two NLRs indicate different kinematical and physical properties. The clear tendency of the NLR1 to have a blue-shifted systemic velocity (though in the case of Mrk 110 and Mrk 841 the blue-shift is marginal and very close to the cosmological redshift) supports the idea of a jet geometry of the NLR (see e.g. Dopita et al. 2003). In this case the receding jet component in the [OIII] lines might be obscured or absorbed by the host galaxy, so one sees only the out flowing gas from the closer part of the jet. Many papers have been devoted to the radial velocity difference between narrow and broad lines (starting e.g. from Gaskell 1982), especially the velocity difference between the [OIII] lines and the Hβ that indicates a jet geometry of the NLR or of part of the NLR (see e.g. Bennert 2002; Zamanov 2002). On the other hand, 2D spectroscopy clearly shows that AGN NLRs have a complex structure (see e.g. Arribas 1997). The tendency for the [OIII] lines to be blue-shifted relative to the Hβ suggests that they are associated with a high-ionization outflow originating in these highly accreting sources (see e.g. Zamanov 2002).

CONCLUSIONS

The NLRs of considered AGNs show complex structure, and we clearly can see at least two phases of the NLR in all observed AGNs. The different intensity ratio of the [OIII] 5007 and [OIII] 4959 also indicates that in the NLR1 and NLR2 exist difference in physical processes. (For more details see Popović et al. 2004).

The clear tendency of NLR1 systemic velocity toward to blue support the idea about jet geometry of NLR (see e.g. Dopita et al. 2003), and in the chosen sample it seems that line-of-sight might be in the jet direction.

ACKNOWLEDGMENTS

This work was supported by the Ministry of Science, Technologies and Development of Serbia through the project P1196 "Astrophysical Spectroscopy of Extragalactic Objects".

REFERENCES

Arribas, S.,Mediavilla, E., Garcia-Lorenzo, B., & del Burgo, C.,ApJ, 490, 227, (1997)
Bennert, N., Falcke, H., Schulz, H.,Wilson, A. S., &Wills, B. J. ApJ, 574, 105 (2002)
Busko, I.C. & Steiner, J. E., MNRAS, 228, 1479-1495 (1989)
Dopita M.A., Bicknell G.V., Sutherland R.S., Saxton C.J., Rev. Mex. A&A 15, 323. (2003)
Gaskell, C. M. Astronomical Society of the Pacific, Publications, vol. 94, p. 891-893 (1982).
Korista, K. T., ApJS, 79, 285 (1992)
Krolik J. H. *Active galactic nuclei: from the central black hole to the galactic environment*: Princeton University Press, (1999).
Osterbrock, D.E. *Astrophysics of gaseous nebulae and active galactic nuclei*, Research supported by the University of California, John Simon Guggenheim Memorial Foundation, University of Minnesota, et al. Mill Valley, CA, University Science Books, 422 (1989).
Popović, L. Č., Stanić, N., Kubicela, A., Bon, E., A&A, 367, 780 (2001)
Popović, L. Č., Mediavlilla, E.G., Kubičela, A., Jovanović, P., A&A, 390, 473 (2002)
Popović, L. Č., ApJ, , 599, 140 (2003)
Popović, L. Č., Mediavlilla, E.G., Bon, E., Ilić D., A&A 423, 909–918 (2004)
Storey, P. J., & Zeippen, C. J., MNRAS, 312, 813 (2000)
Zamanov, R., Marziani, P., Sulentic, J. W., et al., ApJ, 576, L9 (2002)

On phase transitions in Bose gases at constant density and constant pressure

Velin G. Ivanov, Dimo I. Uzunov

CP Laboratory, G. Nadjakov Institute of Solid State Physics, Bulgarian Academy of Sciences, BG-1984 Sofia, Bulgaria
uzun@issp.bas.bg

Abstract. The phase transitions in Bose gases at constant volume and constant pressure are considered. New results for the chemical potential, the effective Landau-Ginzburg free energy and the equation of state of the Bose-Einstein condensate in ideal Bose gases with a general form of the energy spectrum are presented. Unresolved problems are discussed.

INTRODUCTION

The Bose-Einstein condensation (BEC) is a phenomenon due to the quantum statistical correlations (pseudo-interactions) in ideal Bose gases (IBG) of non-interacting bosons. This is a condensation in the momentum space ($\hbar \vec{k}$) but also it possesses some features of the usual (Van der Waals) condensation [1, 2, 3, 4]. This phenomenon strongly depends on the spatial dimensionality d, and the energy spectrum $\varepsilon(k) = (\hbar^2 k^\sigma /2m)$ of the bosons [$0 < \sigma \leq 2$; $k = |\vec{k}|$; $\vec{k} = \{k_j = (2\pi/n_j L_j)\}$, where $j = 1, ..., d$; $n_j = 0, \pm 1, ...$] in IBG in volume $V = (L_1 ... L_d) \sim L^d$ and periodic boundary conditions [3, 4].

The original Bose-Einstein condensate (BEC) is not superfluid but the latter state is possible in many-body systems of interacting bosons (nonideal Bose gas, or, shortly, NBG) [1, 2, 3, 4, 5]. The phenomena of BEC and superfluidity have a number of similar features and both of them are widely discussed in various problems of astrophysics (see, e.g., Refs. [5, 7, 8, 6]. In particular, BEC and superfluidity are relevant to the treatment of the so-called plasma-solid transition in "astrophysical matter" [7], and for the behaviour of the neutron component of the dense matter in the interior of neutron stars [5, 8, 6, 9]; for applications to cosmological models of dark energy and dark matter, see Ref. [10].

The main properties of BEC are known (see, e.g., Ref. [4, 11, 12, 13, 14, 15]. Recently, BEC at constant density and constant pressure has been reviewed in Ref. [16]. Here we shall discuss the equation of state of IBG and BEC for spinless bosons. The treatment can be generalized for bosons with spin [17, 18].

The superfluidity and the effect of interparticle interactions on BEC can be treated by both the mean-field like Gross-Pitaevskii [19] and the microscopic Beliaev-Popov [20] approaches. We shall discuss the latter within the renormalization group theory [3] and for this reason we shall consider the following action of NBG [3, 4, 20]:

$$\mathscr{S}[\phi] = -\sum_q G_0^{-1}(q)|\psi(q)|^2 - \frac{u}{2\beta N} \sum_{q_1,q_2,q_3} \psi^*(q_1)\psi^*(q_2)\psi(q_3)\psi(q_1+q_2-q_3), \quad (1)$$

where $q=(\omega_l,\vec{k})$ is a $(d+1)$-dimensional frequency-momentum vector, $\omega_l = 2\pi l k_B T/\hbar$ is the (Bose-)Matsubara frequency ($l = 0, \pm 1, \ldots$), $\psi(q)$ is a C-number Bose field, u is the interaction constant, N is the number of particle and $\beta = 1/k_B T$. The (bare) correlation (Green) function $G_0(q) = \langle |\psi(q)|^2\rangle_0$ is given by

$$G_0^{-1}(q) = i\omega_l + \varepsilon(k) + r, \tag{2}$$

where $(-r) = \mu \leq 0$ is the chemical potential of IBG ($u = 0$).

Note, that the thermodynamics of NBG is given by the grand canonical thermodynamic potential $\Omega(T,V,\mu) = -\beta^{-1} Z$, where the grand canonical partition function $\mathscr{Z}(T,V,\mu)$ is defined by $\mathscr{Z} = \int \mathscr{D}\phi \exp(\mathscr{S})$ - a functional integral over the possible field configurations. The usual field theoretical investigations of the action (1) are performed in the thermodynamic limit: $N \to \infty$, $V \to \infty$, provided $0 \leq \rho = (N/V) < \infty$.

Using the models of IBG ($v \equiv 0$) and NBG defined by Eqs. (1) – (2) we shall consider the equation of state for BEC. For IBG this can be performed exactly, whereas for NBG we must use the loop expansion (see, e.g., Ref. [3]). The main features of BEC can be revealed within the one-loop approximation [3, 21]. We shall discuss the effect of the inter-particle interaction on BEC by using results from preceding works.

We wish to emphasize that the phase transition to BEC cannot be easily put to the usual classification of phase transitions [3]. Above T_c this phase transition resembles certain features of second order phase transitions but is rather different from the standard notion about these transitions [3]. On the other side, the equation of state of IBG below T_c is quite similar to known equations of (almost) first-order phase transitions and tricritical points [3]. There is a close similarity between the phase transition properties of IBG and the spherical model in the ferromagnetism, and this point will be discussed in the remainder of this report. In our consideration we shall essentially use results from preceding works [16, 11, 12, 13, 14, 15] (for a comprehensive reference to original papers, see, e.g., the reviews [3, 4]). We work with general values of d and σ but our consideration includes the important case of $d = 3$, $\sigma = 2$.

FREE ENERGY AND THERMODYNAMICS

For a convenience we shall introduce a fictitious external field h which is thermodynamically conjugated to the order parameter $\Psi = \langle \psi(0) \rangle$ of uniform BEC. Note, that in case of uniform BEC the ($q = 0$)-mode $\psi(0)$ can be represented by the sum $\psi(0) = \langle \psi(0)\rangle + \delta\psi(0)$, where $\delta\psi(0)$ is the (uniform) fluctuation mode. The consideration of an external uniform field h can be performed by adding a term

$$\mathscr{S}_h = \frac{1}{N}[h\psi^*(0) + c.c.]. \tag{3}$$

The thermodynamic potential of IBG can be written in the form

$$\Omega(T,r,h) = -\beta^{-1} V \lambda_T^{-d} A(d,\sigma) g_{d/\sigma+1}(\beta r) - \frac{1}{N}\frac{hh^*}{r}, \tag{4}$$

where $g_\nu(y)$ is the Bose function [3, 15, 18], the thermal wavelength is given by

$$\lambda_T = \left(\frac{2\pi\hbar^2}{mk_BT}\right)^{1/\sigma}, \tag{5}$$

and

$$A(d,\sigma) = \frac{2^{1-d+2d/\sigma}\Gamma(d/\sigma)}{\sigma\pi^{d(1/2-1/\sigma)}\Gamma(d/2)}, \quad A(d,2)=1. \tag{6}$$

The potential (4) obeys the differential relation $d\Omega = -SdT + Ndr - \Psi dh^* - \Psi^* dh$.

By a suitable Legendre transformation we obtain another thermodynamic potential $\tilde{\Omega}(T,r,\Psi)$, where the natural variable is Ψ:

$$\tilde{\Omega} = -\beta^{-1}V\lambda_T^{-d}A(d,\sigma)g_{d/\sigma+1}(\beta r) + Nr\Psi^2. \tag{7}$$

We can restrict ourselves, without a loss of generality, to real h and Ψ. The susceptibility is then $\chi_T = \partial\Psi/\partial h = 1/Nr \sim t^{-\gamma}$, where γ is the susceptibility critical exponent and $t = (T-T_c)/T_c$ (see,e.g., Ref. [3]).

For small r and energies ($\varepsilon \ll k_BT$), which correspond to the critical regime near the phase transition point, the correlation function $\chi(k) = G_0(0,\vec{k})$ takes the form $\chi(k)^{-1} \sim (ck^\sigma + r)$. This gives us the correlation length $\xi = (c/r)^{1/\sigma} \sim t^{-\nu}$ and $\chi(k) \sim k^{-2+\eta}$ at $r=0$. From $\chi(k) \sim k^\sigma$ we obtain that the Fisher exponent η, defined by $\chi(k) \sim k^{-2+\eta}$, is equal to $(2-\sigma)$ for all dimensional ranges and possible constraints (of constant volume V or constant pressure P).

In order to obtain the correlation length exponent ν and the exponent γ of the susceptibility χ_T, we need to calculate the function $r(t)$. The latter is different for the cases $V = const$ and $P = const$. In the thermodynamic limit, the constraint of constant volume is equivalent to a constraint of constant density ($\rho = const$). The results for the critical exponents corresponding to these cases are summarized in Ref. [16] for various spatial dimensions d. Here we restrict our consideration of the critical exponents and the equation of state to features which demonstrate the difference between the phase transition to BEC and both first and second order phase transitions for the most interesting interval of spatial dimensions $\sigma < d < 2\sigma$ that includes the case $\sigma = 2, d = 3$.

Constant density

To consider the effect of the thermodynamic condition of constant density $\rho = (N/V)$ we need the free energy $F = Vf(T,\rho,\Psi)$. To obtain this thermodynamic potential near the phase transition point to BEC we expand the Bose function in (4) in powers of βr and express r as a function of ρ. This procedure is performed for the potential $\tilde{\Omega}$ given by Eq. (7). Further we obtain the potential F with the help of a Legendre transformation of the form:

$$F = \tilde{\Omega} - Nr|_{r=r(\rho)}, \tag{8}$$

$$\rho = \frac{\partial\tilde{\Omega}}{V\partial r} = \lambda_T^{-d}A(d,\sigma)g_{d/\sigma}(\beta r) + \rho\Psi^2. \tag{9}$$

The lowest temperature T_c, at which the system does not condensate ($\Psi = 0$) is obtained from Eq. (9). When $d \leq \sigma$, the Bose function is divergent for $(\beta r) \to 0$ which means that BEC may occur only at the absolute zero ($T_c = 0$) - a zero temparature BEC [15]. When $d > \sigma$, we obtain that [4, 14]

$$T_c(\rho) = \frac{2\pi\hbar^2}{mk_B} \left[\frac{\rho}{A(d,\sigma)\zeta(d/\sigma)} \right]^{\sigma/d} > 0, \tag{10}$$

where $g_{d/\sigma}(0) = \zeta(d/\sigma)$ is the zeta function. Below the critical temperature T_c, BEC occurs ($\Psi > 0$).

For the most interesting case of dimensions $\sigma < d < 2\sigma$, the result for the free energy density $f(T, \rho, \Psi)$ to the lowest order in $|t| \ll 1$ is given by:

$$f = C_f \left(\Psi^2 + \frac{d}{\sigma} t \right)^{d/(d-\sigma)}, \tag{11}$$

where

$$C_f = \left(\frac{d}{\sigma} - 1 \right) \left[\frac{\zeta(d/\sigma)}{|\Gamma(1-d/\sigma)|} \right]^{\sigma/(d-\sigma)} (k_B T_c) \rho. \tag{12}$$

In our derivation of the energy (11) a Ψ-independent term has been neglected. Such terms can be ignored because they belong to the energy of the disordered phase ($\Psi = 0$). For this reason the net free energy of the BEC can be obtained from (11) by neglecting the Ψ-independent term of type $t^{d/(d-\sigma)}$. The free energy (11) is of Landau-Ginzburg type [3]; one may easily expand $f(\Psi)$ in powers of Ψ.

The external field h is fictitious and can be neglected in studies of the thermodynamics. In this case, the equation of state $h \sim \partial(f/\partial \Psi)$ becomes $(\partial f/\partial \Psi) = 0$. From Eq. (11) we easily obtain the equation of state for $h = 0$:

$$\Psi \left(\Psi^2 + \frac{d}{\sigma} t \right)^{\sigma/(d-\sigma)} = 0. \tag{13}$$

The solutions of Eq. (13) are: $\Psi = 0$ (disordered phase), and $\Psi^2 = -(d/\sigma)t > 0$ (corresponding to BEC for $t < 0$). Note, that the chemical potential ($\mu = -r$) can be obtained in the form

$$\mu = -k_B T \left[\frac{\zeta(d/\sigma)}{|\Gamma(1-d/\sigma)|} \right]^{\sigma/(d-\sigma)} (\Psi^2 + \frac{d}{\sigma} t)^{\sigma/(d-\sigma)}. \tag{14}$$

BEC is possible for $t < 0$ under the condition $\mu = 0$. The latter introduces thermodynamically forbidden (unstable) domains in the phase diagram.

The main conclusion which can be drawn from the Ginzburg-Landau free energy (11) of IBG at constant ρ is that the point $t = 0$ resembles a tricitical point [3]. Usually such multicritical points occur on a phase transition line where the phase transition changes from second order to a symmetry conserving first order phase transition (or vice versa). Here this is not the case but the similarity with the usual tricriticality is in the fact that the coefficients of both the Ψ^2- and the Ψ^4-terms in (11) tend to zero for $t \to 0$. On the other side the phase transition to BEC is a continuous phase transition which exhibits critical exponents identical to the critical exponents known from the Berlin-Kac spherical model in ferromagnetism [22, 23] (see also Refs. [3, 4]). Here the condition of constant density ρ plays the role of the spherical condition for the spins in the spherical model.

Interaction effect. Most of these exceptional properties of the phase transition to BEC in IBG are not present in systems of interacting bosons described by NBG. In this case an additional term of type $u\Psi^4$ will appear in the effective free energy $f(T,\rho,\Psi)$ and the respective coefficient of the Ψ^4- term will remain finite at T_λ – the critical temperature to superfluid state in interacting Bose fluids (gases and liquids). The generalization of our treatment to the case of interacting bosons described by the action (1) may lead to new thermodynamic properties. A similar generalization for the spherical model was made in Ref. [23]. Another way of treatment of NBG may be performed within the loop expansion [21]. It is believed that the λ-transition described by NBG belongs to the so-called XY universality class of standard second order phase transitions [3].

Recent studies of interacting bosons indicated discrepancies in the theoretically predicted values of the critical temperature T_λ [24, 25, 26, 27]. The problem for the calculation of the phase transition temperature of interacting many-body systems is a hard and still unresolved problem of the theory [3]. The fluctuation shifts of the critical temperature that are usually calculated from field models (see, e.g., Ref. [3] are very small and do not include essential contributions due to the large-momentum (high-energy) fluctuations of the order parameter field ψ. Therefore, the correct treatment of the phase transition temperature in many-body systems with interparticle interactions requires new theoretical methods.

Constant pressure

The only papers where the effect of the constraint of constant pressure on the phase transition properties of IBG has been investigated so far are Refs. [13, 14, 4, 16]. This case should be taken in mind in interpretations of real experiments, in particular, in low-temperature experiments on BEC in trapped atomic gases (see. e.g., Ref. [19]), where the density ρ varies but the pressure P is (almost) fixed. In this case BEC occurs at finite temperatures ($T_c > T > 0$) for all spatial dimensions $d > 0$. The critical temperature will be [4]

$$T_c(P) = \left[\frac{\lambda_0^d P}{\zeta(1+d/\sigma)A(d,\sigma)k_B}\right]^{\sigma/(d+\sigma)}, \tag{15}$$

where λ_0 is given by $\lambda_0 = \lambda T^{1/\sigma}$ and Eq. (5). For $d = \sigma = 2$ one obtains the result for T_c known from Ref. [13].

Although a number of results are known from preceding works, in particular, for the two-dimensional case ($d = 2$) [13], the entire picture of the phase transition to BEC at constant pressure is not still clear for all spatial dimensions d. In particular, this is the case of interacting bosons. We have no information about research papers devoted to this problem.

CONCLUDING REMARKS

We have presented a brief discussion of several properties of the phase transition to BEC. New results have been obtained for the effective Landau-Ginzburg free energy (11), the equation of state (13) and the chemical potential (14) of BEC in ING for general values of d and σ.

The effects of the constraints of constant density and constant pressure on BEC in ING and the condensation to a superfluid phase in systems of interacting bosons (NBG) are not yet clarified in a comprehensive way. The renormalization group methods [3, 4] reveal a dimensional (quantum to classical crossover) in interacting Bose systems and other basic universality features of the so-called quantum phase transitions at zero and very low temperatures but the equation of state below the phase transition point is not investigated in details. This is the situation for the whole variety of Bose systems known in condensed matter physics [3, 4] and, in particular, for Bose fluids of real atoms.

For Bose fluids the concept of universality of the quantum critical phenomena seems to be invalid and mainly for this reason the quantum phase transitions of second order can be classified in two groups: as universal and non-universal (see Ref. [4]). However, a number of examples indicate that the quantum phase transitions are often of first order or are described by multicritical points which are different from the critical points of standard second order phase transitions.

In astrophysics, a quantum phase transition ($T_c \sim 0$) may occur in cases of very low density or very low pressure of the respective Bose fluid; see Eqs. (10) and (14). For the matter in the interior of neutron stars one should investigate the so-called classical limit [4] in which the quantum fluctuations are irrelevant. This is the case of $T_c > 0$ discussed in the prevailing part of our report.

BEC of spin bosons [17] can be treated within the framework of a generalization of our treatment. For this aim one may consider a complex vector field $\vec{\psi}(q) = \{\psi_\alpha; \alpha = 1,...n/2\}$. For $n = 2$ one obtains the complex scalar field in Eq. (1). When such a system is placed in an external magnetic field, essentially new results can be obtained [17]. The unresolved problem about the superfluidity of interacting spin bosons in external magnetic field is of special interest in astrophysics.

REFERENCES

1. Huang,K.: 1963, Statistical Physics, Wiley, New York.
2. Landau,L.D. and Lifshits,E.M.: 1980, Statistical physics, Part I, Pergamon, London.
3. Uzunov,D.I.: 1993,Theory of critical phenomena,World Scientific,Singapore.
4. Shopova,D.V. and Uzunov,D.I.: 2003,Phys.Rep.C,**379**,1-67.
5. Tilley,D.R., Tilley.J.: 1974,Superfluidity and Superconductivity, Van Norstrand Reinhold Company,New York.
6. Link,B.: 2003,Phys.Rev.Lett.,**91**,101101.

7. Celebonovic,V. and Daäppen, in: 2000, 20th SPIG Conference, Ed. by Petrovic,Z.Lj.et al, Faculty of Physics and INN.Vinca, Beograd, Yugoslavia; see also:astro-ph/0007337, and astro-ph/0102284.
8. Tsuruta,S.: 1998,Phys.Rep.C,**292**,1.
9. Pèrez Rojas,H.,Pèrez Martìnez,A. and Mosquera Cuuesta,H.J.: 2004,cond-mat/0407047.
10. Nishiyama,M.,Morita,M-a.and Morikawa,M.: 2004,astro-ph/0403571.
11. Gunton,J.D. and Buckingham,M.J.: 1968,Phys.Rev.,**166**,152.
12. Cooper,M.J. and Green,M.S.: 1968,Phys.Rev.,**176**,302.
13. Gunther,L.,Imry,Y. and Bergman,D.J.: 1974,J.Stat.Phys.,**10**,425.
14. Lacour-Gayet,P. and Toulouse,G.: 1974,J.Physique(Paris),**35**,425.
15. Busiello,G.,De Cesare,L. and Uzunov,D.I.: 1985,Physica A,**132**,199.
16. Ivanov,V.G., in: 2004, Meetings in Physics at Sofia University, Ed. by A. Proykova, Heron Press, Sofia; cond-mat/0405537.
17. Yamada,K.: 1982, Progr.Theor.Phys.,**67**,443.
18. da Frota.H.O.,Silva,M.S.,Goulart Rosa Jr,S.: 1984,J. Phys.C:Solid St. Phys.,**17**,1669.
19. Dalfovo,F.,Giorgini,S.,Pitaevskii,L.P. and Stringari,S.: 1999,Rev.Mod.Phys.,**71**,463.
20. Popov,V.N.: 1983, Functional integrals in quantum field theory and statistical physics, Riedel, Dordrecht.
21. Toyoda,T.: 1982,Ann.Phys.(N.Y.),**141**,154.
22. Berlin,T.H. and Kac,M.: 1952,Phys.Rev.,**86**,821.
23. Langer,J.S.: 1963,Phys.Rev.,**137**,No.5A,A1532.
24. Baym,G,.Blaizot,J.-P., Holzmann,M. et al: 1999,Phys.Rev.Lett.,**83**,1703.
25. Huang,K.: 1999,Phys.Rev.Lett.,**83**,3770.
26. Schakel,A.M.J.: 2000,cond-mat/0004142.
27. Schakel,A.M.J.: 2003, J.Phys.Studies,7,140; cond-mat/0301050.

Phase transitions to spin-triplet ferromagnetic superconductivity in neutron stars

Diana V. Shopova, Tsvetomir E. Tsvetkov, Dimo I. Uzunov

*CP Laboratory, G. Nadjakov Institute of Solid State Physics, Bulgarian Academy of Sciences,
BG-1784 Sofia, Bulgaria
uzun@issp.bas.bg*

Abstract. Effects of the anisotropy of Cooper pairs in spin-triplet ferromagnetic superconductors are investigated on the basis of the Ginzburg-Landau theory. A special attention is paid to the triggering of the superconducting state by the ferromagnetic order. The ground states of these superconductors are outlined and discussed. The idea about a possible coexistence of ferromagnetism and spin-triplet superconductivity in neutron stars is introduced.

INTRODUCTION

The cold and dense matter in the interior of neutron stars undergoes phase transitions to both superfluid and superconducting states (see, e.g., Refs. [1, 2, 3]. Both the superfluidity due to neutron Cooper pairing and the superconductivity due to proton Cooper pairing may be of unconventional spin-triplet (p-wave) type [1]. The latter is known from the theory of ^{3}He liquids [4, 5, 6] as well as from the theory of heavy fermion [7, 8]) and high-temperature ([9, 10, 11, 12]) superconductors.

The possible superconducting phases in unconventional superconductors are described in the framework of the general Ginzburg-Landau (GL) effective free energy functional [11] with the help of the symmetry group theory. Thus a variety of possible superconducting orderings were predicted for different crystal structures [13, 14, 15, 16]. A detailed thermodynamic analysis [9, 14] of the homogeneous (Meissner) phases and a renormalization group investigation [9] of the superconducting phase transition up to the two-loop approximation [11] were also performed.

Recent experiments [17] at low temperatures ($T \sim 1$ K) and high pressure ($T \sim 1$ GPa) demonstrated the existence of spin triplet superconducting states in the metallic compound UGe$_2$. This superconductivity is *triggered* by the spontaneous magnetization (M) of the ferromagnetic phase (M-trigger effect [18]). The ferromagnetic order exists at much higher temperatures and coexists with the superconducting phase in the whole domain of existence of the latter below $T \sim 1$ K; see also experiments published in Refs. [19]. Moreover, the same phenomenon of existence of superconductivity at low temperatures and high pressure in the domain of the (T,P) phase diagram where the ferromagnetic order is present has been observed in ZrZn and URhGe, too; for details, see, e.g. Ref. [22, 23, 24, 18]. Note, that the superconductivity in the metallic compounds mentioned above, always coexists with the ferromagnetic order and is enhanced by the latter. Besides, in these systems the superconductivity seems to arise from the same electrons that create the band magnetism.

A similar phenomenon of coexistence of ferromagnetism and spin-triplet superconductivity may exist also in neutron stars. In this case the Cooper pairing of fermions (protons) will be triggered by the spontaneous magnetic moment of the same proton subsystem of the nuclear star matter. The basic features of these phenomena can be described within an extension of the GL theory. The results can be applied to a new and interesting problem of coexistence of superconductivity and ferromagnetism in neutron stars. The coexistence of spin-triplet superconductivity and ferromagnetism may explain the large magnetic field of these stars. To our best knowledge the problem of a possible coexistence of ferromagnetism and superconductivity in neutron stars and asprophysics objects is introduced for the first time in our present report.

The theory allows the description of various types of phase transitions and multicritical points [11, 25]. Following notations in Ref. [18], here we summarize previous studies [22, 23, 24, 18] and present new results on the effect of the Cooper pair anisotropy on the phase diagram of spin-triplet ferromagnetic superconductors [18].

Our consideration is focussed on the ground state, namely, we are interested in uniform phases, where the order parameters (the superconducting order parameter ψ and the magnetization vector $M = \{M_j, j = 1, 2, 3\}$, do not depend on the spatial vector $\vec{x} \in V$ (V is the volume of the system).

GINZBURG-LANDAU FREE ENERGY

Consider the GL free energy $F(\psi, M) = V f(\psi, M)$, where the free energy density $f(\psi, M)$ (for short hereafter called "free energy") of a spin-triplet ferromagnetic superconductor is given by [18]

$$f(\psi, M) = a_s |\psi|^2 + \frac{b_s}{2} |\psi|^4 + \frac{u_s}{2} |\psi^2|^2 + \frac{v_s}{2} \sum_{j=1}^{3} |\psi_j|^4 + a_f M^2 + \frac{b_f}{2} M^4 \quad (1)$$
$$+ i\gamma_0 M \cdot (\psi \times \psi^*) + \delta M^2 |\psi|^2.$$

In Eq. (1), $\psi = \{\psi_j; j = 1, 2, 3\}$ is the three-dimensional complex vector ($\psi_j = \psi'_j + i\psi''_j$) describing the unconventional (spin-triplet) superconducting order and $B = (H + 4\pi M) = \nabla \times A$ is the magnetic induction; $H = \{H_j; j = 1, 2, 3\}$ is the external magnetic field, $A = \{A_j; j = 1, 2, 3\}$ is the magnetic vector potential ($\nabla \cdot A = 0$). In Eq. (1), $b_s > 0$, $b_f > 0$, $a_f = \alpha_f(T - T_f)$ is given by the positive material parameter α_f and the ferromagnetic critical temperature T_f corresponding to a simple superconductor ($M \equiv 0$), and $a_s = \alpha_s(T - T_s)$, where α_s is another positive material parameter and T_s is the critical temperature of a standard second order phase transition which may occur at $|H| = \mathcal{M} = 0$; $\mathcal{M} = |M|$. The parameter u_s describes the anisotropy of the spin-triplet Cooper pair whereas the crystal anisotropy is described by the parameter v_s [9, 14].

The two orders – the magnetization vector $M = \{M_j\}$ and $\psi = \{A_j\}$, interact through the last two terms in (1). The γ_0–term [21] ensures the triggering of the superconductivity by the ferromagnetic order ($\gamma_0 > 0$) whereas the δ–term makes the model more realistic in the strong coupling limit [20]. Both ψM-interaction terms included in (1) are important for a correct description of the temperature-pressure (T,P) phase diagram of the ferromagnetic superconductor [18]. In general, the parameter δ for ferromagnetic superconductors may take both positive and negative values.

The values of the material parameters $(T_s, T_f, \alpha_s, \alpha_f, b_s, u_s, v_s, b_f, K_j, \gamma_0$ and $\delta)$ depend on the choice of the concrete substance and on intensive thermodynamic parameters, such as the temperature T and the pressure P. One may assume that the general form (1) of the free energy may describe the general features of the uniform orders in neutron stars provided one makes a suitable choice of parameters $(T_s, T_f, \alpha_s,...)$.

As we are interested in the ground state properties, we set the external magnetic field equal to zero $(H = 0)$. Besides, we emphasize that the magnetization vector M may produce vortex superconducting phase [1, 11] in case of type II superconductivity. The investigation of nonuniform (vortex) states can be made with the help of gradient terms in the free energy [18] which take into account the spatial variations of the order parameter field ψ. This task is beyond our present consideration. Rather we investigate the basic problem about the possible stable uniform (Meissner) superconducting phases which may coexist with uniform ferromagnetic order. For this aim the free energy (1) is quite convenient.

In case of a strong easy axis type of magnetic anisotropy, as is in UGe$_2$ [17], the overall complexity of mean-field analysis of the free energy $f(\psi,M)$ can be avoided by performing an "Ising-like" description: $M = (0,0,\mathcal{M})$. Further, because of the equivalence of the "up" and "down" physical states $(\pm M)$ the thermodynamic analysis can be performed within the "gauge" $\mathcal{M} \geq 0$. But this stage of consideration can also be achieved without the help of crystal anisotropy arguments. When the magnetic order has a continuous symmetry one may take advantage of the symmetry of the total free energy $f(\psi,M)$ and avoid the consideration of equivalent thermodynamic states that occur as a result of the respective symmetry breaking at the phase transition point but have no effect on thermodynamics of the system. In the isotropic system one may again choose a gauge, in which the magnetization vector has the same direction as z-axis $(|M| = M_z = \mathcal{M})$ and this will not influence the generality of thermodynamic analysis. Here we shall prefer the alternative description within which the ferromagnetic state may appear through two equivalent "up" and "down" domains with magnetization $\mathcal{M}$ and $-\mathcal{M}$, respectively.

For our aims we use notations in which the number of parameters is reduced. Introducing the parameter

$$b = (b_s + u_s + v_s) > 0 \qquad (2)$$

we redefine the order parameters and the other parameters in the following way:

$$\varphi_j = b^{1/4}\psi_j = \phi_j e^{\theta_j}, \quad M = b_f^{1/4}\mathcal{M}, \qquad (3)$$

$$r = \frac{a_s}{\sqrt{b}}, \quad t = \frac{a_f}{\sqrt{b_f}}, \quad w = \frac{u_s}{b}, \quad v = \frac{v_s}{b},$$

$$\gamma = \frac{\gamma_0}{b^{1/2}b_f^{1/4}}, \quad \gamma_1 = \frac{\delta}{(bb_f)^{1/2}}.$$

Having in mind our approximation of uniform ψ and M and the notations (2) - (3), the free energy density $f(\psi, M)$ can be written in the form

$$
\begin{aligned}
f(\psi, M) \;=\; & r\phi^2 + \frac{1}{2}\phi^4 + 2\gamma\phi_1\phi_2 M \sin(\theta_2 - \theta_1) + \gamma_1\phi^2 M^2 + tM^2 + \frac{1}{2}M^4 \\
& - 2w\left[\phi_1^2\phi_2^2 \sin^2(\theta_2 - \theta_1) + \phi_1^2\phi_3^2 \sin^2(\theta_1 - \theta_3) + \phi_2^2\phi_3^2 \sin^2(\theta_2 - \theta_3)\right] \\
& - v[\phi_1^2\phi_2^2 + \phi_1^2\phi_3^2 + \phi_2^2\phi_3^2].
\end{aligned} \quad (4)
$$

In this free energy the order parameters ψ and M are defined per unit volume.

In contrast to the situation in superconducting compounds, for the case of neutron stars, the crystal field anisotropy represented by the v_s–terms in (1) – (4) can be ignored, and for this reason we shall discuss the case $v_s \equiv 0$. We assume that $T_f > T_s$. This is the case when the superconductivity is triggered by the magnetic order. Besides we shall discuss the stable phases in the temperature region $T > T_s$. The case $T_f < T_s$ may also present interest for neutron stars and, hence, it needs a special investigation. As mentioned in Ref. [18], the case $T_s \sim T_f$ allows for a quite simple analytical treatment. All these cases may be of interest to the description of ferromagnetic superconductivity in stellar objects whereas in condensed matter only cases of $T_f \gg T_s$ have been observed so far.

Our consideration is performed within the framework of the standard mean-field analysis [11]. The stable phases correspond to global minima of the GL energy (1). The equilibrium phase transition line separating two phases is defined by the thermodynamic states, where the respective GL free energies are equal.

PHASES AND PHASE DIAGRAM

The calculations show that for temperatures $T > T_s$, i.e., for $r > 0$, we have three stable phases. Two of them are quite simple: the normal (N-) phase ($\psi = M = 0$) with existence and stability domains given by $t > 0$ and $r > 0$, and the ferromagnetic phase (FM) given by $\psi = 0$ and $M^2 = -t$ which has the existence condition $t < 0$, and a stability domain defined by the inequalities $r > \gamma_1 t$ and

$$ r > \gamma_1 t + \gamma\sqrt{-t}. \quad (5) $$

The third stable phase is a phase of coexistence of superconductivity and ferromagnetism (hereafter referred to as FS). It is given the following equations:

$$ \phi_1 = \phi_2 = \frac{\phi}{\sqrt{2}}, \quad \phi_3 = 0, \quad (6) $$

$$ \phi^2 = (\pm\gamma M - r - \gamma_1 M^2), \quad (7) $$

$$ (1 - \gamma_1^2)M^3 \pm \frac{3}{2}\gamma\gamma_1 M^2 + \left(t - \frac{\gamma^2}{2} - \gamma_1 r\right)M \pm \frac{\gamma r}{2} = 0, \quad (8) $$

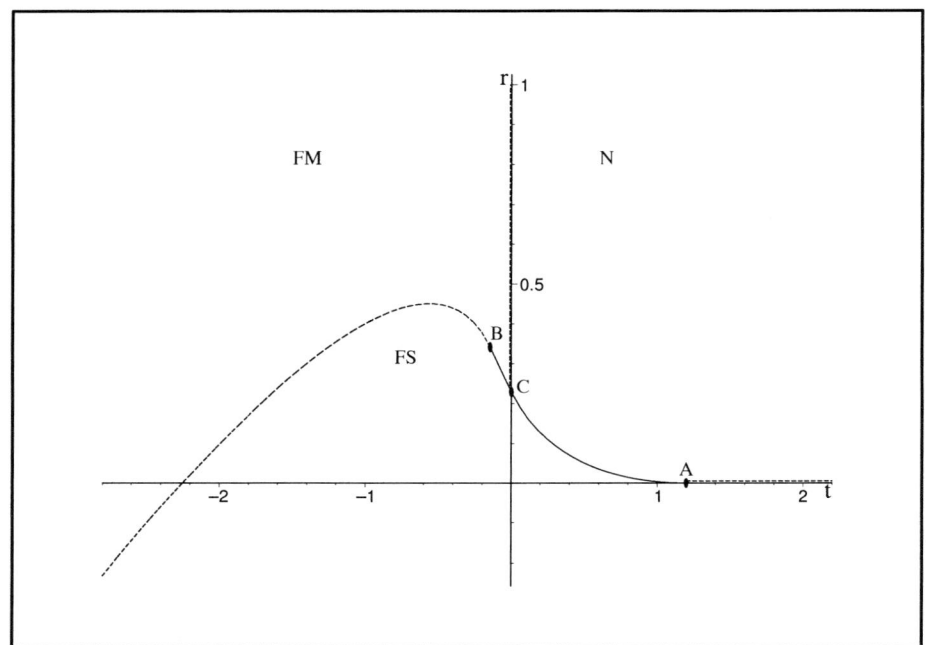

FIGURE 1. Phase diagram in the (t, r) plane for $\gamma = 1.2$, $\gamma_1 = 0.8$, and $w = 0.4$.

and

$$(\theta_2 - \theta_1) = \mp \frac{\pi}{2} + 2\pi k, \qquad (9)$$

($k = 0, \pm 1, \ldots$). The upper sign in Eqs. (7) – (9) corresponds to a domain in which $\sin(\theta_2 - \theta_1) = -1$ and the lower sign corresponds to a second domain which may be referred to as FS*; in the latter, $\sin(\theta_2 - \theta_1) = 1$. These two domains are equivalent and describe the same ordering. We shall focus on the upper sign in (7) – (9), i.e. on FS.

The phase diagram (t,r) is outlined in Figs. 1 and 2 for different values of the anisotropy parameter w. The phase transition lines for $w > 0$ and $w < 0$ shown in Figs. 1 and 2, respectively, have qualitatively the same shape as the phase transition lines corresponding to $w = 0$ [18] but there are essential quantitative differences between these cases. We shall discuss them in the next section.

Note, that the phase diagrams in Figs. 1 and 2 exhibit two types of phase transitions. The dashed curves indicate second order phase transitions of type N-FM, FM-FS and N-FS, whereas the curve AC and the straight line BC indicate the first order phase transitions N-FS and FM-FS, respectively. The points A and B are tricritical whereas C is a triple point, where N, FM and FS coexist [11]. The negative values of the parameter r are restricted by $r(T) \geq r(T=0)$. Having in mind this condition as well as the shape of the phase diagram (Figs. 1 and 2) we easily conclude that both FM and FS ground states (at $T=0$) are possible in systems described by the model (1). This is the case for the itinerant magnets, mentioned above, and one may speculate that this situation may

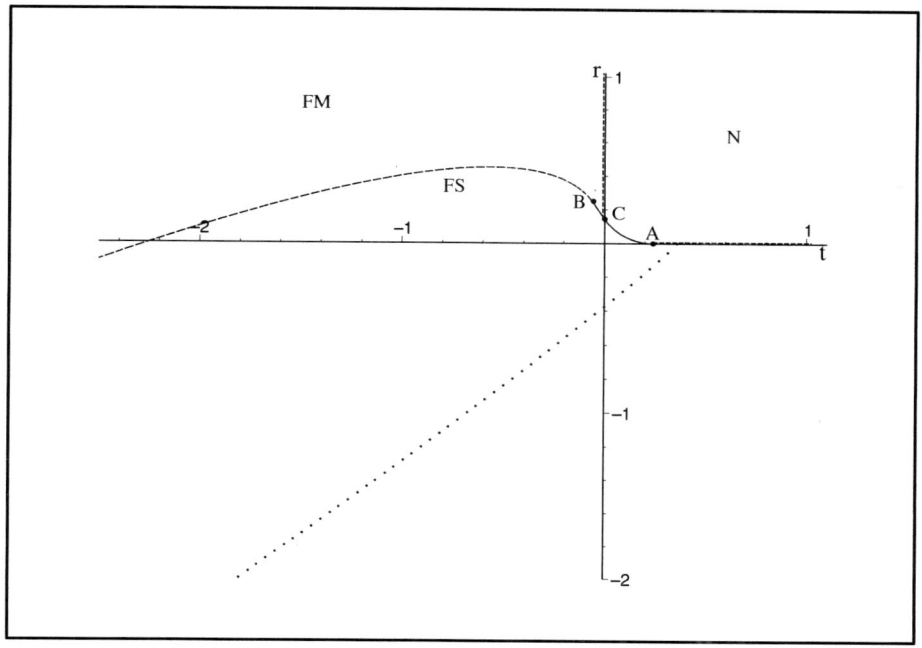

FIGURE 2. Phase diagram in the (t, r) plane for $\gamma = 1.2$, $\gamma_1 = 0.8$, and $w = -2$. The various lines are explained in the text.

occur in neutron stars, too. Whether FM and FS ground state will occur depends on the particular values of the material parameters $(T_s, T_f, \alpha_s, ...)$.

The final aim of the phase diagram investigation is the outline of the (T,P) diagram. Important conclusions about the shape of the (T,P) diagram can be made from the form of the (t,r) diagram without an additional information about the values of the relevant material parameters $(a_s, a_f, ...)$ and their dependence on the pressure P. For example, the equilibrium temperature T_{FS} of the phase transitions to FS phase varies with the variation of the system parameters $(\alpha_s, \alpha_f, ...)$ from values which are much higher than the characteristic temperature T_s up to zero temperature.

ANISOTROPY EFFECTS

Our analysis demonstrates that when the anisotropy of the Cooper pairs is taken in consideration, there will be not drastic changes in the shape the phase diagram for $r > 0$ and the order of the respective phase transitions. Of course, there will be some changes in the size of the phase domains and the formulae for the thermodynamic quantities. This is seen from Figs. 1 and 2 which are shown for the first time in the present report. Besides, it is readily seen from Figs. 1 and 2 that the temperature domain of first order phase transitions and the temperature domain of stability of FS above T_s essentially vary with the variations of the anisotropy parameter w. The parameter w will also insert changes in the values of the thermodynamic quantities like the magnetic susceptibility and the entropy and specific heat jumps at the phase transition points [24].

Besides, and this seems to be the main anisotropy effect, the w- and v-terms in the free energy lead to a stabilization of the order along the main crystal directions which, in other words, means that the degeneration of the possible ground states is considerably reduced. This means also a smaller number of marginally stable states.

The dimensionless anisotropy parameter w can be either positive or negative depending on the sign of u_s. Obviously when $u_s > 0$, the parameter w will be positive too ($0 < w < 1$). We shall illustrate the influence of Cooper-pair anisotropy in this case. The order parameters (M, ϕ_j, θ_j) are given by Eqs. (6), (9),

$$\phi^2 = \frac{\pm \gamma M - r - \gamma_1 M^2}{(1-w)} \geq 0, \tag{10}$$

and

$$(1 - w - \gamma_1^2)M^3 \pm \frac{3}{2}\gamma\gamma_1 M^2 + \left[t(1-w) - \frac{\gamma^2}{2} - \gamma_1 r\right]M \pm \frac{\gamma r}{2} = 0, \tag{11}$$

where the meaning of the upper and lower sign is the same as explained just below Eq. (9). We consider the FS domain corresponding to the upper sign in the Eq. (10) and (11). The stability conditions for FS read,

$$\frac{(2-w)\gamma M - r - \gamma_1 M^2}{1-w} \geq 0, \tag{12}$$

$$\gamma M - wr - w\gamma_1 M^2 > 0, \tag{13}$$

and

$$\frac{1}{1-w}\left[3(1 - w - \gamma_1^2)M^2 + 3\gamma\gamma_1 M + t(1-w) - \frac{\gamma^2}{2} - \gamma_1 r\right] \geq 0. \tag{14}$$

For $M \neq (\gamma/2\gamma_1)$ we can express the function $r(M)$ defined by Eq. (10), substitute the obtained expression for $r(M)$ in the existence and stability conditions (10)-(14) and do the analysis in the same way as for $w = 0$ [18]. The most substantial qualitative difference between the cases $w > 0$ and $w < 0$ is that for $w < 0$ the stability of FS is limited for $r < 0$. This is seen from Fig. 2 where FS is stable above the straight dotted line for $r < 0$ and $t < 0$. This includes into consideration also purely superconducting (Meissner) phases as ground states.

FINAL REMARKS

We have done an investigation of the M-trigger effect in unconventional ferromagnetic superconductors. This effect due to the $M\psi_1\psi_2$-coupling term in the GL free energy consists of bringing into existence of superconductivity in a domain of the phase diagram of the system that is entirely in the region of existence of the ferromagnetic phase.

This form of coexistence of unconventional superconductivity and ferromagnetic order is possible for temperatures above and below the critical temperature T_s, which corresponds to the standard phase transition of second order from normal to Meissner phase – usual uniform superconductivity in a zero external magnetic field, which appears outside the domain of existence of ferromagnetic order. Our investigation has been mainly intended to clarify the thermodynamic behaviour at temperatures $T_s < T < T_f$, where the superconductivity cannot appear without the mechanism of M-triggering. We have described the possible ordered phases (FM and FS) in this most interesting temperature interval.

The Cooper pair and crystal anisotropies have also been investigated and their main effects on the thermodynamics of the triggered phase of coexistence have been established. In discussions of concrete real material one should take into account the respective crystal symmetry but the variation of the essential thermodynamic properties with the change of the type of this symmetry is not substantial when the low symmetry and low order (in both M and ψ) γ-term is present in the free energy.

Below the superconducting critical temperature T_s a variety of pure superconducting and mixed phases of coexistence of superconductivity and ferromagnetism exists and the thermodynamic behavior at these relatively low temperatures is more complex than in known cases of improper ferroelectrics; see. e.g., Ref. [25]. The case $T_f < T_s$ also needs a special investigation.

Our results are referred to the possible uniform superconducting and ferromagnetic states. Vortex and other nonuniform phases need a separate study.

More experimental information about the values of the material parameters $(a_s, a_f, ...)$ included in the free energy (1) is required in order to outline the thermodynamic behavior and the phase diagram in terms of thermodynamic parameters T and P. In particular, a reliable knowledge about the dependence of the parameters a_s and a_f on the pressure P, the value of the characteristic temperature T_s and the ratio a_s/a_f at zero temperature are of primary interest.

The phenomenological GL model (1) is quite general and can be reliably used in considerations of a possible coexistance of ferromagnetism and unconventional superconductivity in stellar objects. Recent investigations [2, 3] of superconductivity in neutron stars can be related with the present consideration.

ACKNOWLEDGMENTS

One of us (T.E.T.) thanks the hospitality of Dr. V. Celebonovic and the other two organizers of the Workshop on Equation of State and Phase Transition Issues in Models of Ordinary Astrophysical Matter (Lorentz Center, Leiden University, 2-11 June 2004). Financial support by SCENET (Parma) and JINR (Dubna) is also acknowledged.

REFERENCES

1. Tilley,D.R., Tilley.J.: 1974,Superfluidity and Superconductivity, Van Norstrand Reinhold Company,New York.
2. Link,B.: 2003,Phys.Rev.Lett.,**91**,101101.
3. Buckey,K.B.W.,Metlitski,M.A.and A. R. Zhitnitasky,A.R.: 2004, Phys.Rev.Lett.,**92**,151102.
4. Leggett,A.J.: 1975, Rev.Mod.Phys.,**47**,331.
5. Vollhardt,D. and Wölfe,P.: 1990,The Superfluid Phases of Helium 3,Taylor&Francis,London.
6. Volovik,G.E.: 2003,The Universe in a Helium Droplet, Oxford University Press,Oxford.
7. Stewart,G.R.: 1984, Rev.Mod.Phys.,**56**,755.
8. Sigrist,M. and Ueda,K.: 1991,Rev.Mod.Phys.,**63**,239.
9. Blagoeva,E.J.,Busiello,G.,De Cesare,L.,Millev,Y.T.,Rabuffo,I. and Uzunov,D.I.: 1990, Phys.Rev.,**B42**, 6124.
10. Uzunov,D.I., in: 1990,Advances in Theoretical Physics, Ed. by Caianiello,E.,World Scientific,Singapore, p. 96.
11. Uzunov,D.I.: 1993,Theory of Critical Phenomena, World Scientific,Singapore.
12. Annett,J.F.: 1995, Contemp.Physics,**36**,323.
13. Volovik,G.E. and Gor'kov,L.P.:1984,JETP Lett.,**39**,674 [1984, Pis'ma Zh.Eksp.Teor.Fiz.,**39**,550].
14. Volovik,G.E. and Gor'kov,L.P.: 1985,Sov.Phys.JETP,**61**,843 [1985, Zh.Eksp.Teor.Fiz.,**88**,1412].
15. Ueda,K. and Rice,T.M.: 1985,Phys.Rev.,**B31**,7114.
16. Blount,E.I.: 1985,Phys.Rev.,**B 32**,2935.
17. Saxena,S.S.,Agarwal,P.,Ahilan,K. et al.: 2000,Nature,**406**,587.
18. Shopova,D.V. and Uzunov,D.I., in: 2004,Progress in Ferromagnetism Research, Nova, New York (in press); see also cond-mat/0404261.
19. Huxley,A.,Sheikin,I.,Ressouche,E. et al.: 2001, Phys.Rev.,**B63**,144519.
20. Machida,K. and Ohmi,T.: 2001,Phys.Rev.Lett.,**86**,850.
21. Walker,M.B. and Samokhin,K.V.: 2002, Phys.Rev.Lett.,**88**,207001.
22. Shopova,D.V. and D. I. Uzunov,D.I.:2003, Phys.Lett.,**A 313**,139.
23. Shopova,D.V. and Uzunov,D.I.:2003, J.Phys.Studies,**7**,426.
24. Shopova,D.V. and D. I. Uzunov,D.I.: 2003, Compt.Rend.Acad.Bulg.Sci.,**56**,35; see also cond-mat/0310016.
25. J-C. Tolédano,J-C. and P. Tolédano,P.: 1987,The Landau Theory of Phase Transitions,World Scientific,Singapore.

Author Index

A

Ayukov, S. V., 147, 173, 178

B

Baturin, V. A., 147, 162, 173, 178
Bon, E., 291
Bonitz, M., 239

C

Čelebonović, V., 269, 280, 288, 291
Chabrier, G., 208
Chigvintsev, A. Y., 255, 261
Christensen-Dalsgaard, J., 18

D

Däppen, W., 3, 219, 230

F

Filinov, V., 239
Fortov, V. E., 147, 239

G

Gough, D., 119
Gryaznov, V. K., 147, 178

H

Houdek, G., 193

I

Ilić, D., 291
Iosilevski, I. L., 147, 178, 255, 261
Ivanov, V. G., 295

K

Kraeft, W.-D., 248

L

Levashov, P., 239
Liang, A., 106
Lin, C.-H., 219, 230

M

Mediavilla, E., 291
Miglio, A., 187

P

Pavičić, G., 288, 291
Perez, A., 208
Popović, L. Č., 291

R

Roerich, V. C., 83

S

Schlanges, M., 248
Shaviv, G., 67, 139
Shaviv, N. J., 139
Shopova, D. V., 302
Starostin, A. N., 83, 147, 178

T

Trampedach, R., 99
Tsvetkov, T. E., 302

U

Uzunov, D. I., 295, 302

V

Vorberger, J., 248
Vorontsov, S. V., 47